PROJECT MANAGEMENT

PROJECT MANAGEMENT

Techniques in Planning and Controlling Construction Projects

Second Edition

HIRA N. AHUJA
University of Toronto

S. P. DOZZI
S. M. ABOURIZK
Dept. of Civil Engineering, University of Alberta

JOHN WILEY & SONS, INC.
New York / Chichester / Toronto / Brisbane / Singapore

This text is printed on acid-free paper.

Library of Congress Cataloging in Publication Data:
Ahuja, H. N.
Project management : techniques in planning and controlling construction projects / Hira N. Ahuja, S. P. Dozzi, S. M. AbouRizk.—2nd ed.
p. cm.
Includes bibliographical references and index.
ISBN 0-471-59168-8 (acid-free paper)
1. Construction industry—Management. I. Dozzi, S. P. II. AbouRizk, S. M. III. Title.
HD9715.A2A418 1994
624'.068—dc20 94-7316

Printed in the United States of America

10 9 8 7 6 5 4 3 2 1

To our wives and families, and many friends in project management and the construction industry.

PREFACE

The need for efficient project management is always continuous. The introduction of quantitative analysis cannot itself solve management problems; it is necessary also to form a synthesis of quantitative facts and the human element to achieve professional project management.

Managing a project involves planning, organizing, executing, and monitoring that project. Planning often has been construed to mean scheduling, but this is not the case. A significant portion of a project manager's time is spent in planning, which is proactive management. A project consists of many elements, all of which require careful planning—time, cost, material, and organization.

In this edition, we have de-emphasized CPM scheduling and have introduced other planning tools. Construction is a dynamic process—the application of resources to accomplish a plan for a project. CPM methods are not dynamic and require that resources be added to CPM networks. This add-on of resources is artificial and is perhaps the foremost reason that CPM methods are not universally applied.

Many contractors use CPM methods to some degree, whereas others resort to simple bar charts with good results.

The scope of this edition is broader than the first edition, because scheduling is treated as only one of several planning areas. Other approaches are also introduced, such as various forms of simulation, which is one of the more promising techniques that deal with the dynamics of a project and resource management.

During the past decade, the construction industry has begun to embrace the total quality management (TQM) philosophy of continuous improvement with a customer focus. TQM requires careful planning and implementation, but, again, a plan is required. Although TQM may not be the perfect approach, it has been proven successful in improving competitiveness and in maintaining a competitive edge.

The organization of this book reflects this broader approach to management of construction projects.

Chapter 1 lays the groundwork for the management of a construction project. The dynamic nature of a project, its life cycle, and the project management model are explained.

Chapter 2 emphasizes the need for planning and the various systems and plans required.

Chapter 3 looks at "people" planning. Different projects require that organizations tailor them to suit their needs. The types of organizations are discussed, as well as the needs of individuals within organizations. Partnering is one approach that is used to improve project performance and to lessen disputes.

Chapter 4 introduces the use of bar charts, because they are the initial type of schedules that are usually produced and are preferred for many contractors. In spite of their limitations, they are readily understood and used.

Chapters 5, 6, and 7 deal with CPM scheduling—both the Activity-on-Arrow (AOA) and the Activity-on-Node (AON) methods.

Chapter 8 looks at additional techniques that optimize planning procedures. Work Breakdown Schedule (WBS), Organizational Resource Chart (ORC), Milestone Networks, Subnets, and Skeltonization are topics that are discussed in this chapter.

Chapter 9 covers resource allocation and planning.

Chapter 10 discusses how to establish the most economical duration of a project.

Chapter 11 looks at estimating and project estimates, because these form the basis for cost control, which is the topic of Chapter 12.

Chapter 12 discusses the cost center and cost breakdown, and explains cost coding.

Chapter 13 remains as it was in the first edition and discusses cash flow, forecasting.

Chapter 14 deals with the remaining procedures for project cost and schedule control, as well as project performance.

Chapters 15, 16, and 17 consider the dynamic aspects of a project, stochastic methods, and simulation. Some of the material in these chapters requires a knowledge of the mathematics of probability and statistics.

Chapter 18 is important for the planning and control of material and equipment resources.

Chapter 19 deals with effective communication.

Planning is useful only if the plans and requirements are communicated accurately and on time. There is a considerable amount of information that is required to execute a project successfully, and it is imperative that there exist a plan for the gathering and dissemination of this information.

Chapter 20 expands on the topic of claims. A plan of action is required to control changes and the resulting claims. Some claims result in disputes, and the mechanisms for handling disputes are discussed.

Chapter 21 looks at the corrective action process and the actions that are required to remedy a situation. Sometimes, in spite of planning, things go off track, and a project manager should be able to recognize the signs of distress and take corrective action.

Chapter 22 discusses the concept of TQM of projects. This is a philosophy that should be applied to all projects in order to create win-win situations.

Chapter 23 examines the concept of constructability, which must be part of the overall plan.

Chapter 24 examines the impact of computers in construction. An understanding of the potential of computers and software is essential for a project manager. In addition, relational databases, expert systems, and neural networks are briefly explored.

Most calculations and examples presented are to be manually computed for the sake of learning and practice only, whereas in reality most planning, scheduling, and cost control would be carried out by the use of computers.

ACKNOWLEDGEMENTS

In the first edition, the following contributors were acknowledged: Victoria Warford, Rodney Pike, Rupinder Rangar, V. Arunachalan, V. Nandakaman, Terry Dwyer, Dave Oliver, Marilyn Warren, Maureen Kenny, Brenda Young, K. C. Thayar, and Bros. Ltd. of New Delhi.

The following people contributed to the completion of this second edition. Word processing was ably performed by Norma Odore at Purdue University and Donna Salvian at the University of Alberta. Word processing and proof-reading assistance also was provided by Christina Dozzi and Agnes Dozzi. Nader N. Chehayeb, Anil Sawhney, Jingshang Shi, and Rod Wales of the University of Alberta provided assistance in many aspects of this work.

The authors also would like to thank their colleagues' contributions to this edition. In particular, we wish to thank Professor James R. Wilson of North Carolina State University for many discussions on the statistical aspects of simulation, many of which are reflected in material presented in Chapter 16. We also wish to express our sincere appreciation to Professor Daniel W. Halpin of Purdue University for his contributions to Chapter 17. Mr. Barry McIntyre of Economic Development, Edmonton, contributed to Chapter 22. We also would like to thank NRC for allowing us to use some material from the book *Productivity in Construction* in Chapter 18.

Finally, we wish to thank the following software developers for providing us with the educational versions of their software that have been used in many illustrations in this book: Pritsker Corp. (SLAMSYSTEM), Learning Systems Inc. (MicroCYCLONE), and Primavera (P3 and PARADE).

CONTENTS

CHAPTER 1

CONCEPTS

Projects are becoming progressively larger and more complex in terms of physical size and cost. In the modern world, the execution of a project requires the management of scarce resources; manpower, materials, money, and machines must be managed throughout the life of the project—from conception to completion.

Numerous endeavors can be categorized as projects. In this text the emphasis is on construction projects, but the management principles apply equally well to any project. Although the objectives are different, the principles of project management apply to construction, manufacturing, social, and personal projects. However, the details and jargon vary according to the type of project. Construction of a building, bridge, or highway requires different trades and knowledge but the management, scheduling, costing, and control of these projects utilize the same tools and techniques, and are subject to constraints of time, cost, and quality.

Except for a few people who report directly to him, the project manager (PM) manages resources indirectly. Essentially the PM is managing information and utilizing systems to achieve the project objectives. There is an hierarchy of systems (Kerzner, 1987), but in project management the systems that a PM utilizes directly are the business, organization, and employee systems. These and others are ongoing entities which facilitate the execution of a project; they provide information for decision making and control. In a business entity, projects come to an end but systems remain for the next project. Flexible systems are the strength of the prominent project management companies; they are the strength of major engineering, procurement, and construction (EPC) contractors. If systems do not exist for the project, they must be created. Systems subsets include financial and accounting, cost and schedule control, marketing, personnel, communications, computing, production, and other lower tier systems. A successful project requires the integration of these systems to achieve the desired results and objectives. Therefore, the PM must have a clear understanding of the systems and their interrelationships. Systems are discussed further in Chapter 2.

As our society becomes more knowledgeable and sophisticated, the demands placed on a project will increase. No longer is it the case that "what's good for General Motors is good for the USA." In some developing countries perhaps the project is justified on the basis of limited economic assessment, that is, the short-term benefits. In the developed countries, especially North America, Europe, and Japan, it is now necessary to consider the cultural and ecological aspects of a project. A few years ago, for example, impact assessments were not required. Today it is inconceivable to initiate larger projects without considering the impacts on the environment and economy.

Formerly, projects were required to comply with local bylaws only. Today, projects experience serious cost and schedule implications due to increased environmental regulations and the need to consider broader rather than just local public opinion. The larger the project, the greater is the environmental impact on the neighborhood and on the project itself.

1.1 DESCRIPTION OF A PROJECT

The emphasis of this book is on construction projects, although the treatment can be very broadly applied. In construction the general scope of a project can be readily described as residential, building (i.e., other than residential), industrial (processing plants, refineries, mills, etc.), and heavy construction, which require a large engineering content. Bridges, tunnels, airports, railways, and harbors are a few that fall into this category. With the increasing concern over the environment, another group can be classed as environmental projects. These broad categories serve to delineate niche markets within the scope of construction projects. Projects can be looked upon in a more generic sense as follows.

A project (construction or otherwise) is a unique undertaking for essentially a single purpose which is defined by scope, quality, time, and cost objectives. The scope could be to provide a facility or a bridge, revise an organization, produce a study, and so on. The cost and quality objectives are met by the use of limited resources.

A construction project is characteristically a capital venture which therefore requires a well-defined start and end point. The next section describes the life cycle within such time constraints of a project.

1.2 PROJECT LIFE CYCLE

A project occurs over an identified period of time during which a changing level of effort is required to complete each stage. The life of the project can be viewed as consisting of pre- and postfund authorization phases. Final approval of funding is the demarcation event which separates these two phases. This event is an all-important milestone. The prefund phase consists largely of planning activities; the postfund phase is the execution phase. Figure 1.1 shows the components of a project. The figure is intended to show relationally an approximate period over which most of the effort for each phase occurs. No scale is intended and each phase indicates an amount of effort. For example, construction is a large effort that occurs during the execution phase. However, some very important construction effort is expended in reviewing constructability problems during the early planning phase, but this level of detail is not shown.

A summation of the component phases is shown in Figure 1.2 (Wideman, 1983) and again, no scales are intended. This figure shows conceptually the level of effort (resources) that is required during the four distinct and essential phases of a project. The four phases are Conceive, Develop, Execute, and Finish. The first letter (capitalized) of each component of the life cycle form the mnemonic **C D E F.**

Conceive is the beginning of the planning phase. The need is identified and a concept is created. Initially project team members begin to establish the feasibility of various alternatives. During this stage, preliminary drawings and block diagrams are produced. As the scope is increasingly defined, budgets and preliminary schedules are used to establish the economic feasibility of the project. Impact assessments are initiated assuming a viable project approval is received and the project moves into the next stage.

In the **Develop** stage, the plan is further evolved with additional engineering and drawing work. Block diagrams are converted into process flow sheets. Additional engineering and economic studies are carried out to develop a better budget, schedule, and cash flow profile. Impact assessments are essentially completed and the work necessary to obtain permits continues. At the end of this stage, the final project brief is complete. A project brief is a document that requests funds for implementation of a project and includes the essentials of a project plan and funding requirements. The scope definition is not necessarily complete, but enough work has been done to reconfirm the economics of the project and final funding approval is received. This is the major "go–no go" decision point. Major or long-lead items are selected in anticipation of approval to continue. The level of effort builds up during the **Develop** stage.

After the funding approval is received, the **Execution** stage follows. Considerably more effort is expended and the project reaches the highest level

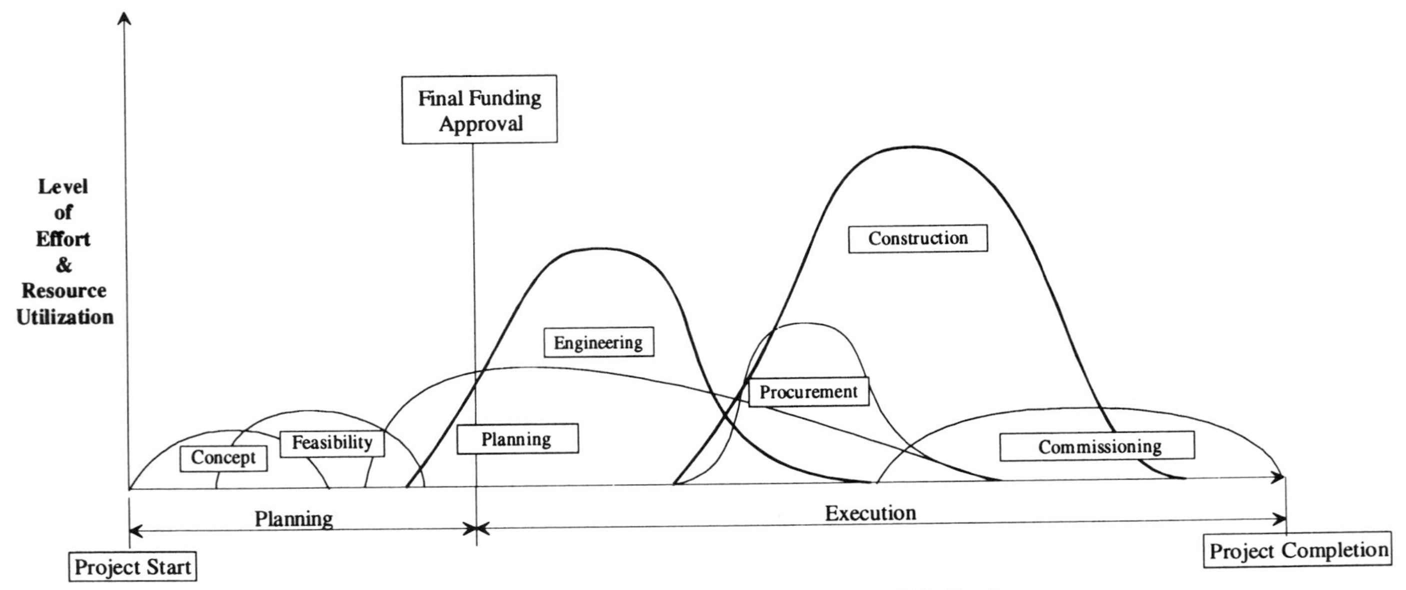

FIGURE 1.1 The Dynamics of the Project Life Cycle

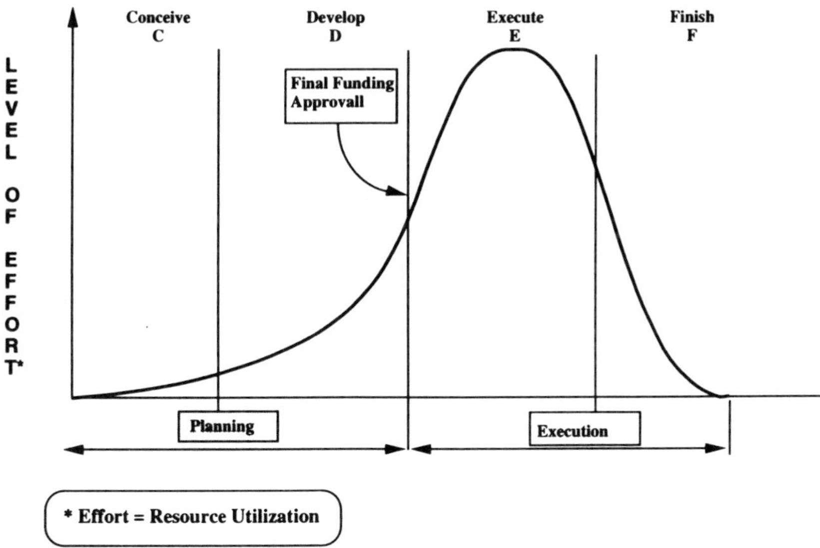

FIGURE 1.2 Project Life Cycle

of expenditures. Most of the engineering, procurement, and construction are completed in this stage.

The project organization is established in order to produce final engineering, drawings, and specifications. Long-lead items are ordered early and the remainder of the procurement effort takes place. The physical project is produced to the scope, cost, time, and quality requirements.

The final procedure is the **Finish** stage, during which the project is terminated. Turnover, or commissioning, and training of operators if necessary are the final steps. The project is then complete, enabling resources to be released. Ownership responsibilities are transferred to the client owner. Receipt of final progress payment constitutes a completed project, notwithstanding any future warranty work. It is important to understand the life cycle concept of a project because the expectations of owners and project personnel must be compatible with what can reasonably be expected during a particular stage. Many owners do not appreciate that a period of time is required for conceptualizing and planning. It is not unheard of to have a contractor in the execution phase while the engineer is still partially in the development stage. The contractor complains about incomplete information and the engineer continues to make changes. Obviously the contractor and engineer are not synchronized, which results in wasted resources and needless friction. For some cases a different form of contract may be required to allow a fast-track method of execution.

Each of these stages requires its allotted time in the project. The durations vary according to the information available. If a facility is an exact duplicate, obviously the planning stages (i.e., **C** and **D**) can be shortened appreciably. At the other extreme, novel projects that utilize cutting edge technology, require more time to develop the design and ensure constructability. A classic example of insufficient planning is the Montreal Olympics (1980) stadium. Experts estimated that the planning stage realistically needed two more years, rather than actual planning period of approximately one year. The results were financially disastrous, with cost overruns exceeding 400%.

An example of a successfully executed project is a pharmaceutical laboratory ($15 million in 1993 costs) that took 30 months from conception to completion. The execution and finish stages (**E** and **F**), which consisted of detail engineering, procurement, and construction, took only 14 months because adequate time was allowed for conceptualizing and planning.

There are no simple rules to determine what is an appropriate time for each stage. The length of stages **E** and **F** are more readily established because of past experiences which have been documented. The duration of planning stages **C** and **D** are not well documented. In fact, there is a lack of documented experiences during these early stages of a project. In summary, project owners and managers need to understand that each and all stages of a project require sufficient time for proper development.

1.3 PROJECT MANAGEMENT CONCEPT

Project management has evolved mainly because of the need to control costs and schedule. Projects have become more complex and demanding for the owners and participants. Hence the risks and potential for losses require better controls. For example, the author worked for a large corporation which was beginning to experience serious losses as projects undertaken were becoming larger. Initially, the projects were run by functional line managers and project coordinators or engineers, both with limited authority. Because of the need for greater accountability and the commensurate authority required, the project manager became a common phenomenon, a natural evolution for better control of projects. The attributes and functions of a project manager are discussed in Chapter 21.

What then is project management? It is the art and science of directing human and material resources to achieve stated objectives within the constraints of time, budget, and quality and to the satisfaction of everyone involved (Wideman, 1983).

As the project environment becomes more complex, so are the requirements of project management. In fact, the development of project management has been driven by these increased demands. The scope of project management will depend on the size and complexity of the project, but it now also requires better knowledge of the physical and social environment of the project.

The systems used will depend on the project. Some simple projects can be successfully completed with a minimal amount of formal systems. The construction of a residential garage does not merit the same attention and control as a nuclear power plant, although each requires planning, execution, and cost control. Project management is the process of applying management techniques and systems to direct and control suitable resources in order to successfully deliver the intended scope of the project.

Figure 1.3 illustrates a project management model (PMI, 1986) as a three-dimensional matrix. This is a framework for project management which the reader can use as a focus for the various topics that are introduced.

It is important to note that the functions, processes, and stages of project management are dynamically linked, that is, each is simultaneously applied to the project. For example, the management processes are applied to the cost function during a particular stage of the project.

The management functions are "what" we manage, which includes scope, time, cost, quality, communications, human resources, and risk. Material and financial resources are subsets of scope and cost respectively. Similarly, risk could be a subset of scope, cost, time, and quality but it is shown as a specific function. The processes are "how" we manage, that is, we plan, organize, execute, monitor, and control. These processes can be remembered by the acronym **POEM/C** from the first letter of each action.

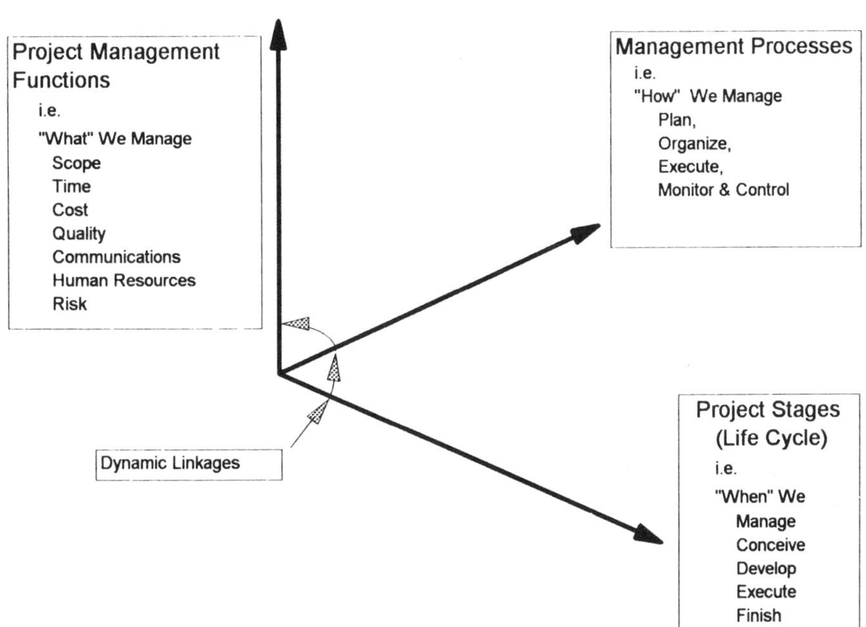

FIGURE 1.3 Project Management Model

In Section 1.2, the four project states were described as **C D E F.** It may be convenient to further subdivide these stages or to use a different set of stages to better describe a contractor's involvement in a project. For an engineering contractor, the **Execution** stage can be categorized under conceptual, preliminary, production engineering, and field support, all of which emphasize the particular company's engineering involvement in the project. For a general contractor, "execution" has a different meaning; likewise for an engineering procurement construction contractor (EPC), or an owner.

Project management is applied in varying degrees and forms, depending on a number of factors. The question that arises is, How much project management do we need? The concepts apply to any project; the extent and types of systems that are applied will vary according to the project needs.

The concept of project management should be applied in the context of total quality management (TQM), that is, on a value-adding basis with the process undergoing continuous improvement. If a system or action does not add some value to the project, it should not be used.

For example, in 1988 a large EPC (Engineering Procurement Construction) contractor was awarded a $10 million refinery revamp project. The contractor had the state of the art scheduling capabilities which could accommodate any project size. Rather than use sophisticated schedule network analyses, the contractor chose to use a gantt (bar) chart schedule. This is all that was needed, and the project was successful. The adaption of all the "bells and whistles" that were available would have added cost without value.

1.4 WHY DO WE NEED PROJECT MANAGEMENT?

Why not use traditional management, that is, the type utilized in the management of any business enterprise? The objectives of traditional management are those for an ongoing business rather than a relatively short-term project. Return on investment, or some other mandated financial measure, is the governing criteria for successful traditional corporate management. These objectives are intended to perpetuate the enterprise and accordingly have a time horizon of one year repeated on an annual basis. However, companies are recognizing that long-term goals are also important for continuing growth and competitiveness.

Personal objectives are major factors in the management of enterprises as well as projects. Corporate objectives are often influenced by personal objectives, which steer and change the course of the corporation.

In a business enterprise which is ongoing, change is relatively gradual or has been so. In the future, shorter product cycles will require that traditional management become more dynamic, running their businesses in almost a project management mode.

Manufacturing and service industries have characteristically traditional management; the roles and responsibilities of each function are usually well understood, each manager having reasonably static responsibilities. The other trait of these organizations is that the tasks of the organization are somewhat repetitive within a known technology.

In a traditional organization several levels of management are imposed on the functional organization. The present trend is to flatten the organizational hierarchy, which includes top, middle, and operational managers and workers. The functions (departments) usually consist of finance, engineering, production, sales, and personnel both for manufacturing and service enterprises. Communications and flow of work move easily up and down within the department (i.e., vertically). They do not move as freely between functional groups (i.e., horizontally between departments), and as a result functional gaps (Kerzner, 1987) develop. Within an organization, these gaps are a natural development, because of departmental allegiance and cost center constraints within departments or functions. Accountability is structured vertically within a department, not across functional lines.

Thus problems can arise when a traditional organization is given the responsibility to carry out a specific project which is not part of their regular functions and output.

Coordination is required to overcome poor communications, resource allocation difficulties, slow response time, and any omissions that fall between the cracks. A project management approach can balance the countervailing needs and make "it" happen.

In summary, project management is required if construction projects are complex and under demanding constraints of time, cost, and environmental regulations. If several activities or disciplines are to be integrated, and there is a need to coordinate the cooperation of various functional departments within the organization, project management is a viable approach. Project management is especially appropriate when the project faces changing environmental needs and other considerations that are external. In contrast to the functional departments' objectives within an ongoing enterprise, project management provides a single focus which is the specific project only.

CHAPTER 2

PLANNING

2.1 CONCEPTS

Planning is one of the key functions (POEM/C) of the management process; it is the project manager's prime activity. Planning is selecting objectives and then establishing programs and procedures for achieving the objectives. It is decision making for the future; it is looking ahead. Decisions are required because a choice must be made from among the alternatives that are available. It may be a choice of systems, equipment, or contract strategy to mention a few. Eventually the plan is accomplished through a structured sequence of events which will lead to a desired set of objectives.

A proactive management style sets a plan and makes it happen. This is the opposite of reactive management, which results from lack of planning. Reactive management is often referred to as crisis management or as a fire-fighting mode. Planning allows leadership to control, whereas reactive management is a lack of leadership and control.

An engineer who is looking for challenging work will soon find himself or herself responsible for a project that demands the utmost in planning skills. In fact, it may be necessary for the engineer to carry out all the planning: to visualize all the operations of the project; to arrange these operations in their proper sequence; to achieve confidence that each operation is understood; to acquire the know-how and means necessary to perform each operation; and to feel convinced that the method thought out for performing each operation is the most economical. Confidence is achieved only through systematic planning. Unfortunately, because of overconfidence many people do not make the necessary effort to develop a plan; they resort to "seat of the pants" management which produces one panic after another and below par performance. Their targets keep shifting from day to day. For our purposes this picture presents a vivid example of how a project should not be carried out.

If, however, one is confident that one knows a project extremely well and it is so small that its progress can be plotted mentally without the need of any formal planning, that individual is justified in saving planning time and using it for other efforts such as organization and coordination. But how many projects are that simple?

Planning is not only scheduling, which is but one of the many plans that can be put in place. Planning results in setting out schedule milestones, which are time objectives. It clearly establishes the work to be done to achieve certain contract scope and cost objectives.

The main purpose of planning is that it reduces the uncertainty that exists before a project or portion of a project is launched. Besides avoiding the crisis management mentioned earlier, planning improves the efficiency of execution; it clarifies the objectives.

Another important product of planning is that it provides a basis for executing, monitoring, and controlling. Thus during the early overall planning the PM *must* be a key participant in the planning process. This planning responsibility should not be delegated; to do so would mean that the PM is abdicating responsibility for planning, requiring that someone else carry this load. This someone else becomes the true project manager.

2.2. DEVELOPMENT OF THE SYSTEMS APPROACH

A system is defined as an assemblage or combination of things or parts forming a complex or unitary whole (PMI, 1986). The future will demand new and better delivery systems. With this in mind, it is useful to review the development of the systems approach, which is a perspective on how to solve complex engineering problems.

The existence of a problem was considered to be a lack in the environment, a need. A solution to remove this lack was viewed as a system that would receive inputs from the environment and transform them into maximum desired outputs. For example, one solution for a lack of transportation facilities was the automobile. For a lack of project cost control, the solution was a project management system. For a lack of confidence in coordinating the operations of a project, the solution was a plan, and so on.

With this lack–solution concept in mind, it is immediately apparent that for any given lack there may be more than one solution. For example, in response to a lack of cost control, a project accounting system presents a reasonable solution, but not if we introduce the constraint that the system must provide a weekly forecast of costs. Knowledge of this constraint requires that a costing system be developed which utilizes committed costs and a forecasting ability.

In order to achieve this solution, the system may be created using problem solving techniques and/or an assemblage of components.

The problem solving procedure that follows is similar to an action sequence utilized in the Kepner–Tregoe system (Kepner–Tregoe, 1981). The following procedure must be adhered to:

1. Analyze the need or problem due to some lack.
2. Define, describe, or specify the problem. It is mandatory that the problem statement and objectives be clearly spelled out before proceeding any further.
3. Develop design criteria that clearly leads to a solution.
4. Generate alternatives, that is, solutions to the problem.
5. Check physical, economic, and financial feasibility.
6. Optimize feasible alternatives.
7. Evaluate optimized alternatives and select best solution. This step requires consideration of the potential problems that could result from the adoption of this system.
8. Implement the solution.
9. Use feedback and control to continuously improve.

When a lack is experienced, a system may already exist. The status quo of the system should be questioned because evolution is inevitable in our ever-changing needs. For example, the space shuttle evolved from the nonreusable rocket booster. It is this facility for innovative design and implementation that makes the system design an ongoing process. In another example, the multistep process of handling data in the accounting and costing systems evolved into a relational database management system where a piece of data is handled only once, reducing the number of processing errors and increasing the efficiency of the system.

The system or "lack–solution" design method is based on the concept that logically related components collectively form subsystems, and linked subsystems combine to create systems. In this way, the system can be regarded as a pyramidal structure. The basic components are synthesized to become part of the system or solution. Figure 2.1 illustrates the pyramid or cascading structure.

Steps 1–4 of the system design process are used in synthesizing a system. Steps 5 and 6 are the steps for analysis. Steps 7 and 8 are for selection, implementation, and modification of the system.

The general idea of the systems approach has been known for years. The successful construction manager has intuitively understood and concentrated on the total system in which he has operated. What is new today is a greatly

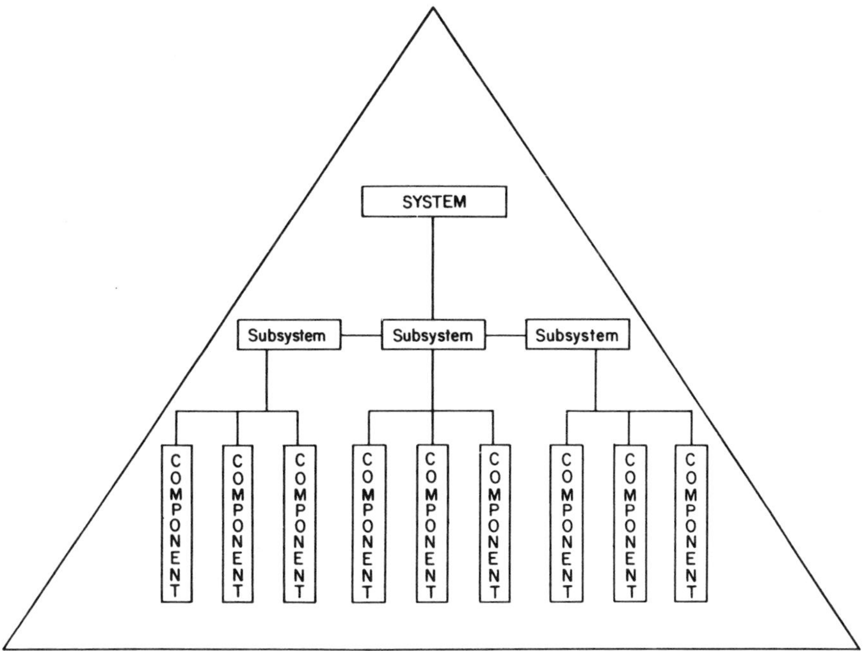

FIGURE 2.1 Pyramidal Structure of a System

increased awareness by professionals at all levels of construction management of the value of systems thinking. Also, with such recently devised tools as networking, simulation, dynamic programming, and computer utilization, the systems approach has become the best method of (1) developing optimum plans for large, complex projects, (2) controlling scope, and (3) administering projects.

The systems approach requires the application of a rational methodology of system design. It demands consideration of a system large enough to evaluate interactions and involves utilization of various tools as dictated by the systems engineering methodology.

Three attributes therefore characterize the systems approach:

1. Emphasizes the importance of the interrelationships that tie the system components into a recognized entity.
2. Utilization of the methodology of system design to develop a project management system.
3. Application of the tools and techniques of system design suitable to the project phase and the needs dictated by project size and complexity. It is evident from the comparison that the methodology for the development and implementation of the project plan is not different from the

methodology for the design and implementation of a system. There is only one major limitation associated with the use of the systems approach in planning projects: the size and complexity of the project must validate the cost of using this rigorous planning technique.

Although it is necessary to go through a similar procedure to develop and select an acceptable project management plan for the design of a system, on certain projects one need not go through all the steps explicitly. Some checks may be made intuitively. Nevertheless, one cannot be sure that a plan is the best, that the scope is adequately controlled, and that project management monitors and controls the task adequately unless all the steps have been taken and survived the tests. Ad hoc and disjointed decisions are not sufficient when the greatest possible level of efficiency in the use of resources has to be achieved. The systems approach must be viewed as a series of logical, interrelated steps that can integrate all the functions to achieve optimum performance. However, on very small projects consisting of standard operations, use of the systems approach may inflict unnecessary cost; a decision about how far to use the systems approach on a certain project depends on the sound judgment of the management.

2.3 SYSTEMS

What do we mean when we speak of systems for a construction project? It is through a system, which is an assemblage of things or parts, that the project manager is able to communicate cross functionally, as we shall see later in this section. A construction project management system consists of a number of subsystems which are put in place to facilitate the execution of the job. The key systems are organization, planning, management information, project control, and techniques and methodologies. These systems support the project team within the project's environment.

An organization is a system of personnel procedures and individuals assembled to perform the work. Organization structures are systems for organizing human efforts and assigning responsibilities. The procedures include personnel policies and all the necessary documentation for hiring, evaluation, and destaffing which, if they did not exist as part of a corporate system, would need to be created for the specific project. Most projects are organized in a matrix format and it is imperative that the functions and responsibilities of each participant be clearly defined.

The planning system consists of all the plans, strategies, goals, and schedules. Planning allocates scarce resources.

The information management system deals with communications and retrieval of information, which provide the intelligence support system for decision making.

The project control system gathers and disseminates data and information on costs, schedules, and technical performance.

The project environment includes these systems, some tools and techniques, and people. Increasingly over the past years and certainly for the future, the make-up of the people is the most important component of any project. The modern PM must consider peoples' values, attitudes, emotions, perceptions, and prejudices. This is the human side of project management.

With this array of available systems it is therefore necessary to carefully plan for the extent of the systems that will be used on a project, that is, which and how much of the systems.

Thus far we have described generic systems, which are the source of more specific project systems. A project uses the corporate systems if they are in place or tailor-made systems are set up for the project. Figure 2.2 illustrates an array of systems which are applicable to most projects. Each is self-explanatory and usually computer based. Additional subsystems can be added, such as the quality control and quality improvement program.

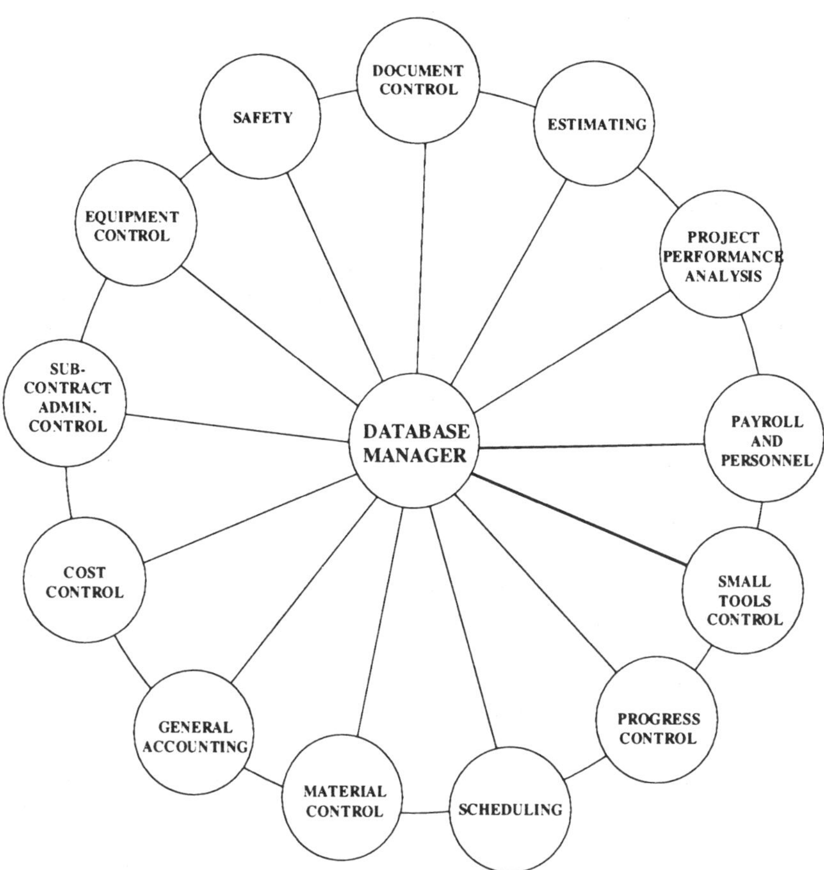

FIGURE 2.2 Project Systems

Systems as shown in Figure 2.2 can process considerable amounts of information. They rely on the computer's ability to perform high-speed calculations or information transfers. More recently a new breed of systems has evolved, namely those classed as knowledge-based expert systems (KBES), which are a subset of artificial intelligence. Expert systems capture the experience and knowledge of experts in a specific area and generate rules for decision making. The result is problem solving methods which copy those of a human expert.

Also on the horizon for construction projects is another problem solving tool called artificial Neural Networks (NN) which is a further extension of expert systems. This technique has the capacity of learning in order to optimize a solution. Artificial intelligence (AI) will provide more advanced delivery systems than have customarily been used in project management.

Two strengths of the most prominent EPC, A/E (Architect/Engineer) companies are their people and their systems. Owners expect that contractors can be operational immediately because proven systems exist and can be utilized for their project. Rarely is there a need to create a new system, although some modification may be required for specific project or owner demands. Computerization has facilitated the creation of systems and the future will bring more relational database systems. These are discussed further in Chapter 24.

2.4. TYPES OF PLANS

Planning is most effective as early in the project as possible. Planning is a process of decision making, and early planning during the C and D stages sets the tone of the project. Figure 2.3 illustrates the effect of timing on the impact of decisions. The impact of a decision made early is usually greater than those made during later stages. It is necessary to make decisions throughout the project, but the earlier decisions have the greatest effect. For example, a decision to use CAD must be made early and that decision has implications throughout the project life. If this decision is made too late, it may not be cost effective.

There are several types of plans. In the C and D stages planning determines the systems to be used, the type of plans to be developed later, and the scope of the physical project. The detailed plans that follow are used primarily during the execution and finish stages (E and F).

Schedules are time plans traditionally associated with the planning process. Other plans include organizational and staffing, procurement, contracting, safety, materials management, total quality management (TQM), communications, and information. The list can be expanded to suit the nature of the project. More specific plans are commissioning, turnover, completions, manufacturing, fabrication, promotional and publicity, and so on. Several of these plans are covered more fully in later sections of this book.

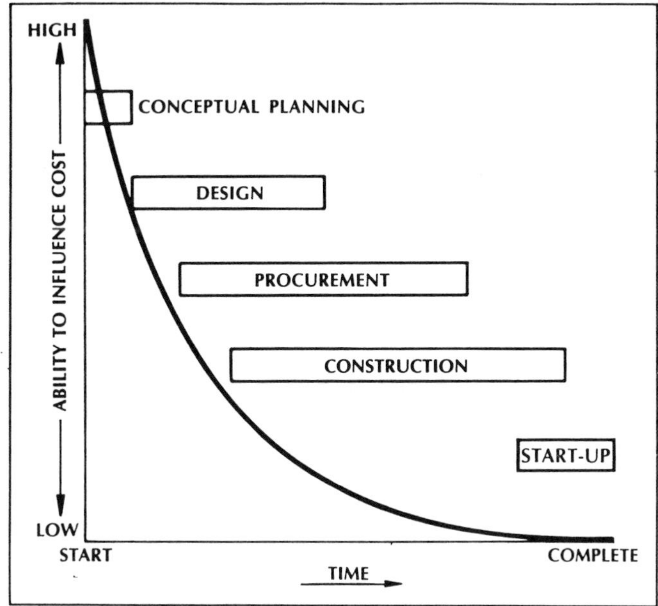

FIGURE 2.3 Relative Ability to Influence Cost. From CII Publication 3-1, Constructability: A Primer, July 1986

2.5. WHY IS PLANNING AND CONTROL NECESSARY?

Today construction is a project-oriented industry. Most work is carried out as a project, which means that facilities to be constructed or objectives to be achieved are defined and that an effort is then made to achieve these within certain time and cost parameters.

One characteristic of projects is their continuous growth in size and complexity as technology advances. Complexity generates the necessity for specialization, and since each specialist has his own jargon, specialization may lead to a breakdown of communications. For example, in a project for the construction of a certain plant, many specialists—structural, construction, mechanical, electrical, heating and ventilating, sanitary, production, and system engineers—may be involved. Each is responsible for a specialized technology. Each has a viewpoint and thinks differently about the project. Some method must be found to facilitate communication so that each person is working toward the same set of project objectives: time, cost, quality, and goodwill. The emphasis that the project manager puts on the project objectives may be different from that of the project team. It is only by using a project plan that the project manager can effectively communicate to the team his or her emphasis on the various project objectives.

Another characteristic of projects is the increasing importance of timely completion. If a company is trying to develop a certain product before a competitor can introduce a similar product into the market, the company must meet its own schedule or it may lose even its present market share. Today engineering design can be completed by modern computers much faster than could be done only a decade ago. Therefore more organizations can quickly place more designs on the market, reducing the life of a design and increasing the necessity to plan and schedule. The need for an early debut to enhance the potential of a design to capture the market makes a sound plan imperative.

In view of present-day resource limitations, optimum use of resources is essential. A project must be efficiently planned to make optimal use of the limited resource. It is thus possible to economize on the present project and to go that much further with other projects in an organization.

The cost of capital is another factor that places a high emphasis on establishing a schedule and working to achieve its targets. No one can afford to lose interest on the money investment in a project while it is unproductive. It is important to make the investment productive as early as possible.

Because of increased outlay, technical complexity, price escalation, and longer time spans, the element of risk in construction projects has increased. Financiers of large projects demand not only technical feasibility reports but also documented schedules and reliable estimates to substantiate the management's ability to execute and control the operations economically. In effect, the financiers require a workable plan.

Shop drawings have to be submitted and approved. Delivery of materials has to be arranged. Many types of material tests by widespread agencies and workmanship inspections by various specialists are required while a technically complex project is in progress. The start of many operations is dependent on the approval of shop drawings, delivery of materials, and completion of certain tests and inspections. Thorough planning of these events can be scheduled so that neither project duration nor project cost increases.

Environment affects a plan. For example, inclement weather may hinder the progress of a project. Similarly, a delay in deliveries of materials may throw a project off schedule. Continuous planning is necessary in these situations. Contingency plans must be developed and solutions must be thought out to predict the consequences of unforeseen exigencies. This is dynamic planning, which implies that the initial plan is amended to incorporate changes and is used to find answers to meet any situation. It is only when a plan exists that modifications can be made and the consequences predicted.

Management must absorb information and make decisions. With the growth in complexity and size of projects, the volume of information has increased tremendously; consequently, in practice it is necessary to manage by exception. A prerequisite for management by exception is the existence of a plan setting the targets against which performance can be measured. When performance matches the standard set by the plan, no exception report to the management is required; only the factors requiring management action are reported. Along with exception reports, accomplishments should be reported also.

The efficiency of a group depends on each individual contributing his share. Planning provides a unity of purpose to the members of a group. The increased efficiency of each individual, which results from this unity, leads to increased production by the group as a whole. Planning is therefore essential for achieving an increase in production.

Another factor that makes planning absolutely essential in projects is the transiency of construction personnel. Individuals familiar with the job often leave before the project is completed. Breaking in a new person is always more costly than working with someone who has knowledge of the job. Basically, a newcomer works much more slowly while he is acquiring the knowledge and experience necessary to work at a proficient rate. The existence of a plan provides continuity that reduces the learning period for a newcomer.

The need for planning can be considered from another point of view. A project is executed only once. Every project manager likes to claim that it was done most efficiently, but there should be evidence to validate this claim. To substantiate a claim, there must not only be a plan, but it must be the best plan possible. If work is done according to the plan, the assertion that the project was done most efficiently can be upheld.

Projects have four common objectives: time, cost, quality, and goodwill. If efficient communications are maintained and the project is built on time, is within cost limits, conforms to the specified quality, and meets the utility, flexibility, service, and aesthetic needs as defined during the design phase, there is no problem maintaining goodwill. Thus goodwill is a variable dependent on the first three variables and the design objectives. Of these, however, the third variable—quality—is controlled by the design engineers through project drawings and specifications. So the only variables amenable to managerial control are time and cost. In this text, cost planning and control is dealt with in a limited sense, that is, as it is affected by planning and control of progress, and not in its broad sense, which would encompass cost reduction, modeling, optimization, and analysis. In physical terms a system is a set of things or parts forming a whole. The time and cost plan comprising a schedule and cost estimates is a system in the physical sense.

Planning for the sake of planning is of no use. The value of a plan lies in its implementation. Progress is measured against the planned targets, and the schedule or cost deviations are corrected. If the corrective action cannot bring the project within the limits, the plan is modified. The project is kept on schedule and within the budget through the control function, which is as important as, if not more important than, the initial plan.

2.6. PLANNING TECHNIQUES

Generally there are a number of things that can be done to improve the likelihood of creating a successful plan. Goals must be understandable and clear. This requires well-documented (i.e., written) objectives. A flexible approach must be maintained; avoid tunnel vision and consider all sensible alternatives. Get the participation of the team members at every level, which includes those above and below. Involving other participants in the planning process will likely result in a better plan. "Two heads are better than one," as the saying goes. Also, a sense of commitment is generated by those contributing because it is "their plan," which is a better strategy for cooperation and motivation than having someone else's plan imposed on team members.

Schedules and estimates must be realistic. Often in the enthusiasm of initiating a plan or an entire project, a plan is formulated that attempts to achieve too much in too little time. Either schedules are missed or cost overruns occur in these scenarios.

Good planning requires that a series of questions be asked. Ask what is to be accomplished and why, how, when, where, and by whom. This somewhat rigid procedure may seem trite, but it is helpful. A procedure that questions to the void (Kepner-Tregoe, 1981) is also effective, which means that the questioning continues on turning over the next stone and the next, as time permits.

An effective technique is the use of prepared checklists which serve as memory joggers and aid in the questioning process because these lists are actually asking questions. Checklists usually categorize the questions into areas such as climatic conditions, geography, systems to be used, personnel, project scope, and many other detailed questions.

2.7 RATIONALE AND STEPS FOR PLANNING METHODS

The planner's primary objective is to develop an instrument that will enable management to exercise control over planning and performance on a project and, by extension, within a multiproject situation. To meet this objective, the planner must consider the project situation in a logical manner and thus form a structured method by which management may expeditiously receive information it requires at various stages of the project in an adequately detailed manner relevant to the required level of management so that decisions may quickly be applied to the project. The method that has evolved requires the use of a number of steps which have been developed for the fulfillment of various objectives:

1. Identify all project elements, such as structure, foundation, electrical, and mechanical.
2. Identify all agencies participating in the project. Agencies includes contractors and owners representative of any contracting party involved in the project.

3. Identify responsibilities for each project element within each participating agency such as engineering, procurement, and construction.
4. Identify key points within the project elements such as start and completion of substructure, structure, masonry, electrical, piping, testing, and so on.
5. Identify separate projects or subprojects between key points.
6. Identify interfaces between projects or subprojects such as concrete work and embedded parts.
7. Identify each event, key point, and interface for which information is required such as completion of concrete pump base and delivery of pump.
8. Identify highest responsibility levels requiring information about each event, key point, or interface. This will be dealt with in more detail in the section on the information plan.

The development of the main steps of planning is considered in the following sections.

2.8 WORK BREAKDOWN STRUCTURE (WBS)

The most fundamental technique used for planning and managing a project is to break down the scope of work into manageable pieces. This breakdown begins early in the project and is from the initial scope statement in a top-down fashion, much like beginning at the top of a pyramid and expanding downward. The owner's project team, including designers, begins this process and develops the initial overall work breakdown.

Later other participants such as contractors and subcontractors develop their own work breakdown suited specifically to their needs. The perspective that follows is that of a single project management team responsible for the entire plan. However, each contractor has a specific plan and work breakdown which represents a subset of the overall or master plan.

As the design progresses, the planner must define the different parts of the project plan. For instance, when the preliminary design is in progress, the general divisions of the work may be apparent. Later, as more detail is known, each of these divisions may be broken down into its respective components. This arrangement into divisions, subdivisions, and further subdivisions is known as the Work Breakdown Structure (WBS). It is the means of dividing a large multiproject program into its component projects, or a complex project into its components, which at the lowest level are called work packages. With the help of a work breakdown structure the planner, instead of trying to grapple with the whole, can tackle one clearly defined part at a time followed by others sequentially, steadily, and comprehensively.

The division of the project into work packages is very useful to the planner in developing his network. It is much easier to plan one work package at a time and to piece the packages together into a project network than it is to develop a complete project network without the WBS. A result of the first attempt at a WBS for a hydropower project is shown in Figure 2.4.

When more information becomes available during the detail design, the planner can add more levels to the WBS. The lowest-level divisions are the work packages. Consider the subdivision "townsite" in Figure 2.5. As the WBS is developed further, this subdivision may be divided as shown in Figure 2.5. The other divisions will be expanded accordingly. The civil work performed on type A residences forms a work package, as do the mechanical and electrical work. These are the end item subdivisions for which either a contractor or a department of the organization is responsible.

There are many ways to develop a WBS and work packages. Subdivisions can be geographic regions, construction site areas, process components, building elements, engineering and process systems, trades, or departments. One or several work packages may form a contract package. Contract packages may depend on bonding limitations, contractor capacity, expertise, and commissioning requirements.

Who is responsible for the WBS? The owner develops his WBS. If, as in Figure 2.4, the owner lets a contract separately for each of the eleven work packages, then that is the owner's WBS. Each contractor will then have a separate WBS. The divisions and subdivisions of the contractor's WBS would include all the subcontracts as well as in-house work by his or her various departments. If a cost-plus contract environment is chosen, whereby the owner and contractor draw the project schedule and cost information from the same database, a scheduling consultant working for the owner may combine both the owner's and contractor's WBSs into the resulting "project WBS" which will show greater detail. Also, if nec-

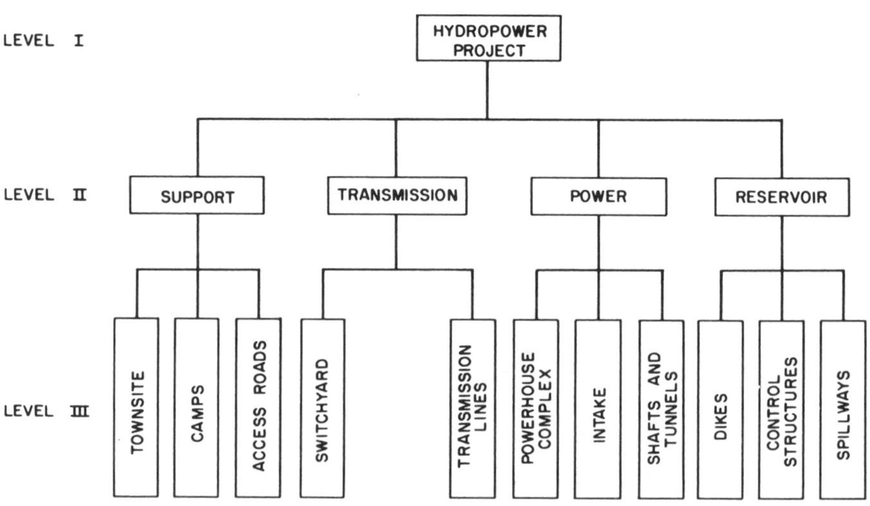

FIGURE 2.4

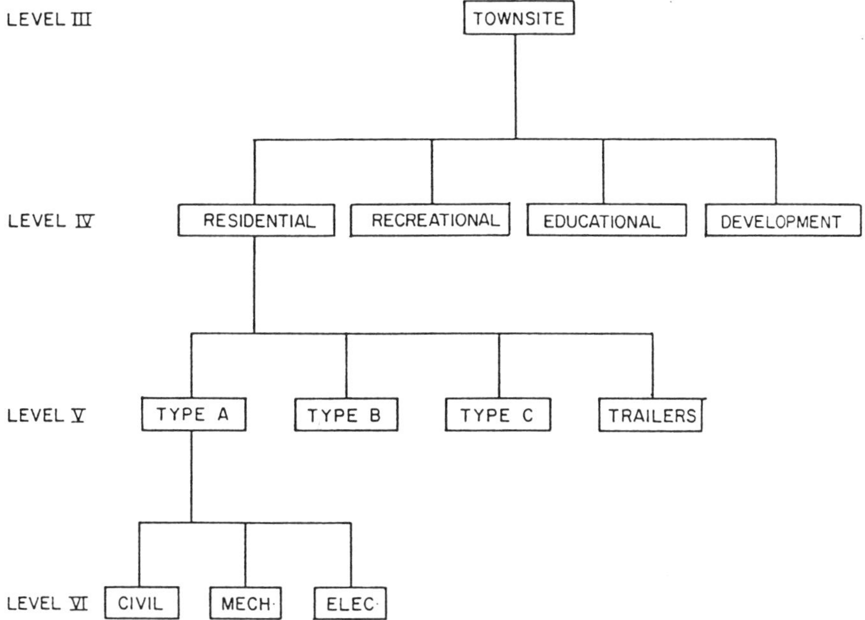

FIGURE 2.5 Example of Work Breakdown Structure-Hydropower Project

essary, WBSs from several projects may be combined into a multiproject WBS.

A work package is also a cost center. Each has a separate estimate budget and expenditure account and is assigned a charge number. A charge number is referred to as a code of accounts, which is discussed in Chapter 12. When a separate WBS is developed for cost control (of which the lowest-level elements are the various cost accounts), it can be integrated with the schedule and cost information of the project WBS. Figure 2.6 illustrates the concept of overlaying work packages. As a result, schedule and cost information can be derived both by the contractor and owner from the same database for all the cost centers of the project WBS. This is a very useful concept and is developed further in Chapter 12.

The work package concept is especially useful for shutdown or turnaround projects in process plants and refineries. In this type of work, the shutdown is planned in very minute detail, and often to the hour. The reason is that when the process has been halted the revenue or income of the owner is negatively affected and therefore the plants are usually shut down for as short a period of time as necessary. Thus it is paramount that when the shutdown occurs, the repair or revision is carried out expeditiously and without delay. To ensure that the package of work can be completed in an uninterrupted manner, the planner must ensure that all of the resources are available, that is, the drawings, materials, equipment, and labor. Only then is the package released so that the work can be started.

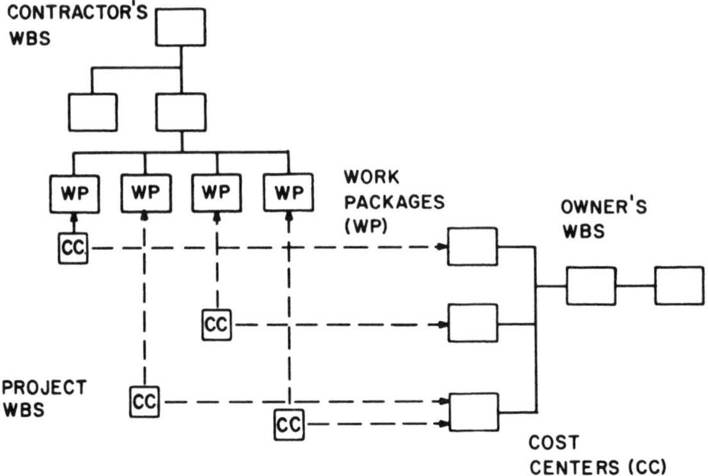

FIGURE 2.6 Overlay of Owner's and Contractor's Work Packages

This package concept works well in all projects because each package is treated as a subproject or a subnet (see Chapter 8). If a package is large, it may delay the project if it is necessary to wait until all of the resources are available. To overcome this problem the package can be broken down further into smaller packages.

The WBS has another important function; it provides the structural framework for managerial control. Division into work packages meets both the planning and control needs of the organization. It is therefore necessary to develop simultaneously an organizational tree and to relate the organizational units to work packages in order to determine the suitability of the WBS.

2.9 ORGANIZATIONAL RESPONSIBILITY CHART (ORC)

The prime responsibility for each work package is assigned to an individual in the project management team. From an owner's or managing contractor's perspective, a project engineer, supervisor, coordinator, or contract manager is the first line of supervision responsible for an assigned package, and often an individual is assigned more than one package.

Depending on size, complexity, and timing, the reason for this responsibility chart is to establish the communication and reporting links as well as the authority structure within the project. Figure 2.7 graphically displays an overlay of the responsibility chart and WBS. Figure 2.7 shows another term used for ORC ie Organization Analysis Table (OAT).

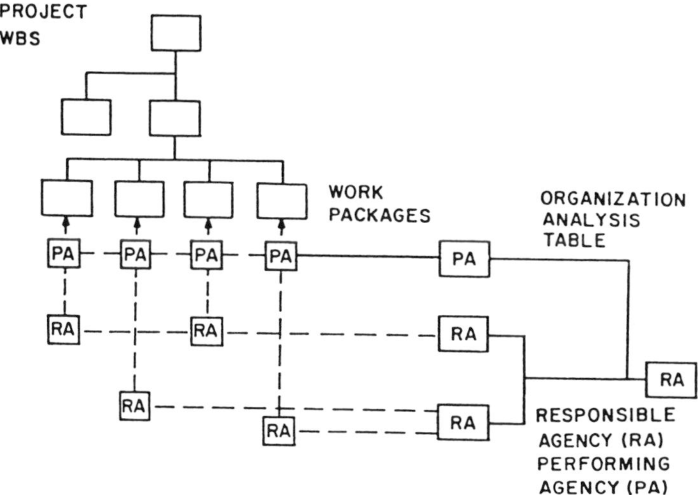

FIGURE 2.7 WBS/Responsibility Structure

The term Responsible Agency (RA) means the first line supervision. Performing Agency (PA) is the contractor's individual responsible for that particular package.

If the performing agency is a contractor, the work package is a contract package, which is a convenient method for establishing each work package. Whether it is called a work package or a contract package, it must have a performing agency.

Responsibility/WBS charts such as that shown in Figure 2.7 are seldom drawn. This figure is included mainly to illustrate the process. It is usually a simple task to allocate the responsibilities for work packages.

It is prudent for a planning engineer to utilize the knowledge and expertise of those individuals responsible for each work package. Each work package has its own schedule which should be produced by or have input from those closest to the work. The planning engineer should elicit input from several levels of supervision, including the foreman level. Involving all key players makes the schedule "their" schedule and this assures their "buy-in," which is in contrast to having a schedule imposed from someone above in the organizational hierarchy. A "buy-in" is the result of participation and contribution and therefore ownership by the key players.

PROBLEMS

2.1 It is your assignment to set up a project management system to complete a new water treatment plant to supply 16,000 USGPM to a small town and large smelter complex. Follow steps 1–8 in Section 2.2 and document the assumptions and decisions.

2.2 Develop a WBS and ORC for a project with which you are familiar. Prepare a list of subnets for the project. (For a discussion on subnets, refer to Chapter 8.)

2.3 Develop a WBS for a roadway project which consists of paved roadway, a bridge, and culverts.

CHAPTER 3

PROJECT ORGANIZATION

A very important function of a project manager is to establish an organization for the execution of the project. Once a plan for the organization is established, staffing is one of the first tasks that is required. The project manager needs specific personnel to execute the various tasks and functions because he must get the work done through others. A team must be assembled that will work in harmony and efficiently. Especially on large projects, considerable thought and effort is required to achieve the right chemistry between the project staff members. There are no established formulas to achieve successful staffing of a project. Casualties and ruined careers often result from staffing failures. A fundamental requirement for the selection of key personnel is that the project manager thoroughly know the candidates and their capabilities.

In this text we present several network-based techniques for project time and cost control. The extent to which these techniques can be applied depends on the size and type of project. The structure of the organization that implements project controls depends on the various project characteristics as well. A difficult task in implementing a project is to coordinate the work of project personnel who are drawn from several disciplines, departments, and even organizations. There are many forms the organization can take to facilitate this coordination, from functional at one extreme to project type at the other. There is also the concept of partnering, which is not an organizational form but rather a method for improving project execution. These techniques are discussed in this chapter. Strategic alliances, which have similar characteristics to partnering, are not discussed.

A new view of organizational structure is emerging—the flat and lean organization which is formed around processes rather than tasks. The trend appears to be away from the functional organization which traditionally has a hierarchical structure based on tasks. Project management is a process, and it appears to be inherently well suited to new organizational structures. Construction organizations tend to have a horizontal structure, mainly because of the method of contracting involves the use of many subtrade contractors.

3.1 PROJECT CHARACTERISTICS

Determining the type of organization that will most effectively meet the needs of a project requires that the characteristics of the project be considered first. The most important characteristics, as shown in Figure 3.1, are:

1. *Objective.* What is the project designed to achieve? The organization must enhance the odds that the project will meet its original objectives without compromise.
2. *Schedule.* The duration of the project and its target dates must be met. The organization must work efficiently enough that its functioning will not hinder the schedule.
3. *Complexity.* The technological requirements determine to a large extent what sort of organization is compatible. A highly specialized and monitored area such as nuclear power would warrant a much more sophisticated organization than a multimillion dollar office complex project.
4. *Size and Nature of the Task.* A project involving thousands of workers and several years duration will have an inherently more complex organizational structure than a six-month, small-scale undertaking.
5. *Resources Required.* Every project requires unique materials and individuals who form the organization to carry it to completion. Even repetitive projects will have variance in these factors.
6. *Information and Control Systems.* According to the peculiarities of the project, each organization will produce suitable information to

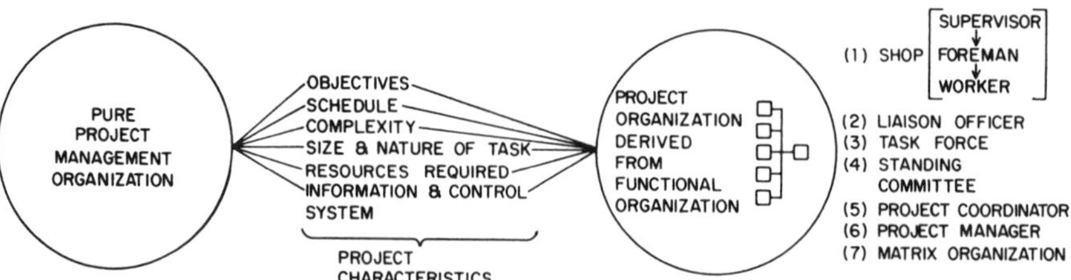

FIGURE 3.1 Project Organization

control effectively the project duration and cost in different ways for the various management personnel.

Depending on the nature of the decisions to be made, management of a project and the organization required to implement it can take different forms. Jobs in a shop may be completed successfully by the engineer in charge and the shop foreman. Small projects may be handled directly by managers. This is project management in its simplest form. The problem, the information, and capable decision makers are all at the same level. When the project requires a large volume of interaction and information flow between any two departments, it is advantageous to appoint liaison personnel to handle the specialized communication between the departments involved. These individuals represent a direct or "horizontal" communication link between departments and avoid the longer, more arduous vertical lines through upper management.

When the project involves several departments and their associated problems and resources, the liaison personnel function is no longer adequate. In such cases task forces are established with representatives from all the involved departments to handle problems on the horizontal basis. The task force lasts only as long as the project; some members are permanent for this duration, while others are called on only for certain periods of the project or as required.

After it is found that these horizontal lines of communication set up by the task force are used over and over again on all of the company's projects, and when the need for these decision makers is experienced repeatedly, a permanent group called a standing committee may be established. The leader of this group is usually the head of the department most involved in or affected by the decision.

When a consensus cannot be reached and the leader of the standing committee is unable to make a decision, the information can be passed up to the executive level for a decision. Since these "failures" of the standing committee are time consuming and expensive, another decision maker separate from the executives is necessary.

In most organizations this role of full-time leader is filled by the project coordinator, project leader, or project manager. The different titles used reflect the importance of the position, which is largely dependent on the project size or complexity. Because of the importance of a project, the project manager may be at the same rank as a department manager and report directly to a division manager. Figure 3.2 shows this relationship. This individual must have enough power to influence the decision process. The project coordinator reports to the executive level and coordinates project work in cooperation with the functional department heads. This cooperation can be achieved only if the project coordinators can effectively use their powers of persuasion rather than depend on their limited authority.

Under the various organizational arrangements for project execution described so far, project work is controlled by a work order system within

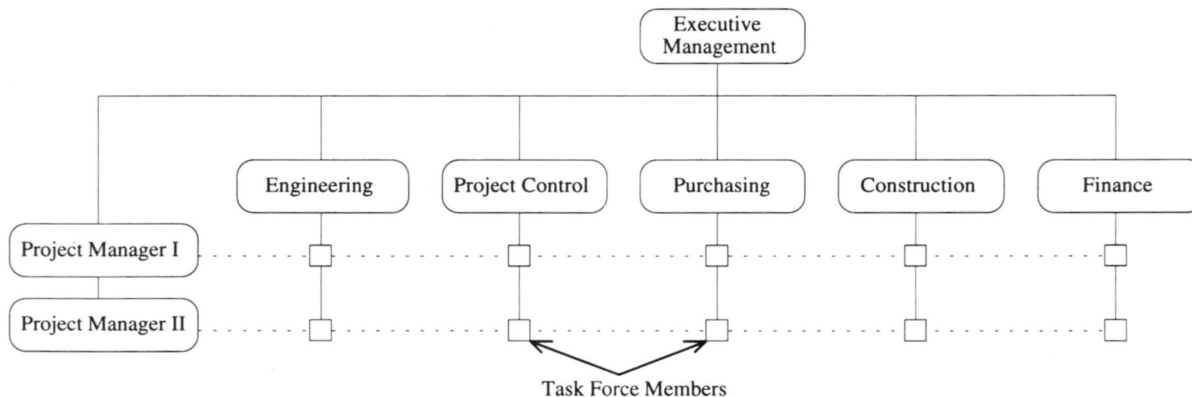

FIGURE 3.2 Functional Matrix Management Organization

the individual functional departments. Problems arise when the work order accounts are not closed on a timely basis. Departments responsible for the work orders may keep charging expenses until 100% of the allotted expenditure for the work order is used up, resulting in distortion of reported figures and consequential loss of control. It is to avoid such situations that a project manager is appointed and given control over the project budget, which is considered a standard tool by most organizations for project cost control. The project manager is given more formal power to coordinate most effectively the interdepartmental decisions. This individual controls the budgets, initiates the budgeting cycle, and buys resources from the functional groups. Even when the project manager has no direct subordinates, he or she is able to exert control through the influential nature of the position.

Whenever the project manager has difficulty in eliciting cooperation from the functional department heads and therefore in integrating the multifarious tasks, there is another possible arrangement. This is the matrix organization, which establishes the "two bosses" relationship between the project manager and the functional manager. Personnel assigned to the project are responsible to both. They receive direction and support from their functional manager and project direction from the project manager. The project manager tells them what to do when and the functional manager tells them how and sets the standards for performance. Usually the project budget is controlled by the project manager, but is structured by the expense items pertaining to the functional departments. In some organizations, each department carries a provision in its budget for a portion of the project budget. It is then easier for the project manager to elicit the cooperation of the functional department heads because of his or her administrative control over a part of their budgets. In a matrix organization the budget is traditionally controlled by the project manager.

3.2 ORGANIZATION AND STAFFING

As described in the preceding section, several factors influence the choice of organizational format. While considering these, another overall decision must be made. Is the new organization to be set up as a derivative of an existing organization, or does the project require hiring new personnel to build up a pure project management type of organization? Figure 3.1 illustrates the difference in the approaches.

3.2.1 Pure Project Management Organization

First, consider the pure project management organization. When an owner has to build one very large, complex project he or she may decide to set up a pure project management organization. This organization is gradually built up and can be gradually disbanded as the project nears completion. This sort of organization is a "one-time" deal; it is set up solely for the purpose of handling that one specific project. Of course there are problems. This

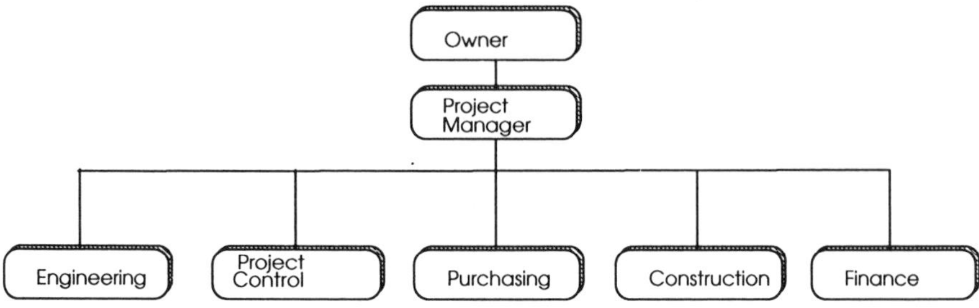

FIGURE 3.3 Pure Project Management Organization

type of organization, because of its exclusivity, is expensive to arrange and run. It requires that specialists be drawn into the team for its duration. Generally, these people are the best in their respective fields and are willing to interrupt their regular jobs only for lucrative salaries. Another difficulty is that since these projects do not offer continued employment, there may be a lack of loyalty toward their employer or the project itself. Figure 3.3 shows this form of organization.

Pure project management is often mandated by the client or customer. Large companies, as well as the government, often prefer this type of organization, especially if it mirrors their own organization. Construction companies will second suitable staff to the pure project organization out of their various functional departments, namely, estimating, data processing, procurement, scheduling, cost control, and construction. Because the seconded staff reports only to the project manager, it remains a pure project organization. This reflected organization makes it easy for members of the owner's organization to communicate with their equivalents in the pure project management organization.

3.2.2 Functional Project Management Organization

The opposite end of the spectrum of organizational form is represented by the existing functional organization, shown on the right-hand side of Figure 3.1, from which many different types of organizations can be derived to manage the project. Each of these seven organizational forms was described briefly in the preceding section and each is tied to the already existing functional organization.

Whereas in the pure project management organization the project manager has full control over the project, neither of the seven types of organizational arrangements derived from the functional organization provides full and ultimate control to any individual. The executive level of management has final say on any problem that cannot be resolved in the lower levels. Let us discuss in some detail the last two organizational arrangements, numbers 6 and 7 of Figure 3.1.

3.2.2.1 Project Manager The project manager occupies a lesser position here than in the pure project management organization. The project manager in a functional organization does have position power, but project participants do not report directly to this individual. Even though the project manager has some power, at times it can be insufficient to get the job done without intervention from higher levels. The control for completion of each function remains with the functional department and the role of the project manager is to coordinate the work, usually with the direct support of the project controls group which provides cost reports and schedules. Figure 3.2 shows this relationship.

3.2.2.2 Matrix Organization The matrix organization provides a blend of functional and project-oriented control. The problem with this approach, however, is that the two organizations remain completely separate. This means that every project participant answers to two bosses. The project manager in the matrix organization has the same level of authority as the functional manager. Figure 3.2 also illustrates the matrix organization, with the functional responsibility indicated by the solid vertical lines and project responsibility shown by the horizontal dashed lines.

The matrix organization has several advantages and disadvantages. Many of these comments also apply to the pure project organization. The pure project organization is a special task force which has been established for a sole purpose. The matrix organization is similar except that it exists within an existing organization whose main purpose is other than the construction of a project, or it may be part of a contracting organization whose main purpose is the execution of projects for owners who are unable to manage their own projects.

The dual reporting of personnel results in a mixed allegiance to the project or functional department. The result is a shifting of allegiance to suit the situation. The lack of permanency of the project organization also results in shifting allegiances, and can be stressful on the individual member of the project.

However, the advantages outweigh the disadvantages. There is a dedication of purpose to the task and accountability is more readily determined. The matrix, and especially the pure project organization, can respond more rapidly to change, conflicts, or other project needs. The project management is allowed better control over costs and schedules. In matrix organizations, individuals can identify with a project, and this can be a motivating influence.

3.3 PROJECT MANAGER'S AUTHORITY

What is the project manager's domain? Figures 3.2 and 3.3 show, respectively, the realms of the project manager in the pure project management organization and the matrix organization.

On any large project, however it is managed, there is an inherent requirement that the resources of various separate departments be pulled together to work harmoniously in the interest of the project. To ensure that project work is performed as smoothly as possible, it is essential that top management's support be continuous and visible. Functional or divisional managers must be convinced that only by cooperating with their respective project managers will the work be completed with relatively little friction. Also, these individuals must be assured that executive support for the project will not threaten their authority or lessen their realm of influence. The project manager's action of gaining support from top management, functional division managers, and all other project personnel is the key to launching a successful team effort.

Top management, before throwing its weight behind the project manager, must be firm on the portion of its own authority it wants to delegate and the extent to which it wants to limit the power of the project manager. In the pure project management organization the top management must expect to give the project manager full, unquestioned authority, otherwise the project manager will be unable to perform the job he or she was hired to do. In the matrix organization, however, top management has much more leeway in the power it assigns to the project manager. Since in a matrix organization management is dealing with an already established group of individuals, it must decide the extent to which functional managers will work well without feeling their power is lessened by the project manager.

In any case, when top management appoints a project manager to be accountable for the success or failure of the project, it must give this individual enough authority to be able to carry out the project. A project manager lacking the power befitting the position is doomed to produce a failure and to become a scapegoat for spineless top management.

As illustrated in Figure 3.3, the heads of each section in the organization report to the project manager in the pure project management organization. The pure project manager must deal with the problems associated with each section. Moreover, the project manager must see to it that each section does its part of the project without unduly hindering the work of any other section. The project manager should have full control over every aspect, giving him the authority and confidence to carry out the job without direct intervention by the owner. There is minimal conflict in pure project management.

3.4 CONFLICT MANAGEMENT

The project manager in a matrix organization, as shown in Figure 3.2, deals with three groups representing different interests. First, the project manager reports to top management. As this person generally has less power than his or her counterpart in the pure project management organization, there must be continuous and visible support of the top management to provide him or her with the authority required to get the job done. The project manager's second interface is with the functional department managers. Though their goals are different from the project manager's, they are all working toward basically the same end—a successful project. The project manager's relationship with these functional managers can be a great source of conflict. Conversely, cordial relations between the functional manager and project manager can make for an easier and less stressful experience.

The individuals put in perhaps the most awkward position by this matrix organization are the project personnel. This third group must deal with not one, but two bosses in positions of balanced, but differing power. Usually the functional department manager has the boss–worker relationship, while the project manager has the position power. The project manager and functional managers must make clear down through the ranks to what degree each individual is responsible to each manager. Their dealings should be separate and distinct, with no confusing overlap or vagueness so they do not cause conflict but, instead, encourage a spirit of cooperation.

3.5 COORDINATION

The function of orchestrating all the skills so that everyone cooperates to carry out the project at hand most efficiently is known as coordination. To ease the coordination of a project (with either type of organization), the project manager can take several key steps to give the project a successful start.

3.5.1 Project Implementation Plan

The project implementation plan, as discussed in Chapter 14, sets out the objectives, defines the project scope, provides the schedule and budget targets, and describes the organization required to implement the project in its various phases. It delineates the involvement of the various functional departments in the organization and the many agencies external to the organization whose work has to be coordinated.

3.5.2 Schedule and Cost Planning

A detailed project work breakdown structure (WBS, see Chapter 2) is next provided to enable development of a schedule and a cost breakdown structure for calculating accurate estimates and budgets for work packages. In developing the project WBS, the project schedule, the cost breakdown structure, and the estimate or budget, all divisions of the project must lend their expertise. Hence this step is also a good way for team members to become accustomed to working together and to identify all the interfaces that will need coordination.

3.5.3 Building the Project Organization

Developing the project organization is the next consideration. In pure project management the project manager has full control in selecting and hiring project personnel. In a matrix organization, however, the project manager has to work within an organization already long established. In such a case the project manager has to trust the judgment of the functional manager as to which individuals in the various departments are best suited to the job at hand.

In either case, once personnel are selected, a formal document should be drafted which outlines in detail the status, authority, and responsibility of all involved in the project, including the functional department and project personnel as well as others external to the organization. Such a document prevents conflicts regarding overstepped authority and makes everyone aware of individual responsibilities, thus facilitating coordination.

3.5.4 Procedures

A key document for coordinating the daily business of many agencies is the Project Procedures Guide. It should cover the following areas:

1. Budget approval
2. Funds appropriation and authorization procedures
3. Purchase requisition authorization procedure
4. Contract award procedure
5. Change order authorization procedure
6. Time-keeping methods
7. Priority determination methods
8. Methods for resolving priority and conflict problems
9. Accounting and audit requirements and procedures
10. Required computer reports and their distribution
11. Formal reporting and review procedures
12. Informal reporting and review procedures
13. Procedure for drawing and vendor data control.

The last two procedures require some further explanation.

Formal reporting procedures for periodically relaying descriptive and computer-generated progress and cost information in a specified format are usually laid down by the owner, top management, or customer, and the project manager has no alternative but to comply. Project managers also usually implement a system of their own to report regularly project status and significant events, thus enabling them to exercise effective control over the project. Since both types of reports are derived from the same database, this dichotomy does not present too much of a problem, although it must be minimized in the interest of economy.

Informal reports and reviews are not regularly scheduled, and may have varying formats. They may be casual verbal exchanges or brief interoffice memos. The project manager must do the utmost to encourage these types of reports so that he or she can stay close to the action and be aware of what is happening at all times. In this way the project manager will be cognizant of potential problem areas and never be caught unprepared.

3.5.5 Drawing and Vendor Data Control

An often overlooked potential problem is that of coordinating the flow of information and documents within a project. Chapter 19 deals with the flow of information. This section covers drawings, specifications, and equipment data sheets.

A critical function is the coordination of drawings between disciplines. Some drawings have input from more than one discipline and therefore require information for each discipline. This requires timely transmittal of information between disciplines, and oftentimes with suppliers. Approvals and signatures are required, and time will be lost unless the flow of documents is properly coordinated.

Information is often required from vendors of equipment. Certified drawings are necessary to locate anchor bolts, to determine sizes of foundations, and other such requirements. Serious delays may result if the delivery dates for vendor data are not specified in the purchase order. This information must reach the appropriate person in a timely manner.

On some projects the responsibility for the coordination of information and drawing flow is assigned to specific individuals.

3.5.6 Project Headquarters

To develop a sense of loyalty and dedication to the project, project personnel should be housed in one central location, preferably in proximity to the project. The existence of such an office reduces, for the project manager, conflicts arising out of dual responsibility.

An identifiable separate geographic location for the project task force also improves communications within the project and with the client.

3.6 OVERCOMING RESISTANCE TO PLAN IMPLEMENTATION

As stated in Chapter 14, it is necessary to have a plan to implement the project plan, as well as an organized group of individuals for implementation. Yet the project manager's intentions to implement the plan and the existence of a group organized as a project services control department do not guarantee that the project personnel will enthusiastically follow the plan. A project services control department is responsible for the costing and scheduling functions, and for reporting on these matters. As a rule, implementation of a plan meets resistance, particularly in a weak matrix organization. In a sound plan, objections to its implementation are anticipated and means to overcome them are included. The causes of resistance and means to overcome them are now considered.

One environmental factor that pervasively affects the ease with which a plan is implemented is an understanding among the personnel of the need for the plan and confidence in its effectiveness. Without the realization that the plan is needed, the effort required to make the plan work may not be forthcoming.

There is some natural unfavorable reaction to the discipline needed to work with the network plans from those who are unfamiliar with them. Foremen and supervisors, who are unfamiliar with or who do not believe in project control methods, believe that the allocation of costs to wrong items of work is not serious because, it may be argued, it is all part of the total job cost. Although it is easy to cheat a schedule and cost control system, for efficient control bad news should be reported objectively as well as good news. One useful method of dealing with such reactions is an in-house indoctrination course, which replaces unfamiliarity with confidence. This is the education approach.

The schedule and cost control plan is not usually a popular subject with construction personnel other than scheduling and cost engineers. For every member of an organization who realizes its value, there will probably be several others who do not—some because they suspect that it is a waste of time and money and will only lead to more paper work and larger overhead and others because they do not want the additional responsibility.

A systematic implementation of a schedule and cost plan undoubtedly creates additional work, costs money, and must be done by technical staff. Some construction personnel fight it because of the demands it makes on their time. They can be won over if the plan is simple and easily adaptable to the needs of the organization, and the benefits of planning are evident.

Sometimes project personnel resist a plan because they consider it a constraint on their operational freedom. They would prefer to work without a commitment of time for which they will be held responsible and would rather not have someone knowing as much as they do about the costs of the jobs they are supervising. This resistance can be overcome in several ways. First, it is necessary to evolve a plan to which each participant has a commitment. This is achieved by developing a plan of the work including schedule, methods, and budget for which that individual is responsible. These plans are fitted into the master plan. It is also possible that, because of unforeseen turnover, the association of those who are initially involved in planning ceases. The new personnel are requested to review the existing plan within their jurisdiction, and any innovative ideas proposed by them are considered. One of the objectives of the master plan is that it should always be an integrated plan for all persons responsible for the project, giving each a chance to put in his or her own method of performing an activity. Another method is to convince members of the project team that they have more freedom since management concerns itself with only about 20%* of the activities, which are the critical ones in a project. They are free to manipulate all other activities on which there is float, so long as these activities do not become critical.

It may also be useful on some projects to restrict the rigorous methods of network planning, including feasibility analysis and optimization, to the level of planners only. For other personnel networks can be translated into critical path method (CPM) bar charts, which are not resisted to the same extent as networks. This arrangement has all the advantages of rigorous planning, as in networks, combined with the simplicity and familiarity of bar charts.

Despite all efforts, some contractors are reluctant to follow a plan. They believe in doing things by the "seat of their pants." They accept network planning because the cost of networking is always borne by the owner, whether he or she pays for the planning consultant directly or the contractor makes a provision for the consultant's fee in the bid. No one except the consultant stands to benefit financially in such circumstances. To ensure that the owner and the contractor benefit from the use of the network plan, it is essential that the contractor's progress payments be linked with the network. By so doing, the contractor is encouraged to make his or her progress claims on the basis of the activities completed as well as the activities in progress. This clearly can increase the contractor's commitment to network plans. Having obtained a price breakdown by work items, the costs can be distributed over activities. Progress payments to the contractor become payable subject to performance over the activities of the plan. Consequently, the contractor's commitment to the plan is enhanced. In the case of in-house work, cash authorizations against departmental budgets can also be made dependent on the progress of activities, thus providing motivation for following the plan.

Another means of encouraging project personnel to use systematic project control techniques is the continued and visible support of management. If the project manager is insistent, the individuals reporting to him do not take long to get the message. Nevertheless, such insistence fails to win willing cooperation. The project manager has to demonstrate his commitment by following the schedules, particularly for the operations that are his responsibility. This means decisions have to be made faster and without vacillation, approvals have to be arranged, and resources have to be obtained as scheduled. Such a leadership example can break down resistance and create an environment of mutual confidence. This is possible if the project manager believes in the plan and ardently strives for its achievement for the success of the project and organization.

*Pareto's law—see glossary.

3.7 WINNING AND SUSTAINING THE ENTHUSIASTIC SUPPORT OF PROJECT PERSONNEL

Regardless of the size or type of a project organization, the important point to keep in mind is that it is composed of individuals, each of whom has a part to contribute toward completion of the project. Of course individual performance varies. Nevertheless, these individuals are a resource and, as such, must be utilized to the fullest if the organization is to realize the maximum return on its investment.

Therefore let us begin by examining the project participant, and how management can motivate this individual to contribute the most productive output he or she has to offer.

Project management influences project team members' attitudes and becomes a major element in their motivation, determining to a large degree how much will be accomplished. A worker who is positively motivated and enthusiastic about what he or she is doing will put effort into it. This person will work longer and harder, resulting in increased production. The quality of work will improve, resulting in time, cost, and labor savings.

Team members who are motivated and working toward a common goal, will be more amenable and cooperative when it comes to working with others. Team discipline and achievement should greatly improve as interpersonal problems and conflicts are alleviated or resolved.

Much of this teamwork is dependent on the project manager. The project manager must ensure that adequate horizontal and vertical communications flow throughout the organization not only to promote maximum coordination with and between disciplines but also to spot possible interpersonal problems and to resolve them as quickly as possible.

All project team members have certain basic needs that must be at least partially satisfied if any productive work is to be expected. Among these are physiological needs, safety or security needs, social needs, ego needs, and self-fulfillment needs. These needs have been defined as follows: (1) physiological—food, shelter, rest, and exercise; (2) safety or security—protection from danger, threat, and deprivation; (3) social—belonging, association, acceptance, and giving and receiving of friendship; (4) ego—self-esteem, self-confidence, independence, pride, competence, reputation, status, recognition, respect, and appreciation; and (5) self-fulfillment—realization of one's own potential for continued self-development.

Since the days of the industrial revolution most of the basic needs, namely, physiological and safety needs, have been satisfied. Hence, if management attempts to realize its objectives by integrating the personal goals of workers with the project goals, then it must appeal to the secondary needs, the social, ego, and self-fulfillment needs.

3.7.1 Social Needs

Among the elements that comprise job satisfaction, interpersonal relationships between team members ranks high. Working with compatible individuals is important, as it allows personnel to cooperate and work productively. Management may also encourage leisure-time activities for team members to reinforce interpersonal relationships and develop natural trust. It is a worthwhile effort to build trust and confidence among the members of the team. To further enhance the team spirit, workers should be allowed to have some input in choosing their co-workers. In this way friends or people who recognize that they work well together can be assigned to the same activities. The project manager should also keep an eye on teams that have above-average productivity and initiative so that he or she can make use of the same group on other projects. The project manager should also find out why these are exemplar performers and encourage others to learn from the above average performers.

3.7.2 Ego Needs

For individuals' ego needs to be satisfied they must be given adequate responsibility and the chance to make decisions that affect them. Every team member must be able to claim some achievement or task as "his or her own" in order for appreciation, advancement, and fulfillment to be effective. Recognition, such as a mention in the company newsletter, can heighten the individual's sense of pride and responsibility as well.

Some organizations set up incentive programs whereby the individual or team is rewarded if it surpasses the set standards or beats a deadline. Functional managers, especially, in the matrix organization, can do much good by taking the time to attend to each of their divisions' members personally. This sort of one-on-one recognition enhances the worker's sense of belonging and importance. Praise for good work by the project manager in the presence of the individual's functional manager inspires the individual to produce consistent, high-quality workmanship.

3.7.3 Self-Fulfillment Needs

Personal fulfillment is basically derived from accomplishment, productive output, and high-quality work. Management's role is to provide a steady flow of work by means of good scheduling, planning, and coordinating. It must also anticipate and solve any problems, labor disputes, change orders, and other dilemmas so that there will be fewer reasons for the individual to withhold his or her willing contribution. An added benefit is that the project team members will be motivated to work more productively when they see that others depend the timely completion of their work for which they are responsible. Also management must be perceived as doing its best in coordinating all work so that the satisfaction that the workers derive from a job well done is not thwarted by delayed or late deliveries.

3.7.4 The Project Manager's Role

The project manager is ultimately responsible for the productivity of the people in the project team; it is therefore the project manager's job to maintain cohesion and cooperation among all those involved on the project. The project manager must be a leader, one who can inspire and motivate people who have ties both to the project as well as to the functional organization. Honesty and integrity are desirable qualities since the project manager is entrusted with the delegated authority of general management. Good communication skills are essential since it is the project manager's job to communicate upward with the client and the management, and downward with the key functional managers and technical professionals assigned to the project. An understanding of the basic managerial skills of planning, staffing, directing, coordinating, and controlling is required and the project manager must be alert, tough, quick to make decisions, energetic, imaginative, versatile, and genuinely flexible.

Since project managers are dependent on their capacity to get things done through other people, they must hone their "people" skills. They must take extra care that they are not perceived as procrastinators, but as leaders who quickly assimilate conflicting viewpoints, weigh pros and cons, and make timely, well-considered decisions which they seldom have to retract. They must maintain a good humor in dealing with the project personnel. By acting this way project managers enhance not only the chances for the project's and their organization's success but also the chances of their own success.

3.8 PARTNERING

The construction industry has been forced into a continued search for improved business methods. Several issues such as inflation, quality, new technologies, high-risk investments, and foreign competition have required new business strategies, with an emphasis on cost effectiveness and total quality. The traditional contracting strategies have produced adversarial and unrewarding relationships because of the disproportionate amount of risk and the threat of liability. The evolution of contract documents has focused on punitive measures to enforce performance. Consequently, there has been a dramatic increase in litigation, which is expensive and counterproductive to everyone's efforts to deliver quality projects within budget and on time. In this costly and intolerable situation, a vehicle for improvement in the business environment is required to create a win–win attitude among all team players. The partnering concept provides a vehicle to reach improved performance in which both customers and suppliers of goods and services can achieve more satisfaction and trust in their relationships.

Partnering is not a panacea. It is not a legal partnership. Partnering requires total commitment, the appropriate conditions, and the right chemistry between organizations to succeed. Partnering is a long-term commitment between two or more organizations for the purpose of achieving specific business objectives by maximizing the effectiveness of each participant's resources. This requires changing traditional relationships to a shared culture without regard to organization boundaries.*

The partnering concept is not a new way of doing business. It is going back to the way people used to do business when a person's word was their bond and people accepted responsibility.* Partnering recognizes that every contract includes an implied covenant of good faith. Over the years the industry has allowed much of the control to pass to the legal profession and lawyers. This has created the aforementioned fear of liability and distrust between the contracting parties.

The key elements of partnering are as follows.

Long Term Commitment. Experience has shown that the benefits of partnering are not achieved quickly. Time is required to solve the problems of this association and to constantly improve to achieve a competitive advantage.

Partnering requires that all stakeholders "buy into" the project.

A partnering agreement may grow from a single project to multiple projects. A long-term commitment permits "lessons learned" to be passed on to later projects.

Trust. Trust enables the resources and knowledge of each partner to be combined, eliminating an adversarial relationship. Teamwork is not possible where there is cynicism about others' motives. A better understanding of each stakeholder's risks and goals creates trust and with trust comes the possibility for a synergistic relationship. The "not invented here" syndrome is avoided. The partnering process empowers project personnel, which encourages decision making and problem solving at the lowest possible level of authority.

Shared Vision. Each partner must understand the need for a shared vision and common mission for the partnering relationship. A candid and open atmosphere must be established to promote a mutual exchange of ideas and agreement on the expectations for the partnership. Goals and objectives are jointly developed and mutually agreed upon. These goals may include achieving value engineering savings, limiting cost growth for the project, limiting and reducing the time period required for contract submittals by elimination of a formal bidding process, minimizing the paperwork required for protective or defensive reasons, no litigation, and other specific goals.

Partnering and Total Quality Management (TQM). Partnering provides an environment for total quality management because it focuses on the long-term approach of continuous improvement of construction processes. It

*Much of the information in this section has been drawn from CII Special Publication 17-1, *In Search of Partnering Excellence* and the Association of General Contractors of America publication *Partnering, A Concept for Success*.

provides a culture of continuous improvement which leads to performance improvement.

These concepts are similar because both require a major cultural change in the method of operation of the partners. Both require long-term commitment and support from every level of the organization. For success, both concepts require that people understand and are motivated to work in the new culture.

Benefits of Partnering. Benefits accrue to owners, contractors, architect/engineers, suppliers, and subcontractors. Giving lip service to the concept is not enough. A formal periodic evaluation is required to ensure the effectiveness of the process. Evaluations should be in written form and include recognition of positive behavior and not just deficiencies. A numerical scoring system can be used to evaluate communications, problem solving, trust/candor, and progress on goals.

A Construction Industry Institute task force identified several benefits from partnering and they are summarized as follows:

1. Improved ability to respond to changing business conditions. Over 80% of the survey respondents claimed that there was increased openness and trust. The relationship was less adversarial than the traditional one, and it permitted improved resource planning.
2. Improved quality and safety and fewer errors.
3. Reduced cost and time and improved profits. Respondents claimed that total project costs were reduced by 8%, contractor profitability improved 21%, and there was a 7% improvement in schedules.
4. More effective utilization of resources. Engineering cost reductions of 21% and administrative cost reductions of 6% were reported.

3.9 ORGANIZATIONAL DESIGN CONSIDERATIONS: SUMMARY

A number of considerations will assist in determining which type of project organization will suit the purpose. If projects are conducted as a part of an ongoing business, there are four considerations:

1. The volume of project work executed by the company is a factor in considering whether to opt for a special task force or departmental project management in which the project is managed by a specific department or division.
2. If projects are technologically complex, it may be preferable to separate this project from the management of the ongoing business.
3. Similar to considerations of technological complexity is that of uniqueness of the individual project. Many ongoing businesses are not skilled

in the management of projects that are complex or require unique or specialized knowledge and skills.

4. Some projects are internal to the company; others have an important interface with outside agencies.

Environmental impact and legislative requirements often require special attention. Today the public interest must be properly addressed. Many projects are stalled at this early stage and the result is an extended schedule.

In addition to the preceding four considerations, the following are important factors that could influence the type and design of a project organization:

5. The degree of integration of resources that is required and the availability of resources such as personnel and workload capacity.
6. Size of the project. Mega projects obviously require a different organization structure than that required for a small school.
7. The level of authority required by the project manager will dictate to whom he should report.
8. The pressures of time and cost may dictate the degree of dedication and concentration of effort required. The project may need a special task force.
9. Financial controls stipulated by lending agencies may require a separate or special reporting system, and a dedicated project organization.
10. Different tax implications may dictate a certain type of organization.

CHAPTER 4

BAR CHARTS (GANTT)

Bar charts are the most frequently used schedules in construction because they are simple and easy to understand. Their simplicity makes them very useful for milestone and summary schedules, which are used for global control at the project management and executive level. At the working level superintendents, foreman, and design supervisors use special function bar charts for one- or two-week daily schedules.

4.1 PREPARING BAR CHARTS

The anticipated start and completion date of each major activity is depicted as a horizontal bar. The length of bar represents the duration of the activity. Figure 4.1 shows a simple bar chart. The amount of information shown on each bar is very limited. The level of labor intensity during each activity, which may vary considerably, is now shown. An interrupted bar is used to show intermittent activity.

The first step in preparing a bar chart is to develop a list of activities that will be shown on the schedule. This list can be as extensive and detailed as required, but usually is less than 20 items, representing a major group of activities such as superstructure frame, floor slabs, roof, and so on. Next, the durations of activities are estimated. On the horizontal axis select an appropriate time scale which can be in days, weeks, months, or calendar dates.

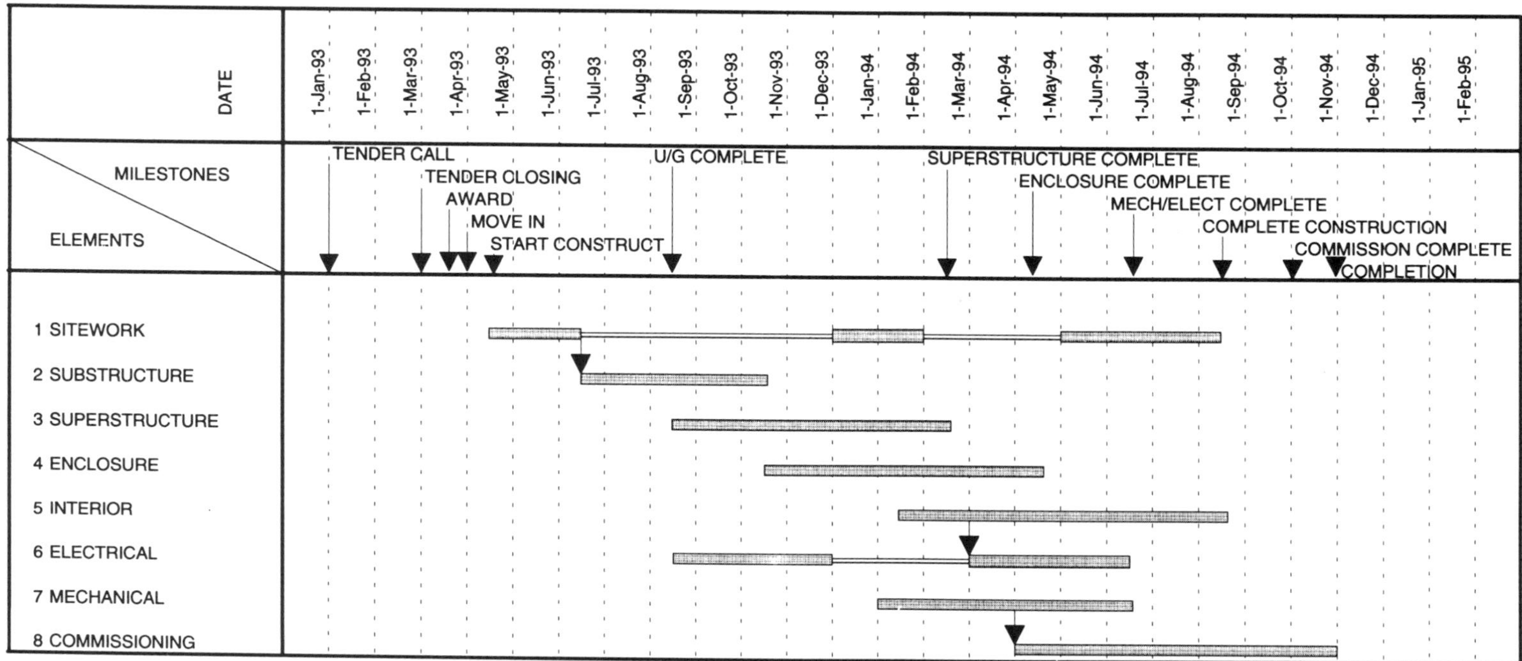

FIGURE 4.1 Simple Bar Chart

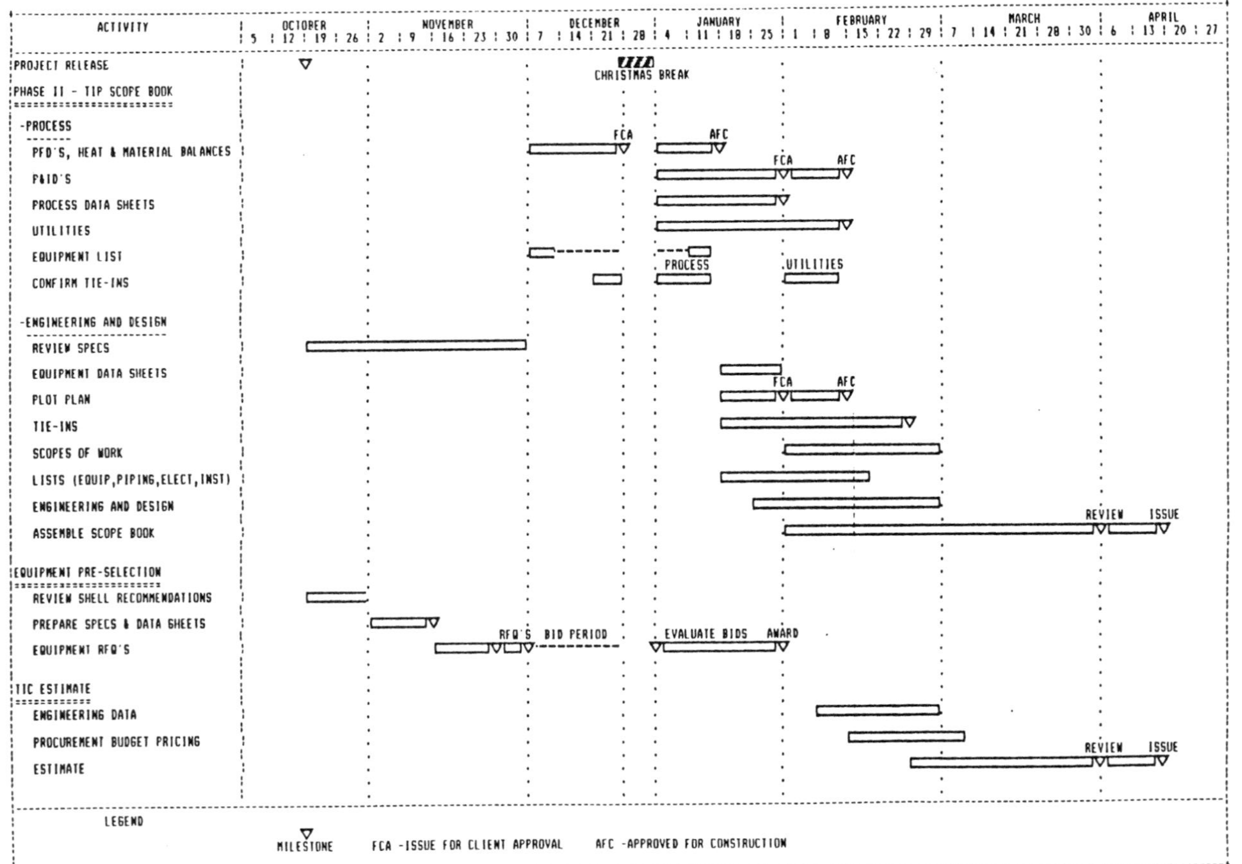

FIGURE 4.2 Milestone Schedule—Modification Project

Figure 4.1 also shows vertical arrows which are intended to show logic sequence ties. Some computer software packages are capable of showing these logic ties.

Float can also be shown by a narrow horizontal bar, a dotted line, or a different color. Many other refinements can be made, but the inclusion of too many features will nullify the simplicity of the bar chart. If considerably more detailed information is desired, it is recommended that a CPM network or other type of schedule be used.

Figure 9.11 (Chapter 9) illustrates a bar chart schedule with the critical path highlighted with dark bars. A table is added to facilitate reading activity information.

4.2 MILESTONE SCHEDULES

The purpose of a milestone schedule is to highlight the dates of significant events. Two examples of significant events are the start or finish of an activity or project and submission of a drawing, specification, or purchase order for client approval (FCA). Figure 4.2 is an example of a milestone schedule produced for the preconstruction phase of a process plant modification. Additional milestones such as approved for construction (AFC), review, or issued for construction are especially useful for this type of project. Other projects have their own specific needs and therefore milestones are selected accordingly.

A useful set of symbols are:

▽—milestone start
⧸▽—missed milestone
▼ or △—milestone completion

Milestones should be critical events which have serious schedule implications. If a milestone is missed and it is acceptable because this missed event does not have a major impact on the schedule, it may be advisable to eliminate this particular milestone.

CHAPTER 5

CPM: ACTIVITY ON ARROW (AOA)

A CPM—critical path method—network is a model of the project execution. There are two types of networks: activity on arrow (AOA) and activity on node (AON). This chapter introduces CPM planning using the AOA method, which is the more cumbersome of the two but is easier to understand for novices to scheduling. In the next chapter the AON method (precedence diagramming) is discussed.

This chapter introduces the logic of AOA and general CPM terminology. Although there is some variation in terminology between AOA and AON, it is essentially the same, and the differences are ignored in this book.

It is not necessary to always use a CPM network for scheduling. Often a bar chart will suffice, but this practice has the potential for serious difficulties. The authors recommend that bar charts be complemented with the use of CPM schedules. The extent of the scheduling effort and type of schedule will depend largely on the experience level of those concerned. For example, on a project with repetitive operations, it may be preferable to use a line of balance chart or a bar chart. A major national contractor uses only a bar chart for high-rise building projects. Another uses a work package concept and bar charts for revamp and maintenance operations in refinery work. However, for most planners it is advisable to avoid short cuts. There is considerable benefit and knowledge obtained from developing the logic of a CPM.

There are numerous computer software packages available to produce CPM networks or schedules. Most packages use AON methods and can provide a bar chart version of the same schedule. Some can also provide an AOA network as a derivative of the AON schedule. It is likely that a computer will be used if a CPM network is desired. For larger projects the calculation and graphics requirements mandate the use of a computer. However, for instructional purposes, a manual approach is used in this book.

As the name implies, a CPM is used to determine the critical path which is the shortest duration for the project. Other useful information can be derived from a CPM network, such as activity times, event times, and float.

5.1 ACTIVITIES AND SEQUENCING

In developing a CPM network, it is first necessary to break the project down into component operations or activities. How fine the breakdown will be may vary from project to project, depending on factors such as the nature of the work involved, the location of the work, the class of labor involved, and the overall sequence of the project. At the initial stages the breakdown listing trades, location, or other criteria might prove useful.

The first step in developing a schedule is to visualize the project as a group of interrelated components. In other words, the project is broken down into components and this is known as the work breakdown structure (WBS) of the project. This is the starting point for initiating the milestone schedule. Initially the scope of the project is understood in terms of the larger components. For example, a coal-fired generating station would have a boilerhouse, coal-handling system, coal storage, cooling ponds, water treatment, ash handling, and so on. As the scope of each of these components is developed, the breakdown is refined further. For example, the coal-handling system would consist of a run of mine (ROM) hopper, ROM crusher conveyors, stackers, stockpile reclaim system, frozen coal crusher, and storage bins. Each of these components is broken down further. The ROM crusher would consist of a grillage, foundation excavation, concrete foundation, backfill, crusher, crusher electrical power feed, and so on.

The preceding discussion illustrates how the schedule activities for a schedule are generated using a work breakdown concept. The authors have chosen this simple introduction of WBS so that a student can begin to study the mechanics of scheduling activities using bar charts and CPM networks. More detailed discussions on WBS, organizational responsibility charts (ORC), and cost codes follow in later chapters.

After the activity list has been prepared, it is necessary to establish the relationship between these activities. The following questions must be asked of each activity:

1. What activities precede this activity?
2. What activities follow this activity?
3. What activities are concurrent with this activity?

By asking these questions the effects of any physical, safety, resource, crew, or technological constraints can be assessed and the network of activities arranged so as to lessen their impact.

A CPM network is a graphical representation of a project. The tasks or activities that comprise the project are represented by arrows. The arrow tail and head represent the start and finish of an activity, respectively. The estimated time to be spent on an activity is called its duration. For network clarity a concise description and the duration is given on each activity. Figure 5.1 shows a conventional method. If the unit of time (day or week) used on a given network is consistent throughout, only a number is needed to denote the duration. Activity arrows are not drawn to scale.

Figure 5-2 gives an example of a small network. Notice that:

B precedes *D*
A precedes *B*
A precedes *C*
B precedes *E*
E precedes *G*
C precedes *F*
D precedes *F*
F precedes *H*
G is a terminal activity
H is a terminal activity

Activity D, for example, can start only when B is completed. Activity E can be performed along with D, whereas F must wait until the completion of D and C.

Sometimes relationships between activities cannot be shown using regular activity arrows. The dummy represented by a dotted arrow is used for this purpose. A dummy has zero duration and involves no work. In Figure 5.3*a* the dummy *X* shows the dependency of *H* on *E*. Figure 5.3*b* shows how *F* and *G*, though independent of each other, are dependent on the completion of *D*.

A network should have a consistent flow of activities from left to right. It should be drawn more than once to improve clarity and reduce crossovers

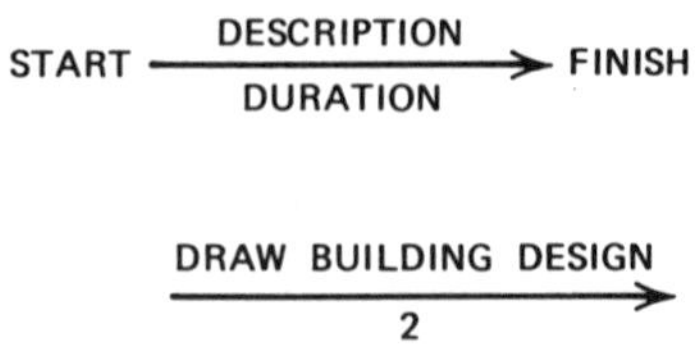

START —DESCRIPTION / DURATION→ FINISH

DRAW BUILDING DESIGN / 2 →

FIGURE 5.1 Activity Meaning

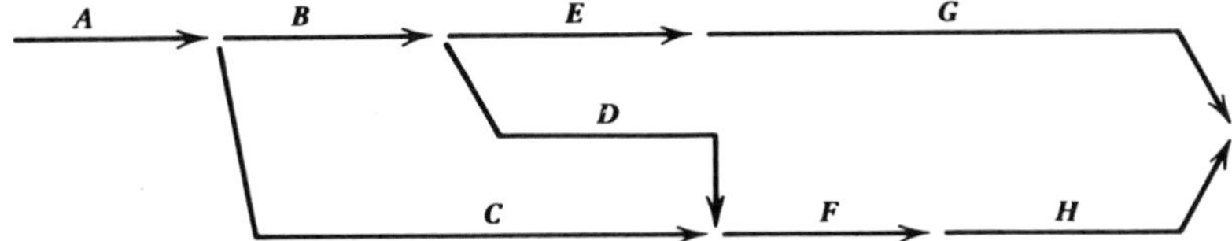

FIGURE 5.2 Example of a Small Network

to a minimum. When it is impossible to eliminate all crossovers, they should be shown in one of the two ways illustrated in Figure 5.4. Although any type of crossover is suitable, only one type should be used in a given network.

5.2 LOCATION AND IDENTIFICATION OF ACTIVITIES

For a large project, a network may have thousands of activities. A method for easily identifying and finding an activity is necessary. Each activity is defined by two nodes or events—a tail node and a head node. The head node is referred to as the *j* node, and the tail node is called the *i* node, as shown in Figure 5.5. The *i* node is the start event, and the *j* node is the finish event; they define those points in time when an activity is started and completed. A node is commonly known as an event because it depicts that moment when all the activities entering that node are finished and, therefore, all the activities leaving that node can begin.

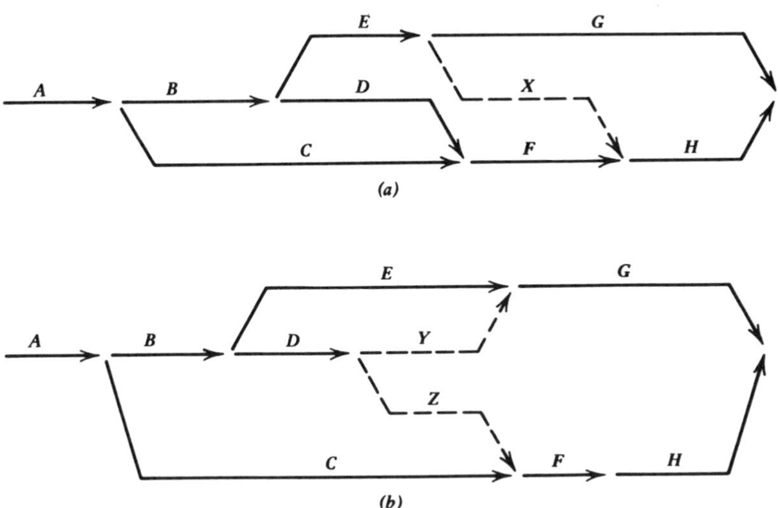

FIGURE 5.3 Networks Illustrating the Use of Dummy Activities

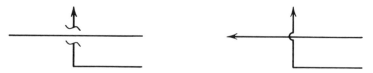

FIGURE 5.4 Crossovers

For node numbering it is convenient, although not necessary, to use the *i–j* rule, where the *i* node number is always less than the *j* node number. To differentiate one node from another, each is given a unique identification number. Hence the activities in Figure 5.6 are easily distinguished from each other as activity 1–2, activity 2–3, activity 2–4, activity 3–5, and activity 4–5. When two or more activities join two nodes, as shown in Figure 5.7, there is an identification problem. In Figure 5.8 a dummy activity is used to clarify the situation. Instead of two activities 15–16, there are 15–16 and 15–17.

The events are sometimes given block numbers 10, 20, 30, 40, and so on, instead of serial numbers 1, 2, 3, 4, 5, and so on. This practice facilitates the addition of activities to networks without renumbering the nodes. The node of activities that fall between such blocks can then be numbered 11–12, 12–14, 14–17, and so on, between block numbers 10 and 20.

There is another, almost completely different, way of identifying and locating activities that is useful in large networks made up of a number of separate sheets. Each sheet is numbered and divided up by a grid. The node is identified by a five-digit number. The first two digits give the sheet number, the next the column number, and the next the row number; the final digit distinguishes the node from other nodes in the same grid section. An activity is defined by its tail and head node. Figure 5.9 illustrates the use of this technique. On a network with 60 sheets, the activity "dig ditch," 14211–14212, is immediately known to be on sheet 14, starting and ending in column 2, row 1, and has the distinguishing digit 1 for the *i* node and 2 for the *j* node.

For the effective use of this method, no node should be drawn on a grid line, nor should more than nine nodes be drawn in one grid section.

A summary of instructions is listed next. If these are followed, CPM networking will be much easier:

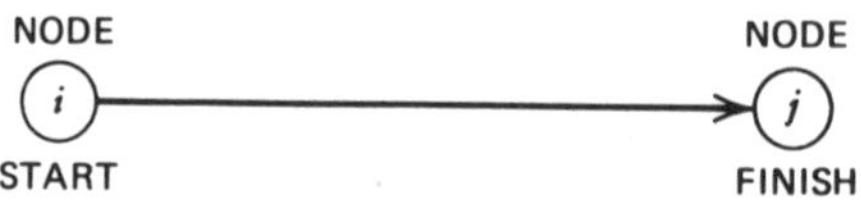

FIGURE 5.5 Nodes Using *i–j* Notation

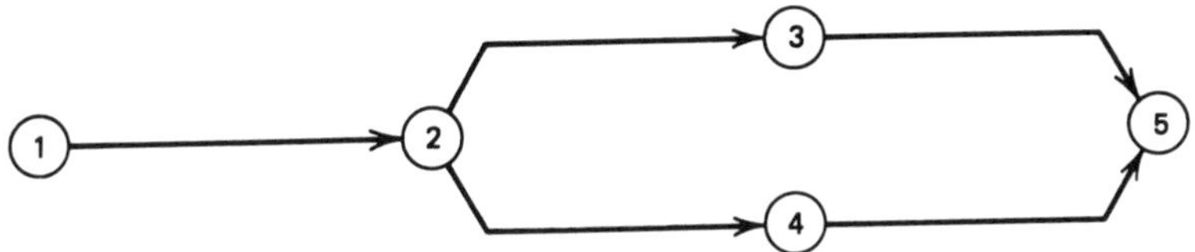

FIGURE 5.6 Sequential Node Numbering

1. When constructing a network, develop the logic one step at a time. Always ask the questions:
   a. Which activities must be finished before this activity may begin?
   b. Which activities are carried out simultaneously with this activity?
   c. Which activities are dependent on the completion of this activity?
2. An arrow can be any length as required for the sake of clarity.
3. Dummies may be used freely, but those that are redundant should be eliminated before proceeding further.
4. Always draw an arrow, with at least a portion of it drawn horizontally. The descriptions and durations can then be written on a horizontal plane, enabling them to be read with ease.
5. Write the description above the horizontal portion and the duration below so that the two are never confused.
6. For the sake of clarity avoid crossovers whenever possible. This might require rearranging small sections of a network.
7. Always number the nodes after the diagram is complete. This removes the hassle of changing the node number during planning. Brief descriptions suffice for identification purposes in the earlier stages.
8. Draw the arrows from left to right (this aids the network mechanics immensely).
9. To avoid confusion, always use a consistent method of presentation.
10. Always include a key and title block with a network diagram.

5.3 ACTIVITY DURATION

Obtaining reasonable activity duration estimates is important when constructing a network. The construction superintendent, being experienced, is

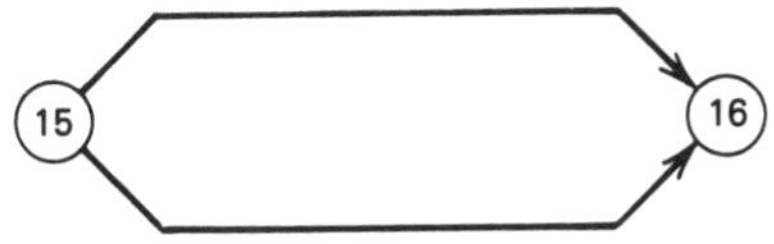

FIGURE 5.7 Incorrect Node Numbering

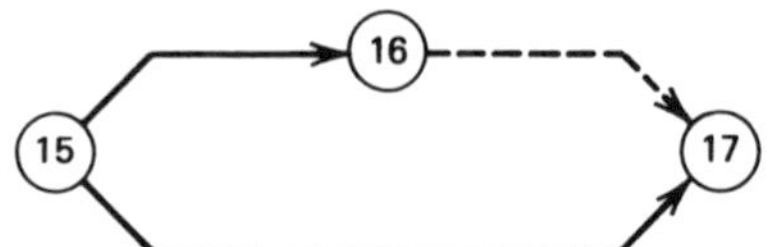

FIGURE 5.8 Use of Dummy Activity

probably best suited to estimate activity durations. Solicit information from those closest to the execution of the activity. Tables showing crew sizes and their productivity, historical data for the different trades, and consultations with crew supervisors are helpful in this task.

Given a defined quantity of work (cubic meters) to be performed for an activity (as described in the plans and specifications) and the resources available, the time required to do the work can be determined by dividing the quantity of work by the labor productivity (manhours/cubic meter). This is illustrated in Figure 5.10. Productivity implies the average or expected rate of work performed and can be obtained from one of three sources: published productivity records, the construction organization's own productivity charts derived from past projects, or the accumulated experience of the construction supervisors.

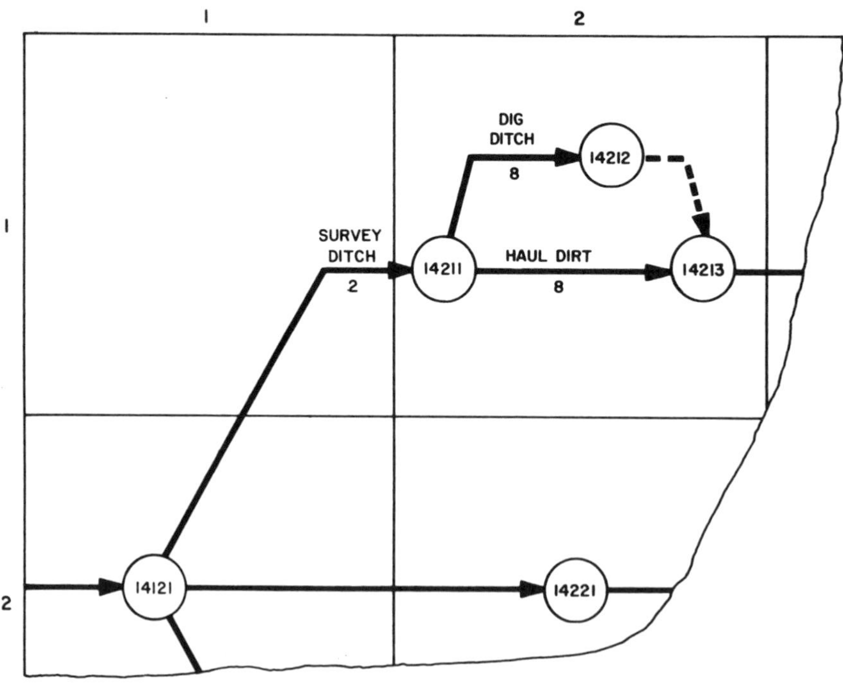

FIGURE 5.9 Grid Numbering of Nodes

FIGURE 5.10 Process for Determining Activity Duration

When published productivity data is used, the planner may find that 125-mm thick, 400 m^2 slab on grade can be laid by 1 foreman, 3 building laborers, 2 cement finishers, and 1 concrete pump operator in 1 day. If there are no space constraints, the concrete pump has no capacity constraints, and the workers can be provided other work later, the crew can be doubled to pour 800 m^2 a day. Thus the planner, knowing the quantity of work in each activity must decide on the crew size to determine the activity's duration.

Unfortunately, the figures obtained from productivity records published in most journals, textbooks, or manuals refer to the average productivities taken from several locations and from several different projects. This average may not represent the productivity of a given organizations' workers. Likewise the data derived from past projects may not reflect present rates of work. When using published data for a certain item of work, the contractor, planner, or estimator must always supplement the available information with historical data from his own organization, if available, and his own experience and judgment. The planner should also consult the crew supervisors of the project. They should be encouraged, but not pressured, to give duration estimates.

To avoid building inefficiency into the plan, usually no provision is made in the duration estimate for overtime, extended work weeks, strikes, late deliveries, authorization delays, abnormal weather, and acts of God. Except for the latter, time can be allocated for these occurrences by including an activity at the end of the network. Its duration depends on the number of days expected to be lost on the project because of any or all of these misfortunes. No thought is to be given to the project duration while estimating the duration of activities, otherwise the latter may be biased by the former.

Overtime and planned extended work weeks can be accounted for in activities although this has a tendency to build inefficiency into the plan. It is pointless to schedule for catastrophic or abnormal events such as strikes, tornadoes, and other abnormal weather.

Extra time needed during seasonal cold weather should be built into the activity or allocated at the end of the affected series of events. For example, cold weather may cause a 25% inefficiency during December, January, and February. The length of an activity is affected by the rest of the progress that is possible during a period of time, and the rate may or may not include the effects of overtime or extended work weeks. The inefficiency due to cold weather can be compensated for by overtime, increased labor input, or an increased work week. The duration of an activity that will be executed during cold weather must include these factors.

Durations can be estimated in hours, shifts, days, weeks, or months, depending on the industry practice. The estimates, which are initially only rough approximations, are most accurate at implementation, when more details about the project are available. An activity with an original estimated duration in months may eventually have a duration in days.

Absolute accuracy is not necessary in estimating durations. Small errors, which usually cancel each other, have little effect on project duration. Is it worse to be 5% wrong on a 50-week job than 50% wrong on a 2-week activity. The sensitivity of errors in durations should be evaluated; only a few key activities need checking.

5.4 EVENT TIMES

An event, synonymous with a node, is that point in time when all preceding activities have been completed and all immediately succeeding activities can begin. In CPM an early and a late event time are associated with each event.

The early event time for a given node or event is defined as the earliest time an activity leaving that node may begin. A computational process known as the forward pass is used to determine the early event time (EET) for each and every event. Either of the symbols shown in Figure 5.11 can be used to distinguish early times from any other type of time. In the forward pass, calculations begin at the first node and travel toward the end until the early event times have been computed for every event on the CPM network.

It is usually assumed in normal calculations that the early event time for the first node is zero. The early event time for the following event is found by simply adding the duration of the succeeding activity to the early event time of the preceding event.

When more than one activity enters a node, the finish time of the last activity to finish becomes the EET of the node. Figure 5.12 shows a network with early event times calculated. At node 5, the early event time is 8, not 7, since 8 is larger than 7. At node 12, the early event time is 15, not 14, since 15 is larger than 14.

The late event time (LET) for a given node or event is defined as the latest time an activity entering that node can finish. A computational process known as the backward pass is used to determine the late event times for each and every event. Either of the symbols shown in Figure 5.13 can be used to distinguish late event times from any other type of time.

□ = ▷ = *E* = EET = early event time

FIGURE 5.11 EET Symbols

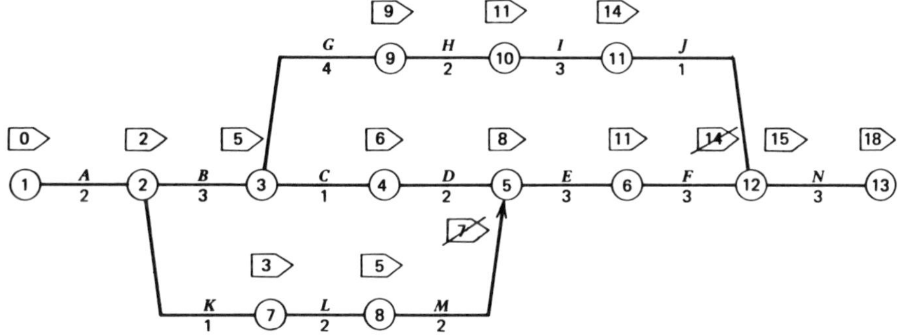

FIGURE 5.12 Sample AOA Network

In the backward pass calculations begin at the last node and travel toward the start until the late event times have been computed for each and every event on the CPM network. The EET at the last event is also taken as the LET of the last event. Obviously, if a job can be finished at the earliest by a certain time, say, time 28, then the latest time one would like to have a job completed is that same time. There is no apparent advantage to delaying a project by more than the time required to do it. The LET for the preceding event is found by simply subtracting the duration of the activity between the two nodes from the LET of the succeeding event.

When more than one activity leaves a node, the start time of each activity that leaves that node must be considered before the LET can be determined. Then, of course, the start time of the activity that is the first to begin is taken as the LET. In Figure 5.14 the LET of node 3 is 5, not 6, since 5 is smaller than 6. At node 2 the late event time is 2, not 4, since 2 is smaller than 4.

Various symbols can be used to distinguish event times. An example of a notation is shown in Figure 5.15. There i is the preceding node number, j the succeeding node number, E the early event time, and L the late event time. For clarity one type of symbol should be chosen and used consistently on any given network.

5.5 ACTIVITY TIMES

In AOA, activity times are computed from event times and are usually presented not on the network but in tabular form. The earliest start (ES_{ij})

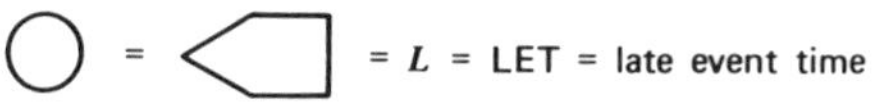

= L = LET = late event time

FIGURE 5.13 LET Symbols

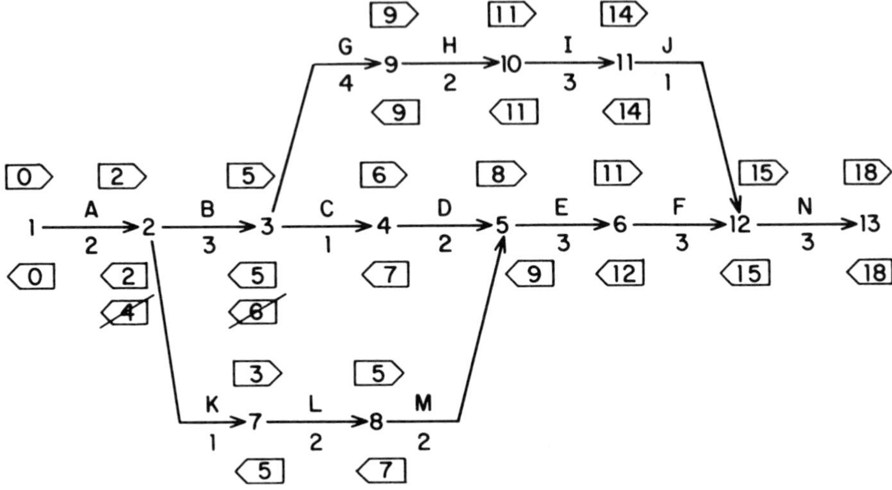

FIGURE 5.14 Sample AOA Network (continued)

of any activity is the early event time of the node that it leaves (i.e., the *i* node). The latest finish (LF_{ij}) of any activity is the late event time of the node that it enters (i.e., the *j* node). The latest start (LS_{ij}) of an activity is its LF_{ij} minus its duration. The earliest finish (EF_{ij}) of an activity is its ES_{ij} plus its duration. The LS_{ij} of an activity is always greater than or equal to the LET_i of the preceding node. The EF_{ij} of an activity is always less than or equal to the EET_j of the following node.

Figure 5.16 shows the relative status of EF, EET, LS, and LET in an activity.

Once the network, with early and late event times noted, is available (Fig. 5.17), the following six-step procedure can be used to compute and tabulate activity times (see Table 5.1).

Step 1. List all activities in column 1 according to the *i* major, *j* minor sorting sequence (list activities in ascending order by *i* node first, and then for each *i* node arrange the activities in ascending order of *j* node).

Step 2. Record the description of each activity in column 2 and its duration in column 3.

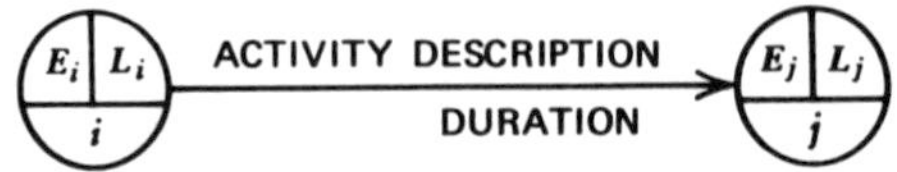

FIGURE 5.15 AOA Node Symbols

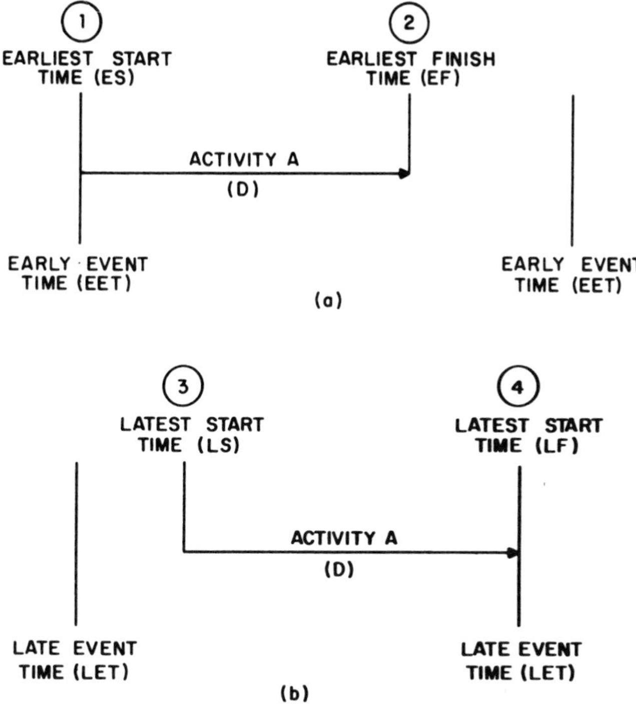

FIGURE 5.16 Activity Relationships

Step 3. Record the earliest start time for each activity in column 4. These are the early event times at the i nodes of activities.

Step 4. Compute the earliest finish for each activity by adding its duration to its earliest start and place in column 5.

Step 5. Record the latest finish time for each activity in column 7. These are the late event times at the j nodes of the activities.

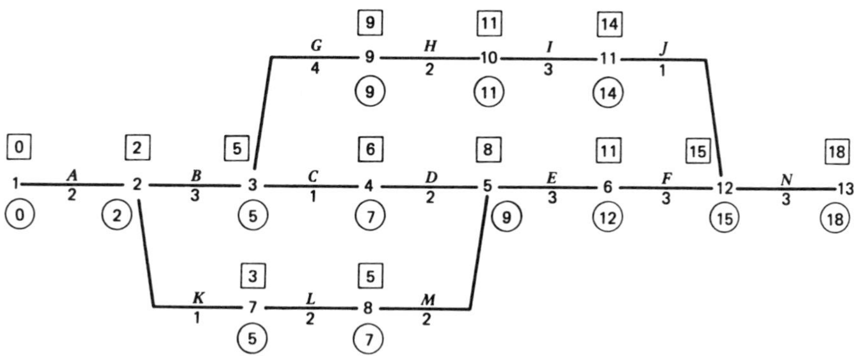

FIGURE 5.17 Sample Project

TABLE 5.1 Example Project

(1)	(2)	(3)	(4)	(5)	(6)	(7)
			Earliest		Latest	
Activity i–j	Description	Duration (D)	Start (E_{ij})	Finish ($E_{ij} + D_{ij}$)	Start ($L_{ij} - D_{ij}$)	Finish (L_{ij})
1–2	*A*	2	0	2	0	2
2–3	*B*	3	2	5	2	5
2–7	*K*	1	2	3	4	5
3–4	*C*	1	5	6	6	7
3–9	*G*	4	5	9	5	9
4–5	*D*	2	6	8	7	9
5–6	*E*	3	8	11	9	12
6–12	*F*	3	11	14	12	15
7–8	*L*	2	3	5	5	7
8–5	*M*	2	5	7	7	9
9–10	*H*	2	9	11	9	11
10–11	*I*	3	11	14	11	14
11–12	*J*	1	14	15	14	15
12–13	*N*	3	15	18	15	18

Step 6. Compute the latest start of each activity by subtracting its duration from its latest finish and place in column 6.

5.6 FLOAT: START, FINISH, TOTAL, FREE, AND INDEPENDENT

Each activity in a project should be completed within the thresholds of the earliest start time and the latest finish time. As long as the activities are completed within these limits, the project schedule will not be extended. When the time difference between these two limits exceeds the duration required for an activity, there is obviously some spare time available either before the start or after the finish of the activity. This spare time is known as float.

The time difference between an activity's earliest and latest start time is known as start float; the time difference between an activity's earliest and latest finish time is known as finish float:

$$\text{Start float} = \text{latest start} - \text{earliest start}$$

$$\text{Finish float} = \text{latest finish} - \text{earliest finish}$$

Both types of float may be associated with an activity. It can be seen in Figure 5.18 that the start float is equal in duration to the finish float. It can also be seen that the start float and finish float are not the difference between the event times but the differences between the activity times.

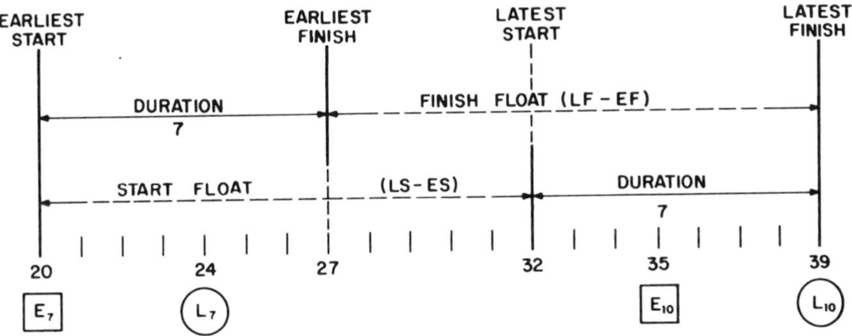

FIGURE 5.18 Floats

The most significant is total float (TF). It indicates the number of time units an activity may be prolonged without extending the project's scheduled completion date. This delay in the start of an activity may cause delays in some of the activities that follow, but it will not extend the project duration. A serious delay can thus be distinguished from a delay that is not detrimental to the timely completion of the project. Total float is defined as the latest finish time of an activity minus the earliest start time of that activity minus its duration:

$$\text{Total float} = \text{latest finish} - \text{earliest start} - \text{duration}$$

$$\text{TF} = \text{LF} - \text{ES} - \text{D}$$

As shown in the following equation, total float can be either start float or finish float:

$$\text{TF} = (\text{LF} - \text{D}) - \text{ES} = \text{LS} - \text{ES} = \text{start float}$$

$$\text{TF} = \text{LF} - (\text{ES} + \text{D}) = \text{LF} - \text{EF} = \text{finish float}$$

For simplicity the subscripts ij have been omitted in the preceding three equations.

The main difference between these is that, whereas start and finish floats refer to float either before the latest start or after the earliest finish, total float may refer to either of the two. Total float can also be split. Part of it may be used before the latest start of activity, and the balance after the earliest completion of the activity. This may be seen in Figure 5.19. Notice that four units of total float are used before starting the activity and the balance of eight units of float are available on completion.

Total float of an activity may also be defined as the late event time L_j of the succeeding node minus the early event time E_i of the preceding node minus the duration of the activity identified by these nodes:

$$\text{TF}_{ij} = L_j - E_i - D_{ij}$$

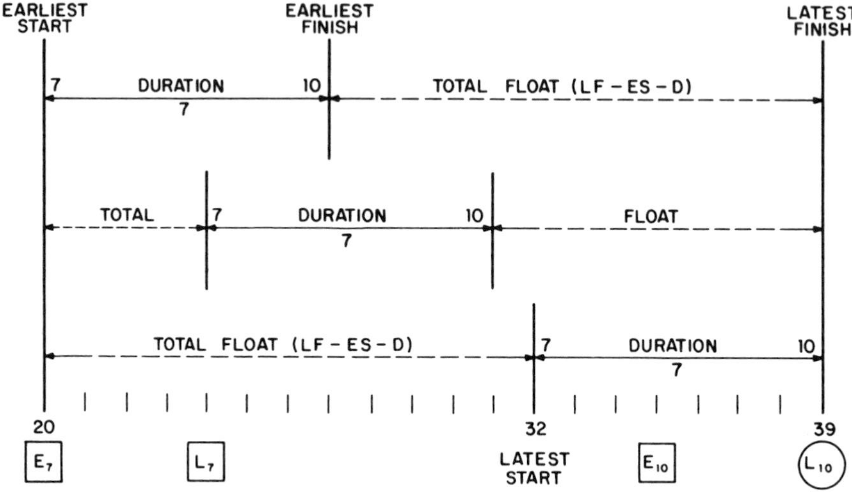

FIGURE 5.19 Floats—Split

The main use of total float is to establish priorities, depending upon the float associated with the activities.

Free float (FF) is defined as the early event time E_j of the following node minus the early event time E_i of the preceding node minus the duration of the activity identified by these nodes:

$$FF_{ij} = E_j - E_i - D_{ij}$$

Free float is mainly used to identify the activities that can be delayed without affecting the total float on the succeeding activities.

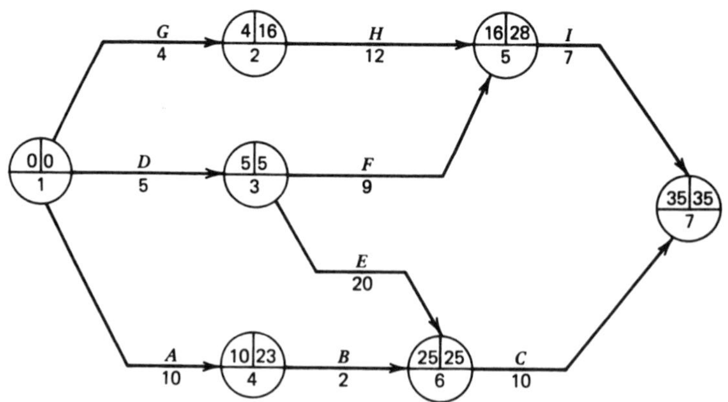

FIGURE 5.20 Example Network

TABLE 5.2 Example Network

			Earliest		Latest		Floats		
Activity	Description	Duration	Start	Finish	Start	Finish	Total	Free	Independent
1–2	*G*	4	0	4	12	16	12	0	0
1–3	*D*	5	0	5	0	5	0	0	0
1–4	*A*	10	0	10	13	23	13	0	0
2–5	*H*	12	4	16	16	28	12	0	0*
3–5	*F*	9	5	14	19	28	14	2	2
3–6	*E*	20	5	25	5	25	0	0	0
4–6	*B*	2	10	12	23	25	13	13	0
5–7	*I*	7	16	23	28	35	12	12	0
6–7	*C*	10	25	35	25	35	0	0	0

* See text for explanation.

The standard definition of independent float (IF) is the early event time E_j of the following node minus the late event time L_i of the preceding node minus the duration of the activity identified by these nodes:

$$IF_{ij} = E_j - L_i - D_{ij}$$

Independent float identifies the activities that, even if delayed, will not affect the total float of either the preceding or the succeeding activities. It is noteworthy that in some instances, as in activity 2–5 in Figure 5.20 the calculation yields a negative value for the independent float (16 − 16 − 12 = −12). In that case the float would be taken as zero. Replacement of this negative value with a zero is highlighted with an asterisk in Table 5.2.

Figure 5.21 graphically illustrates total, free, and independent float:

$$TF_{ij} = L_j - E_i - D_{ij} = \text{total float}$$

$$FF_{ij} = E_j - E_i - D_{ij} = \text{free float}$$

$$IF_{ij} = E_j - L_i - D_{ij} = \text{independent float}$$

Total, free, and independent floats are computed and tabulated with activity time. Table 5.2 is an example based on the CPM network shown in Figure 5.20. Note that whenever an activity has zero total float, there is also zero free float and zero independent float. Therefore, whenever a calculation yields no total float and some other float has a positive value, it is clear that an error has been made in the computations.

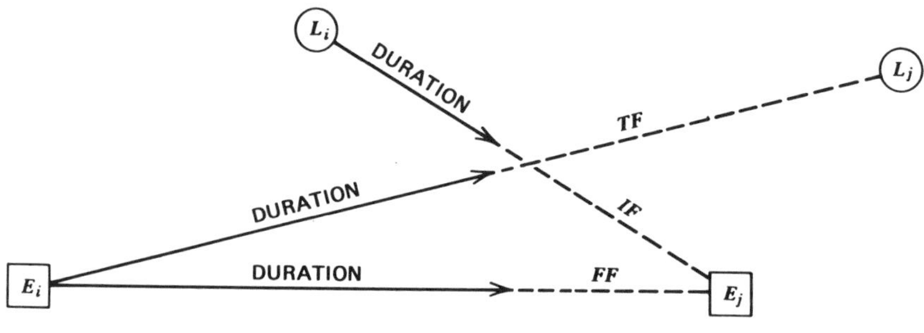

FIGURE 5.21 Float Relationships

PROBLEMS

5.1 In the expansion of a telephone service the following activities are involved:

1. Receive circuit request from traffic department.
2. CNT checks facilities.

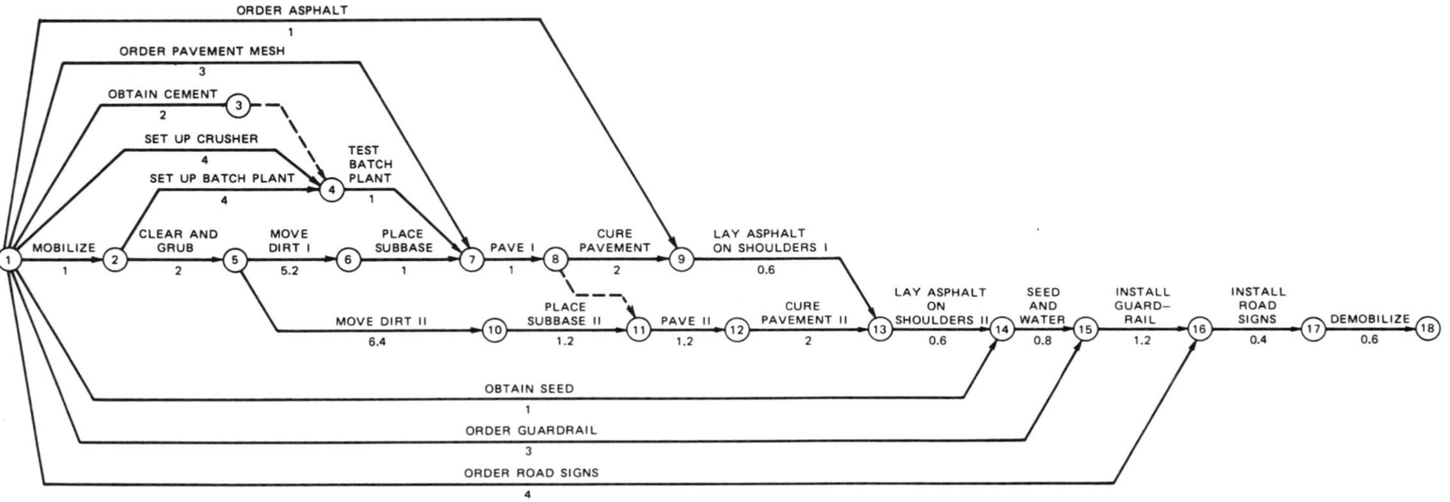
ORDER ASPHALT
1
ORDER PAVEMENT MESH
3
OBTAIN CEMENT
2
SET UP CRUSHER
4
SET UP BATCH PLANT
4
TEST BATCH PLANT
1
MOBILIZE
1
CLEAR AND GRUB
2
MOVE DIRT I
5.2
PLACE SUBBASE
1
PAVE I
1
CURE PAVEMENT
2
LAY ASPHALT ON SHOULDERS I
0.6
MOVE DIRT II
6.4
PLACE SUBBASE II
1.2
PAVE II
1.2
CURE PAVEMENT II
2
LAY ASPHALT ON SHOULDERS II
0.6
SEED AND WATER
0.8
INSTALL GUARD-RAIL
1.2
INSTALL ROAD SIGNS
0.4
DEMOBILIZE
0.6
OBTAIN SEED
1
ORDER GUARDRAIL
3
ORDER ROAD SIGNS
4
1
2
3
4
5
6
7
8
9
10
11
12
13
14
15
16
17
18

PROBLEM 5.3

3. Transmission engineering checks facilities.
4. Issue trunk work order.
5. Issue circuit work order.
6. Obtain plug-in equipment.
7. Office wiring.
8. Test circuit.
9. Turn up for service.
10. Install equipment in local office.
11. Adjust records

Activities 2 and 3 follow 1; 4–6 follow 2 and 3; 7 follows 5 but precedes 8; 8 precedes 9; 6 precedes 10; 11 follows 10; 9 and 11 are terminal activities; 4 precedes 8. Draw a CPM network to show these operations. Number the nodes.

5.2 Given the following information, construct the network and compute the early and late event times:

a. *U* and *R* can be performed concurrently and are the start of the project.
b. *K* must follow *E*.
c. *X* is independent of both *Q* and *K*.
d. Neither *F* nor *G* can start before *R* is finished, but *F* and *G* can be performed concurrently.
e. *U* must precede *E* and *Q*.
f. *Q* must precede *J*.
g. *C* is independent of both *F* and *G* but follows *K*.
h. *E* and *Q* can be done concurrently.
i. *H* can begin only after *C*, *X*, and *J* are completed.
j. *H* is the last operation.
k. *X* is dependent on *F* and *G*.

5.3 Compute the activity times for the network shown and arrange them in a table in early start sequence.

5.4 Write notes on the following:

a. What questions should a planner ask of the project personnel to establish the logical sequence of the project activities?
b. Describe and compare the various types of float.
c. What allowances for weather must be made in a network plan?

CHAPTER 6

AOA CRITICAL PATH ANALYSIS

In this chapter activities with and without float and their relative importance are discussed. The replacement of activity times by calendar dates is explained, restraints that may be imposed on a CPM network are investigated, and finally CPM bar charts are discussed.

6.1 CRITICAL ACTIVITIES

The chain of activities that takes the longest time to complete determines the earliest time by which a project can be completed. This time is often known as the project time or project duration, but more commonly as the critical path. A critical path always begins at the very first event and proceeds from there through the network until it terminates at the last event. Each activity in the critical chain is known as a critical activity. It is therefore very important to identify each and every critical activity. Before it can be accepted as being critical, an activity must meet the following three criteria:

1. The early and late event time at the i node must be equal:

$$E_i = L_i$$

2. The early and late event times at the j node must also be equal:

$$E_j = L_j$$

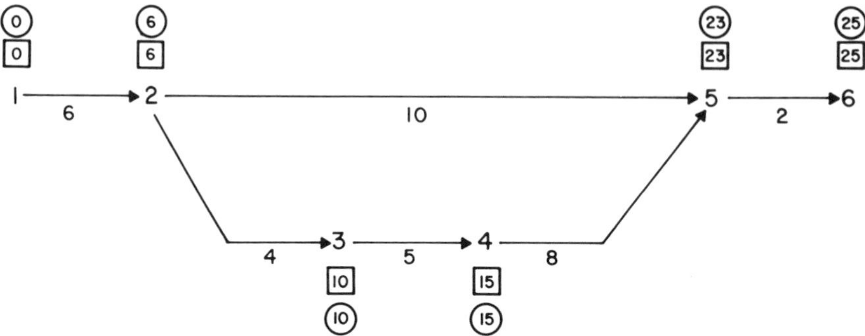

FIGURE 6.1 AOA Example Network #2

3. The duration for the activity must be equal to the difference between the late event time at the j node and the early event time at the i node:

$$L_j - E_i - D_{ij} = O$$

Criteria 3 indicates that for an activity to be critical it must have no float of any kind. If total float is zero, then all other floats are equal to zero. Total float is therefore a useful tool because, whenever the total float for an activity is zero, it indicates that the activity is critical. It is important to note that a critical activity must satisfy all three conditions. For example, in Figure 6.1 activity 2–5 is noncritical because it does not satisfy Condition 3.

There can often be more than one critical path. Even though the critical path, as a whole, goes from start to finish in the network, sometimes short chains of critical activities leave and return to the critical path. See chain 3–4–16 in Figure 6.2.

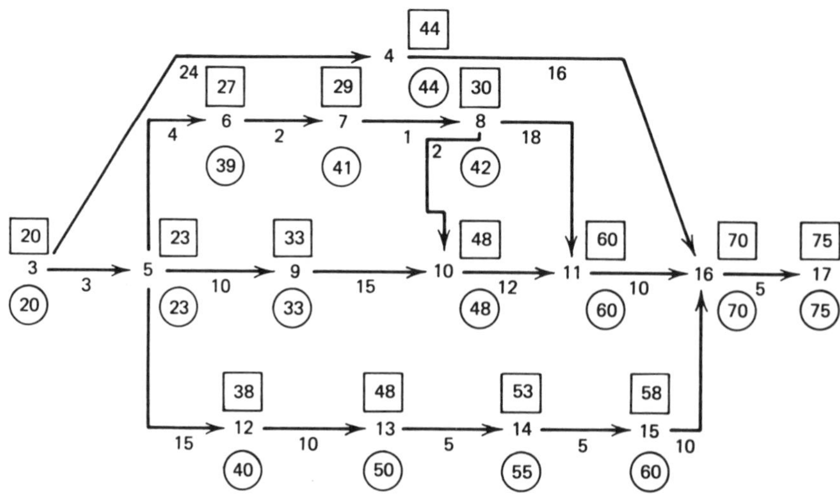

FIGURE 6.2 AOA Example Network #3

Different methods of construction may require completely different sets of activities, which would generate dissimilar critical paths. The CPM may be used to help decide which of two different methods of carrying out a project would be better in time and cost.

The critical path method establishes priorities. Critical activities must be completed on time or the project will be delayed. Activities with only a little float are subcritical, and activities with an appreciable amount of float may be considered as noncritical. The more float an activity has, the less critical that activity is in comparison to the others.

Typically critical activities make up a small percentage, perhaps 10–15%, of a project and warrant the majority of management effort. The CPM facilitates management by exception, especially on large complex projects.

6.2 THE USE OF FLOAT

Total float is shared by all activities in a particular chain. When one activity uses total float, the total float available to the other activities is reduced by that amount. Hence all activities in a chain are affected by any use of the available total float.

In Figure 6.2 the chain of activities 3–5–9–10–11–16–17 is critical and has no float. The chain of activities 5–6–7–8–11 has a total float of 12; the chain 5–12–13–14–15–16 only has a float of 2. Because the last chain has less float, the activities in this chain are more critical than those of the first.

Free float is shared only with the preceding activities in a chain, and its use therefore has no effect on succeeding activities. Free float is useful in network analysis during the planning stage. In Figure 6.2 activity 15–16 has a free float of $70 - 58 - 10 = 2$. It is the only activity in the chain with free float. Its use affects the preceding activities. If the free float of this activity is reduced by increasing the activity duration, the float of the preceding activities will be reduced by the reduction of free float in this activity.

Unlike total or free float, the use of independent float does not affect any other activity. As indicated by its name, it is independent of any other activity, affecting only the particular activity involved. An example is activity 8–10 in Figure 5.2.

One use of independent float is to show which activities might have their durations increased without causing any delay. Consider the following situation where the planner uses independent float to determine how much an activity could be delayed without any detriment to the project schedule or any additional cost. Given one week to complete a 50-hour plumbing job the cost will be $1100 (40 hours of regular time at $20/hr plus 10 hours of overtime at time and a half). However, if the job has an independent float of 0.25 weeks, then the 50 hours may be spread over 1.25 weeks. This means that no overtime will be incurred, and the project will cost only $1000 (50 hours at $20/hr). Utilizing independent float not only avoids overtime work and

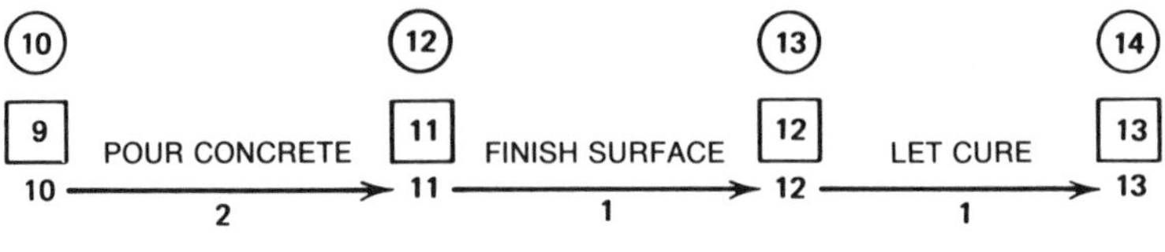

FIGURE 6.3 Chain with Unusable Float

the associated extra costs but achieves this without sacrificing the total project schedule.

When the float available to certain activities is used, these activities can become critical or near critical, resulting in a new critical path. Thus for any activity, exclusion from or inclusion in the critical path is directly affected by the use of floats.

Sometimes, the available float cannot be used. Consider the straight-line network shown in Figure 6.3, illustrating a concrete pour for a slab on grade. Activity 12–13 has a total float of 1 ($14 - 12 - 1 = 1$), which cannot be used. As soon as the concrete is spread on the grade, its surface must be finished before it sets and hardens. The physical nature of the work determines whether available float can be used.

6.3 THE USE OF DUMMIES IN OVERLAPPING ACTIVITIES

Dummies can have float, and can also be critical and form part of the critical path. In a network where many operations overlap, dummies may be critical. Normally, in a CPM network the preceding job or activity must always be finished before the next one can begin. This is not so in industry, where activities often overlap. To illustrate how CPM may be used to represent such a situation, consider the unloading of a truck at the grouting plant for pressure grouting of a dam rock base. The sand is stored in 2-ton containers which are lifted off by a crane. After the sand is unloaded, there are five more operations before it is ready to be delivered as grout to a mixer for a high-pressure pump. All six activities are illustrated in Figure 6.4.

Figure 6.4, however, does not show that these activities overlap one another. For example, the total load of sand need not be unloaded before the cleaning begins. The term package means blending and adding cement.

Consider a truck with a capacity of 8 tons or 4 containers. The sand from container 1 is being cleaned while container 2 is being unloaded. There is one crew available for each of the six operations, and on any container each

FIGURE 6.4 Sequence of Activities

operation must be finished before the next crew can begin its work on that container.

Figure 6.5 illustrates this repetitive sequence of operations. Dummies are used to make the logic correct. Notice that the dummies are critical because the second container load cannot be cleaned until it is unloaded, as indicated by dummy 3–6. The importance of the dummies is even more apparent when it is considered that the operation times may vary. In this network certain dummies are critical, and others are not. Four dummies are part of the critical path chain in Figure 6.5. Critical activities are highlighted by means of two slash marks across the activity arrow.

6.4 ACTIVITY SIZE

The superiority of a plan depends on the innovative skill of the planning team, as evidenced by the methods of construction selected and the way the job is broken down into various operations. For instance, if a building is to be finished floor by floor, it could be broken down into different floors and then into different sections.

Each section can be further broken down into operations for each trade. These operations could include items such as pouring of concrete, placing reinforcing steel, and forming of deck slab.

A criterion for determining the size of an activity is that it must be identifiable on the site, so that its progress can be monitored. If a roofing system in a building is one activity, it is difficult to monitor its progress on site when membrane waterproofing, insulation, asphalt, metal flashing, gutters, skylights, hatches, caulking, and sealants are all at different stages of progress. The activities should be small, clear, identifiable tasks.

Yet another criterion is to make the relationships among all the activities explicit so that some unknown relationship does not appear later, which may jeopardize the project schedule. This situation may occur when too much is included in an activity, perhaps warranting a further breakdown. For example, "basement walls" includes formwork, placing reinforcing steel, pouring concrete, curing, and stripping formwork. This activity should be broken down because of the different trades employed and rates of progress for the component operations.

The start and end points of activities should be clearly defined. If "pour slab" is an activity, does it start with the installation of a steel deck followed by rebar and end with the concrete pour? If the encasement of steel beams in concrete and the slab pour occur simultaneously, is the steel beam encasement part of "pour slab" or not?

The identification of activities is facilitated and ambiguities are avoided if a list is prepared and a detailed description is given for each activity. The planner's objective is to provide a network plan for control of the project and this will be compromised by an overwhelming number of small activities.

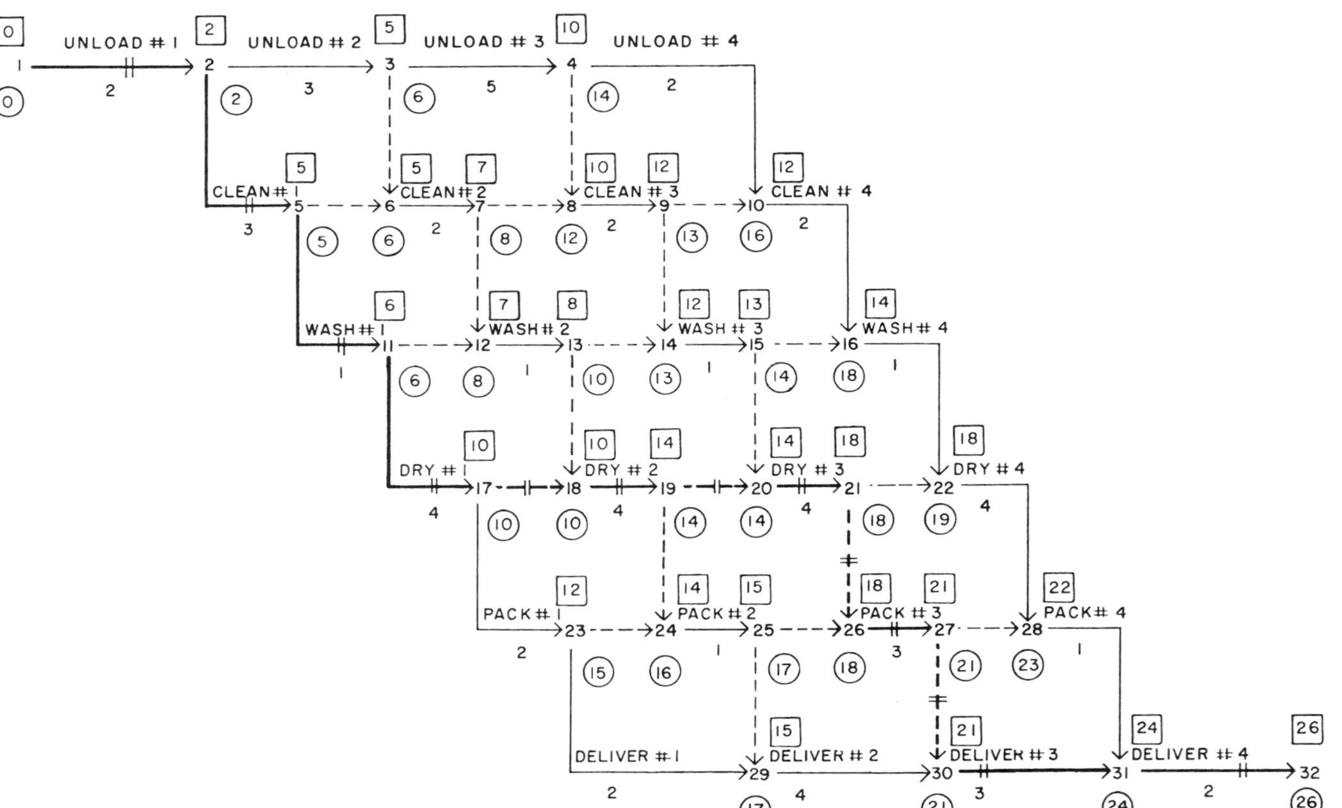

FIGURE 6.5 AOA Network of a Repetitive Operation

For example, formwork would be obscured by describing this activity as layout panels, assemble, and erect formwork.

6.5. COMPUTING CALENDAR DATES

In CPM planning calculations are done in workdays. Some special application programs such as Primavera Finest Hour permit scheduling to the hour, which is necessary in refinery turnaround projects as an example. However, the plan must be implemented according to the calendar. A means of converting workdays to calendar dates is shown in Table 6.1. This calendar is based on a 5-day workweek. In the row allotted for workdays, the number of working days since the project start is given, corresponding to the calendar date immediately above it. A quick glance at the calendar shows that day 20 of the project is March 28 or that day 9 is March 11. Though a 5-day week is shown, any size workweek can obviously be accommodated by this calendar method.

If, on a project, several contractors are responsible for work and they observe different workweeks, say a 5- and 6-day week, a single project calendar may be set up with two or more rows to record separately the respective workday numbers for the two contractors.

In previous examples it was assumed that projects start at time 0. Time 0 refers to the afternoon of the day before the project officially starts. The project then actually begins next morning on day 1 and finishes in the afternoon of the last workday. (Note that some software programs such as P3 differ from this approach. In P3, if an activity has a 1-day duration, the calendar start and finish dates are on the same day). Activities *A, B,* and *C* are shown in the form of a bar graph in Figure 6.6, where each unit represents a day. When the event times in a CPM network are computed, the time when a preceding activity finishes and a succeeding activity can start is determined. There is no gap in between. Working time only is considered but not rest time. When activity *A* is finished in the afternoon of day 5, the next activity commences. Thus the end of day 5 can be conceived as the beginning of day 6. This is true if resting time is ignored. Activity *A* starts on the morning of June 1 and is finished on the morning of June 8 (considering two days of the rest time). Activity *B* takes 3 days (i.e., June 8–10) and finishes on the morning of June 11, not taking up any time on June 11. This is the start time for activity *C,* which finishes on June 16 at 5 P.M., which is the same as the morning of June 17.

Presentation of information in a schedule, with real start and finish times indicated, is achieved by subtracting 1 from all completion times, as shown in Table 6.2. The method illustrated in Table 6.2 provides a realistic way of preparing the schedule for the activities. One precaution, however, must be observed. In computing the completion time for a dummy, do not subtract 1 or the schedule will indicate completion of the dummy before it actually

TABLE 6.1 Project Calendar[a]

	March						
	Sunday	Monday	Tuesday	Wednesday	Thursday	Friday	Saturday
Calendar dates			1	2	3	4	5
Workdays			1	2	3	4	*
Calendar dates	6	7	8	9	10	11	12
Workdays	*	5	6	7	8	9	*
Calendar dates	13	14	15	16	17	18	19
Workdays	*	10	11	12	13	14	*
Calendar dates	20	21	22	23	24	25	26
Workdays	*	15	16	17	18	19	H
Calendar dates	27	28	29	30	31		
Workdays	*	20	21	22	H		

[a] Asterisks denote weekends and an H is a holiday. These are not added to the total workdays.

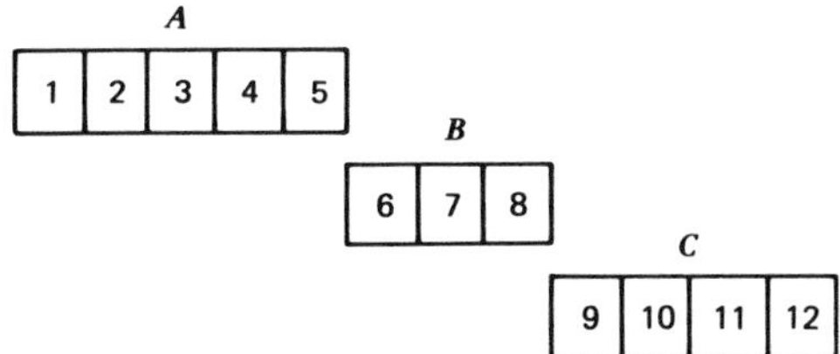

FIGURE 6.6 Bar Graph for Example

begins. In Table 6.2 the earliest finish times have been adjusted, and the corresponding dates have been read off the calendar.

In the previous example, if work had started on day 0 instead of day 1, a similar calendar could be obtained by adding day 1 to the start times. Depending on the situation, 1 is added or subtracted to achieve the real start and finish times for the activities, but the dummies stay the same.

TABLE 6.2 Calculation of Calendar Dates

	Earliest Start		Earliest Finish	
Activity	Day	Date	Day	Date
A	1	01 June	6 − 1 = 5	07 June
B	6	08 June	9 − 1 = 8	10 June
C	9	11 June	13 − 1 = 12	16 June

6.6 DATES FROM MULTIPLE CALENDARS

In Table 6.3, consider three activities with 5-, 6-, and 7-day workweeks, and with different start times from the previous example.

Assuming that the activities take place in June and July, we want to give the dates for the activity times. We can use the calendars in Figure 6.7 for June and July which show 5-, 6-, and 7-day workweeks. Since activities *A*, *B*, and *C* each use a different workweek, the activity times can be located on the calendar for the corresponding workweek—5, 6, or 7. Then the date can be read off the calendar from the activity time. For example, if we want the late start date for activity *A*, we take the late start time, which is 20, and read off the corresponding date from the 6-day workweek calendar. The date is June 23. The other results are as listed in Table 6.4.

Similarly, if we want the early start date for activity *C*, we take the early start time, which is 7, and read off the corresponding date from the 5-day calendar. The date is June 9.

TABLE 6.3 Example of Network Calculations

Activity	Duration	ES	EF	LS	LF	Number of Days in Work Week
A	2	3	4	20	21	6
B	4	8	11	22	25	7
C	3	7	9	23	25	5

Consider the same three activities, but with start times adjusted for activities not dependent on each other, as shown in Figure 6.8 and Table 6.5.

Assuming that the activities take place in June and July, we want to give the dates for the activity times. We can use the calendars for June and July shown in Figure 6.7(c).

For activity *A*, which is a 6-day workweek activity, the activity times can be matched to the dates by reading off the date against the activity time occurring in the 6-day workweek row. The results are:

Activity	Duration	ES	EF	LS	LF
A	2	June 3	June 4	June 23	June 24

S	M	T	W	T	F	S	S	M	T	W	T	F	S
JUNE 1983							JULY 1983						
			1	2	3	4						1	2
			1 1 1	2 2 2	3 3 3	 4 4						23 27 31	2 28 32
5	6	7	8	9	10	11	3	4	5	6	7	8	9
  5	4 5 6	5 6 7	6 7 8	7 8 9	8 9 10	 10 11	 33	24 29 34	25 30 35	26 31 36	27 32 37	28 33 38	 34 39
12	13	14	15	16	17	18	10	11	12	13	14	15	16
  12	9 11 13	10 12 14	11 13 15	12 14 16	13 15 17	 16 18	 40	29 35 41	30 36 42	31 37 43	32 38 44	33 39 45	 40 46
19	20	21	22	23	24	25	17	18	19	20	21	22	23
  19	14 17 20	15 18 21	16 19 22	17 20 23	18 21 24	 22 25							
26	27	28	29	30									
  26	19 23 27	20 24 28	21 25 29	22 26 30									

(*a*) (*b*)

S	M	T	W	T	F	S	S	M	T	W	T	F	S
JUNE 1983							JULY 1983						
			1	2	3	4						1	2
			1	2	3	4						28	
5	6	7	8	9	10	11	3	4	5	6	7	8	9
5	5 6	6 7	7 8	9 8	10 9	10							
12	13	14	15	16	17	18	10	11	12	13	14	15	16
	11 11	12	13	14	15	16							
19	20	21	22	23	24	25							
	17	18	19	20	21	22							
26	27	28	29	30									
23	24	25	26	27									

x — 5-day WORKWEEK
y — 6-day WORKWEEK
z — 7-day WORKWEEK

FIGURE 6.7 Multiple Calendars

TABLE 6.4

Activity	Duration	ES	EF	LS	LF	Days in Workweek
A	2	June 3	June 4	June 21	June 24	6
B	4	June 8	June 11	June 22	June 25	7
C	3	June 9	June 13	July 1	July 5	5

Since activity *B* succeeds activity *A*, the schedule dates of activity *B* must follow the schedule dates of activity *A*.

Using the same procedure for finding the schedule dates as in the previous cases, one would expect the results to be as follows for the 7-day workweek of activity B.

Activity	Duration	ES	EF	LS	LF
B	4	June 5	June 8	June 22	June 25

A problem arises here though. The late start date for activity *B* is before the late finish date for activity *A*, which is not possible. In this case, instead of using June 1 for the starting position, we use June 24 (from the LF date of activity *A*) on the 6-day workweek calendar. Then we start counting off the workday 22 on the 7-day workweek calendar. The date corresponding to workday 22 is June 25. This procedure also has to be done for the LF time. Using June 24 as the starting point (from the LF time of activity *A*), we count off to workday 25 for the 7-day workweek calendar. The date corresponding to 25 in table 6.5 is June 28. The ES and EF dates for activity *B* now become:

Activity	Duration	ES	EF	LS	LF
B	4	June 5	June 8	June 25	June 28

Since activity *C* succeeds activity *B*, the schedule dates of activity *C* must be later than the schedule dates of activity *B*. Therefore the workday corresponding to June 8 (from the EF of activity *B*) on the 7-day workweek calendar is used as the starting position for the early start time of activity *C*, and we start counting off to workday 9 and workday 11 on the 5-day workweek calendar. Hence the ES and EF dates of activity *C* are June 9 and June 13, respectively.

A → B → C
2 4 3

FIGURE 6.8 Example of an Activity Chain

TABLE 6.5 Example of Network Calculations

Activity	Duration	ES	EF	LS	LF	Days in Workweek
A	2	3	4	20	21	6
B	4	5	8	22	25	7
C	3	9	11	26	28	5

In the same manner, the workday corresponding to June 28 (from the LF of activity *B*) on the 7-day workweek calendar is used as the starting position for the late start time of activity *C*. Then we start counting off to workday 26 and 28 on the 5-day workweek calendar. Therefore the LS and LF dates of activity *C* are June 29 and July 1, respectively.

It should be noted that for relatively large projects, computers are used to calculate the schedule dates on multiple calendars, but the user of these schedules must know how these calculations are performed, which is the purpose of this section. Some software packages use a MERGE command to consolidate three schedules or calendars into one.

6.7 SCHEDULING ACTIVITIES

In this section the influence of restraints in CPM is investigated. The three types considered are the following:

1. Delayed starts
2. Deadlines
3. Required completion dates

In a project certain activities cannot begin until certain time-restricting events have occurred. For example, in northern regions the greater protion of the construction of an earthen dike cannot begin until the frost has left the ground. In other cases certain construction operations cannot begin until the necessary materials have arrived. Thus the start of immediately succeeding activities may be delayed. Figure 6.9 shows a section of a network where, because of financial problems, activities *B* and *C* cannot start until day 30. This restraint is shown in Figure 6.10 by an inverted triangle positioned at node 8. On a given chain of activities only the early event times of the preceding nodes remain unchanged by the inclusion of such a restraint.

In many cases activities have to be completed by certain dates to meet deadlines. In Figure 6.9, on completion of activities *R*, *S*, and *T*, a distinct part of the project is finished. The owner plans to use this part, starting on day 24. In Figure 6.10 this deadline is shown in an inverted triangle positioned at node 7. The inclusion of such a restraint affects only the late event times of the node involved and the preceding nodes in a given chain of activities.

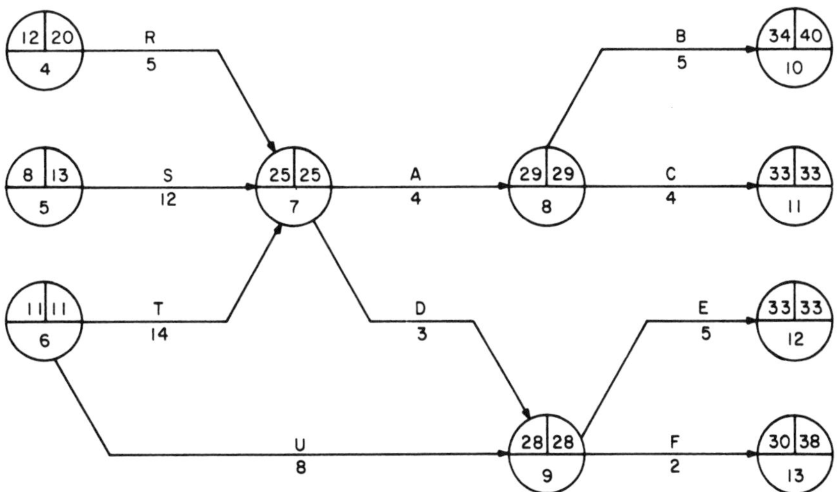

FIGURE 6.9 Example of a Network with No Restraints

In Figure 6.9 activity *F* is an inspection, but the inspector is available only on day 31. Hence activities *D* and *U* must finish by day 32, and activities *E* and *F* can start only on or after day 31. In Figure 6.10 the required completion date is shown in an inverted triangle positioned on node 9, which must occur on day 31. Only the early event times of the succeeding nodes

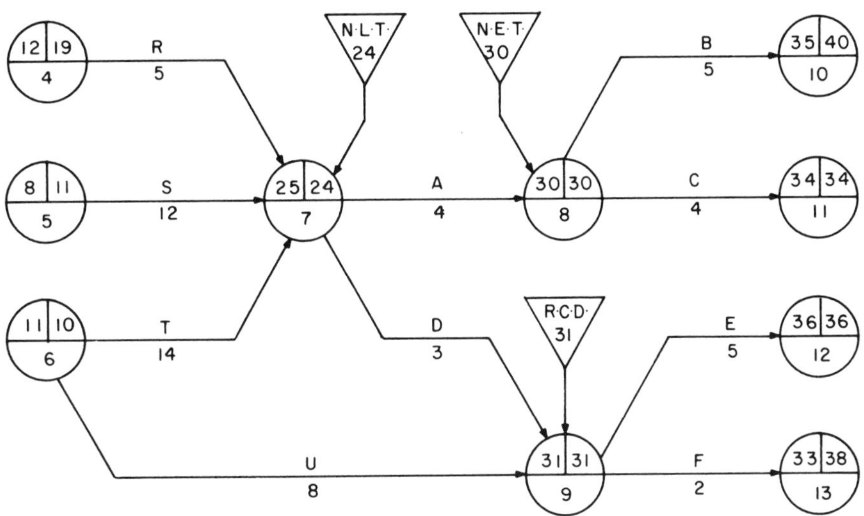

FIGURE 6.10 Example of a Network with Restraints

and the late event times of the preceding nodes in a given chain of activities are affected by the inclusion of this type of restraint.

For a better understanding of the effects of restraints, the reader should carefully compare the event times of Figures 6.9 and 6.10. It is noteworthy that event 7 has a negative slack. A decision will have to be made before passing on the schedule to field engineers for implementation as to how this negative float will be eliminated. Possibly the duration of activity *T* can be reduced to 13 days by using overtime.

Unlike the examples just discussed, in many practical situations only one or two activities about a node are restrained. More detailed representation may be needed on the network in these cases. For the network in Figure 6.10 it is now determined that only *B* and not *B* and *C* must be delayed to day 30. In Figure 6.11 dummy 8–10*A* clearly illustrates the new situation. Referring again to Figure 6.10, it is found that the day 24 deadline need be met only by activity *R*. In Figure 6.11 dummy 7*A*–7 shows that only *R* must be completed by day 24. Finally the situation at node 9 in Figure 6.10 changes in Figure 6.11. Although *E* and *F* still must start on day 32, with the aid of dummy 9*A*–9 it is clear that only *D* and not *U* has to finish on exactly day 31.

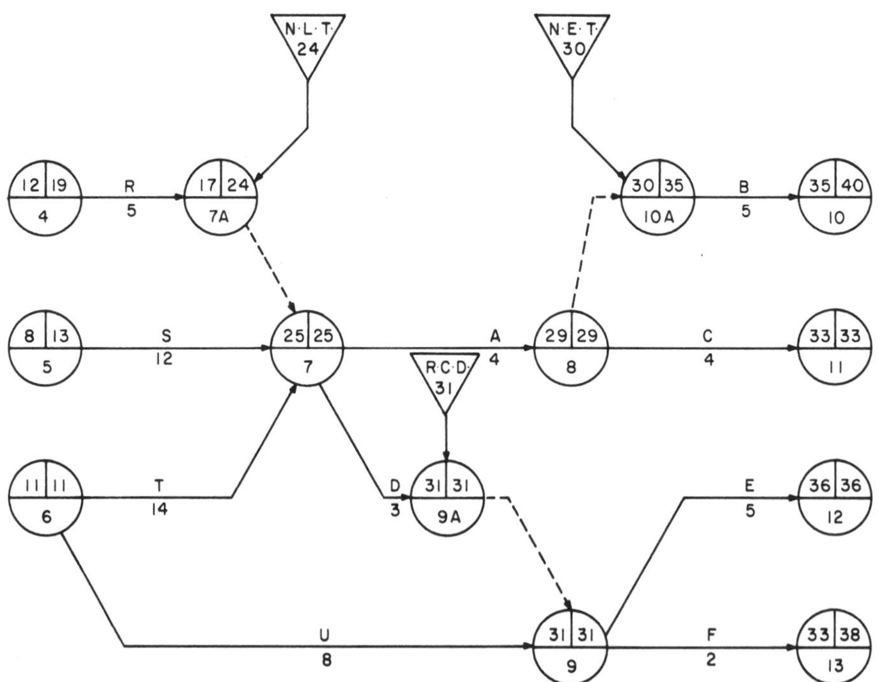

FIGURE 6.11 Example of a Final Network with Additional Restraints

6.8 STEPS TO DETERMINE A SCHEDULE

The preceding sections have dealt comprehensively with the critical path method. As a review, these are the steps to follow in determining a schedule:

1. Complete the CPM network diagram.
2. Label each activity with its proper description.
3. Check network logic.
4. Estimate activity durations.
5. Identify each event with a unique node number.
6. Determine the early event times by implementing the forward pass computations.
7. Determine the late event times by implementing the backward pass computations.
8. Compute total, free, and independent float, and identify the critical activities on the network.
9. Calculate calendar dates for event and activity times.
10. If project duration is not suitable, either revise the network or reschedule the start/completions of the event/activities, and once again go through Steps 5–8.
11. If the project duration, starts, and finishes of activities are suitable, arrange the activities in the desired sequence, and tabulate activity identification, description, duration, early and late starts, early and late finishes, and floats.
12. Finally, identify the critical activities.

6.9 CPM BAR CHARTS

This section is an extension of Chapter 3 and illustrates how an AOA Network is converted to a bar chart.

Construction personnel at the site sometimes prefer to use bar charts instead of CPM. A bar chart that does not show relationships between the activities is archaic. However, it is not difficult to draw a bar chart based on CPM computations.

The immediate prerequisite of a CPM bar chart is a network, which provides the times associated with the activities as well as the relationships of the activities. Bar charts utilize bars or thick lines to represent activities. These bars are drawn horizontally according to a time scale laid out across the chart. The length of each bar therefore represents the estimated duration of the activity. Each bar has two identification numbers, as in CPM, one at the beginning of the bar and the other at the end of the bar. These are the node numbers used to identify an activity on the CPM network.

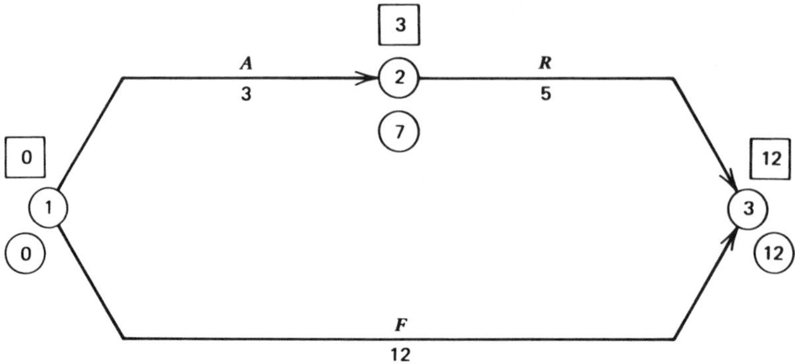

FIGURE 6.12 Example Project—AOA Network

The bar chart can illustrate an activity, its duration, its identification and thus its relation to other activities. The next thing that must be shown by a bar chart computed by CPM is float. For example, in Figure 6.12 activity *A* has a float of 4, which activity *R* shares. On a bar chart the float is illustrated by a dotted line. This float, however, can be used in three different ways for activity *A*. It can precede the activity, succeed it, or partly succeed and partly precede it. This necessitates a decision on the use of float. Is one activity to be assigned this float, or is it to be distributed over several activities? Where float is used, and how it is to be used, is very important and must be determined before starting the CPM bar chart. In the rest of the illustrations in this section all activities will begin at their earliest start times.

Figure 6.13 illustrates a CPM bar chart for the CPM network shown in Figure 6.14. The identification numbers on each bar, as shown in Figure 6.13, indicate the relationship among activities. In addition bars are sequenced in ascending order of *j* major, *i* minor. The CPM bar chart has been drawn by going through the following step-by-step procedure:

		Months									
Activity		1	2	3	4	5	6	7	8	9	10
1–2	(*A*)	1			2						
2–3	(*B*)				2				3		
1–4	(*E*)	1				4					
2–4	(*D*)				2	4					
3–5	(*C*)								3	5	
4–5	(*F*)					4				5	

FIGURE 6.13 Example Bar Chart Based on AOA Network

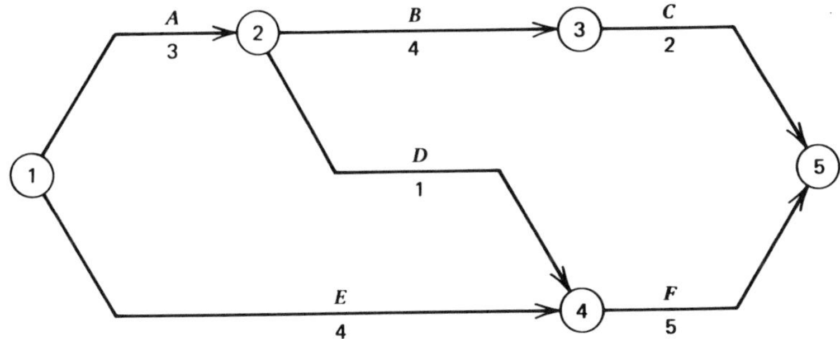

FIGURE 6.14 Example AOA Network

1. Complete the CPM network diagram before beginning.
2. Construct or obtain a blank table for drawing the bar chart and fill in the time units.
3. List all the activities in the leftmost column.
4. Draw each activity in ascending order, *j–i*.
5. Represent the duration of every activity by the length on the activity bar.
6. Begin every activity at its scheduled start time.
7. Identify each bar by putting the start and finish node numbers obtained from the CPM network at the start and finish of the respective activity bars.
8. Indicate any float by dotted lines.

All activities and their identification can be shown on a bar chart along with float.

Note that the CPM bar chart identifies the activities, illustrates the logic, indicates float, is useful in monitoring progress, and because of familiarity, is more acceptable to construction personnel. One disadvantage it does have, and which the CPM network does not, is rigidity. As mentioned earlier, the earliest start or latest finish must be decided before the activity bars can be placed on the bar chart. If the project falls behind or gets ahead of schedule, and certain changes are required in the timing of certain activities, the CPM bar chart has to be redrawn because of the rigid sequencing that is followed. However, with computerized scheduling packages, this is not a serious problem.

PROBLEMS

6.1 List the criteria to determine a critical activity. Is it possible to have a discontinuous critical path? Indicate the critical path on the following network.

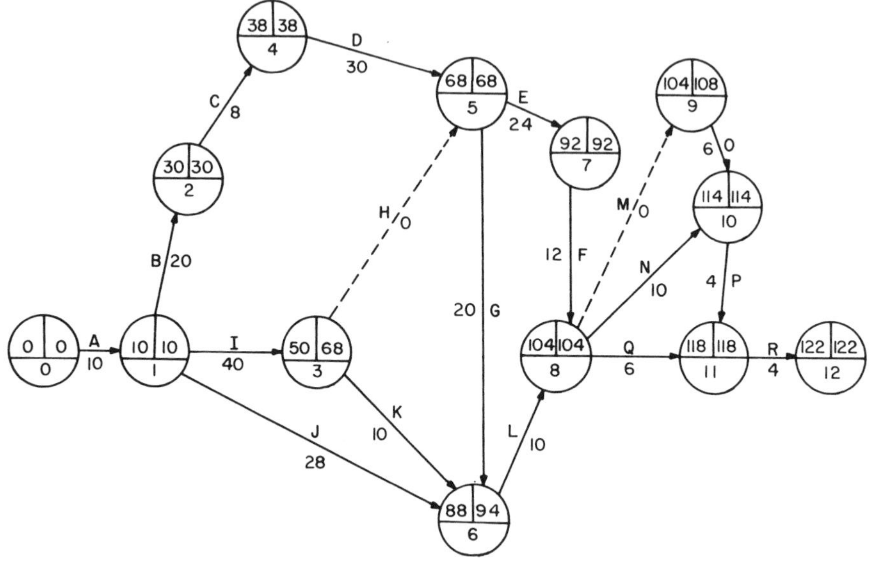

PROBLEM 6.1 Network

6.2. The activities in a land assembly project involving three streets of approximately one-half mile each are:

1. Clear and grout
2. Excavate trenches
3. Lay sewer pipes
4. Install manholes
5. Lay water mains
6. Provide hydrants
7. Backfill
8. Provide curb and gutter
9. Subgrade
10. Grade
11. Pave

Draw up a CPM network when there is one crew available for each operation, and the work of a crew has to be finished on one street before it can be sent on to the next street. Estimate durations for the activities. Make a schedule. Identify the critical path. Select the operations you can delay without affecting the schedule. If you can expedite certain operations, revise their estimated durations. Draw up another schedule and compare the critical path in the two networks. Which schedule do you prefer? Give reasons.

6.3 Draw the CPM network for the following activities and determine the critical path. Tabulate the earliest and latest start and finish dates when the project starts on June 1, 1983.

Sequence	Activity	Duration	Type of Activity (Days)
1–2	*A*	1	5
2–3	*B*	5	6
2–4	*C*	3	6
2–5	*D*	4	5
3–5	*E*	0	5
4–5	*F*	2	Done on Wednesday and Thursday
4–6	*G*	5	7
4–7	*H*	9	6
5–7	*I*	4	5
6–7	*J*	2	Sundays only
7–8	*K*	2	5

Type of activity given refers to the number of days in a week an activity can be worked on.

6.4 (a) The activities in the following diagram form part of a network. Complete the event times for the network so that the deadline is met. (b) Assume only activity *A* is to be influenced by this restraint; redraw

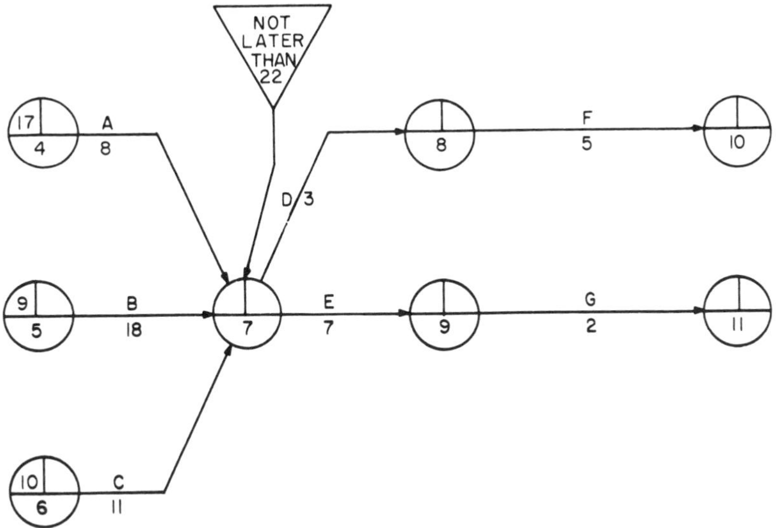

PROBLEM 6.4 Network

the network to represent this situation and compute the event times. Assume activities 8–10 and 9–11 are critical in both cases.

6.5 Construct a project schedule for the gas bar described when the project starts on December 1, 1983. Assume (a) a 7-day week for supply items and (b) a 6-day week for remaining items.

The gas bar is proposed to be built on an already developed site. It will consist essentially of the sales outlet and the office block. The cash office and pumps comprise the sales outlet. The manager's office building, which also houses public washrooms and an air compressor, is called the office block. Adjacent to the pumps will be a concrete pit that will house the gasoline tanks.

The entire area, excluding the office and pumps site, is covered with a concrete slab, and there is a low perimeter wall in the rear. The utility company has undertaken to install an electric meter on the site and connect it to the mains. Gasoline pumps must be obtained from the manufacturers, and after being installed, they are to be connected to the gasoline tanks and the power supply. Before use they must be inspected by the local authority to ensure safety and compliance with regulations. Gasoline tanks are housed in concrete pits and covered by the concrete slabs. Before they are covered, however, the tanks and the associated pipework have to be inspected by the local authority. The sales outlet base is excavated first, the pipework and tanks second, the office block third, and the trench for underground services last.

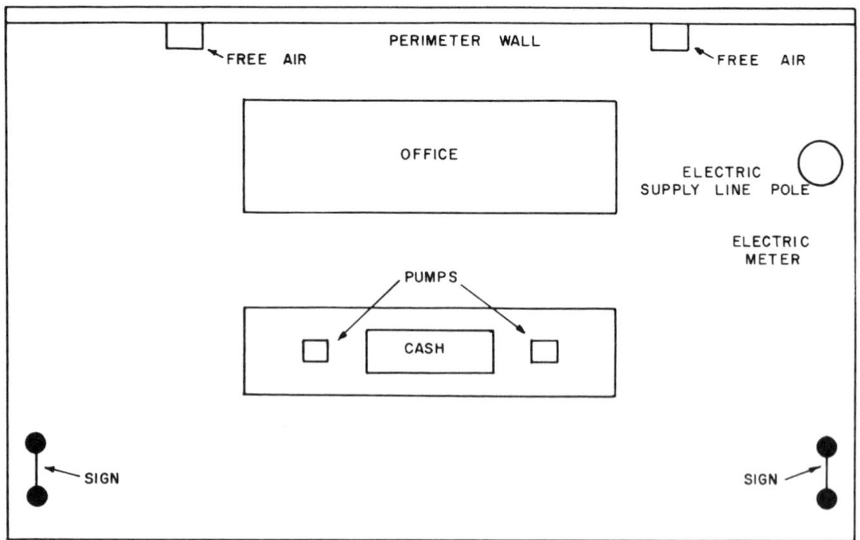

PROBLEM 6.5 Network

After the excavation for the tanks and pipework is completed, work can proceed on the construction of the perimeter wall and air points.

The insurance company insists that the manager's office have an efficient burglar alarm. Once these alarms are installed, the insurance company will inspect them. The alarms are powered by the main electricity. All furnishings for the office and washrooms must be specially obtained and, when delivered, must be stored under cover to prevent weather damage. Some furnishings will require painting after being placed in position.

Compressed air for inflated tires will be supplied by an electrically driven compressor, which must be inspected by a competent person before the compressor is put into use. The air lines to the "free air" points are installed with the general underground services, and the points themselves are mounted on the perimeter wall. The air points can be hooked up after the concrete slab has been poured.

To advertise the gas bar, signs will be erected on the approach road. Sites for these signs have been earmarked generally, but actual site selection will not be made until construction of the project is started. It is expected that the signs will be in position by the time the gas bar is ready for use by the public.

Mobilization to start work comprises, among other preparations, the moving of a trailer to the site to store tools, furnishings, and any weatherprone parts and to serve as the site office. Similarly, when work at the site is completed, the trailer will be removed, and all scaffolding and construction equipment taken away. This is known as "demobilization and cleanup of site."

List of Activities	Days Required
1. Excavate for sales outlet	1
2. Construct sales outlet base	1
3. Construct cash office	2
4. Obtain pumps	16
5. Install pumps	2
6. Connect pumps	1
7. Inspector approves pump installation	2
8. Obtain office furnishings	8
9. Paint and furnish office and washrooms	2
10. Connect office and toilet lighting	1
11. Excavate for office block	1
12. Construct office block	1
13. Build offices and washrooms including all services	15
14. Install burglar alarm	1
15. Connect burglar alarm	2
16. Insurance company inspects burglar alarm	2
17. Utility company installs elctric meter	14
18. Connect main cable to meter	1

List of Activities	Days Required
19. Install area lighting	4
20. Mobile site	1
21. Set out and level site	1
22. Excavate trench and lay all underground services	1
23. Excavate for pipework and tanks	1
24. Construct concrete pit	2
25. Obtain pipework and tanks	2
26. Install pipework and tanks	3
27. Obtain compressor	10
28. Install compressor	1
29. Connect power to compressor	1
30. Inspection of compressor	2
31. Backfill and cover tanks	1
32. Pour concrete slab	2
33. Construct perimeter wall including air points	2
34. Connect air points	1
35. Demobilize and clean site	1
36. Obtain approach road signs	4
37. Select site for approach road signs	2
38. Erect approach road signs	1
39. Inspection of pipework and tanks	2

6.6 It is proposed to construct a wharf for the docking of large fishing vessels. The wharf consists of a rock-fill causeway with topping and armor stone extending to approximately one-half the length of the total structure. The remainder of the wharf consists of four rock-filled concrete cribs over which are placed concrete box girders, deck beams, and a poured in place, reinforced concrete deck.

The rock fill, armor stone, and causeway topping material are available locally. The box girders and deck beams are precast and delivered to the site ready for placement. Reinforcing steel and a concrete supply are readily available, and all concrete framework is prefabricated on the site.

The concrete cribs are constructed in the following manner. When the causeway is rock filled to one-half its length, a crib launchway consisting of a tipping structural steel platform, prefabricated on shore, is erected on the causeway. On this platform the crib base is poured, slipforms are installed, and the sides are poured to a height of approximately 10 ft. After a short curing period the platform is tipped, and the crib is launched. It is then towed to deeper water and poured to the required height. Two sets of slipforms are used, and concrete is supplied from a barge.

Before the cribs are placed, the area is dredged and a rock-fill mattress with a layer of topping is prepared for each crib. After placing, the cribs are rock filled to one-half their height before the box girders are placed from the barge. Deck beams are placed from the barge over the box girders, and the concrete for the deck is pumped from the causeway. After curing, the wheelguards and capstans are placed, and the wharf is ready for operation.

Draw a CPM network for the wharf. Compute and tabulate activity times and the total float of each activity.

List of Activities	Weeks Required
1. Fabricate crib launchway	6.0
2. Erect crib launchway on causeway	1.0
3. Pour crib base	0.2[a]
4. Formwork for crib base	2.0[b]
5. Pour concrete to 10-ft height	0.4[a]
6. Cure crib to 10-ft height	1.0[a]
7. Launch crib	0.2[a]
8. Tow to position	0.2[a]
9. Dredge	1.0[a]
10. Place rock mattress	1.0[a]
11. Place topping on rock mattress	0.4[a]
12. Place crib	0.4[a]
13. Half fill crib with rock	1.0[a]
14. Place box girders	0.6
15. Place deck beams	1.4
16. Form edges of deck	0.8
17. Place rebar on deck	3.0
18. Set anchors for wheelguards	0.8
19. Pour concrete for deck	1.4
20. Install prefabricated wheelguards	1.2
21. Install capstans	1.6
22. Pour concrete above 10-ft height	2.0[a]
23. Place rebar for crib	1.0[a]
24. Install slipforms	0.4[a]
25. Formwork for deck	1.5
26. Rockfill for causeway (first half)	1.5
27. Armor stone	3.0
28. Causeway topping	1.0
29. Strip forms	0.2[a]
30. Prefab box girders	2.0
31. Prefab deck beams	2.0
32. Prefab wheelguards	1.0
33. Obtain capstans	6.0
34. Rockfill for causeway (second half)	2.5

[a] The duration is given for one crib or span only.
[b] Durations with this sign will be reduced to one-third of the original duration when these operations are repeated.

6.7 Write on (a) the scheduling restraints and sorting parameters for schedule reports and (b) conversion of a workday schedule into a calendar day schedule.

6.8 For the network data given below, draw a CPM bar chart and plot a cumulative curve for expenditure outlay. All activities will start on their earliest start time.

i–i	Duration	ES	EF	LS	LF	$/Activity
1–2	4	0	4	2	6	800
1–5	7	0	7	0	7	3500
1–6	3	0	3	9	12	300
2–3	3	4	7	6	9	3000
3–4	2	7	9	9	11	200
4–8	6	9	15	11	17	3600
5–6	4	7	11	8	11	3600
5–7	2	7	9	7	9	1000
6–8	5	11	16	12	17	1000
7–8	8	9	17	9	17	5600

CHAPTER 7

PRECEDENCE NETWORKS: ACTIVITY-ON-NODE (AON)

Precedence diagraming method (PDM) is another graphical representation technique for project scheduling. It too determines the critical activities and is sometimes also known as AOA. In this text, however, AOA and AON are treated as two separate methods of planning. AON models repetitive and dovetailed work more efficiently than AOA, and it is therefore preferred to AOA when the work is repetitive and many activities are undertaken concurrently.

7.1 WHEN PRECEDENCE IS USED

When project activities overlap, they must be split, to establish their logical sequence in AOA. Figures 7.1 and 7.2 model the same situation through AOA and AON, respectively. The superiority of precedence in this case is demonstrated by the relative simplicity of its diagram. Two sets of four activities are each reduced to one, and no dummies are required. Activities are represented by rectangles. The layout of activity information within the rectangle is user defined. Several conventions are followed, as shown in Figure 7.3.

The left edge of the rectangle represents the start of the activity, and the right edge the finish. The activity, totally presented on a node, is called a nodal.

Spaces indicated by ES, EF, LS, and LF are provided for the early start, early finish, late start, and late finish times, respectively. Other symbols often used to show AON nodes are shown in Figure 7.3a, b, and c.

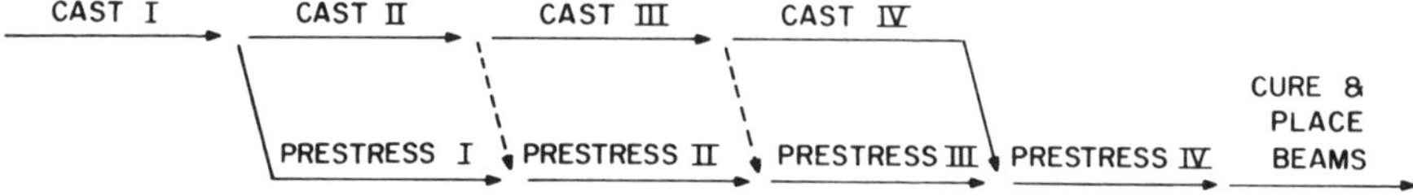

FIGURE 7.1 CPM Model for Production of Beams

7.2 PRECEDENCE RELATIONSHIPS, LEAD AND LAG

Relationships among activities are shown by connecting arrows. The network flows from left to right. The bottom arrow in Figure 7.2 shows the start-to-start relationship (start of the succeeding after the start of the preceding). The upper arrow shows the finish-to-finish relationship (finish of the succeeding after the finish of the preceding). The arrow between the last two activities in the given precedence network shows the finish-to-start relationship (start of the succeeding after the finish of the preceding). It should be noted that a fourth relationship is sometimes called for in AON networks. It is the start-to-finish relationship (finish of the succeeding after the start of the preceding). Its usage, however, is limited, and generally, the three relationships explained here are sufficient to represent all the relationships of construction activities.

The relatively concise representation of a precedence network is possible because the arrows show not only the sequence of activities but also the lead and lag times for the start and finish of each activity. The lead and lag times, often referred to a lead and lag factors, are indicated by pairs of subscripts, the first representing the preceding activity and the second the following activity. When activity *i* precedes activity *j*:

FF_{ij} = lag time for a finish-to-finish relationship. (The succeeding activity finishes this amount of time after the completion of the preceding activity.)

SS_{ij} = lead time for a start-to-start relationship. (The preceding activity starts this much earlier than the start of the succeeding activity.)

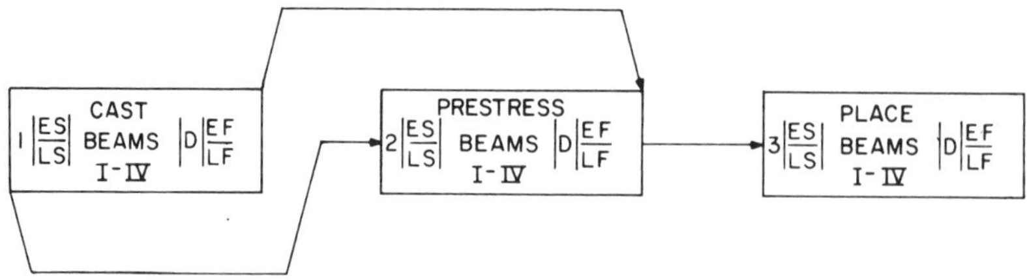

FIGURE 7.2 PDM Model for Production on Beams

(a)

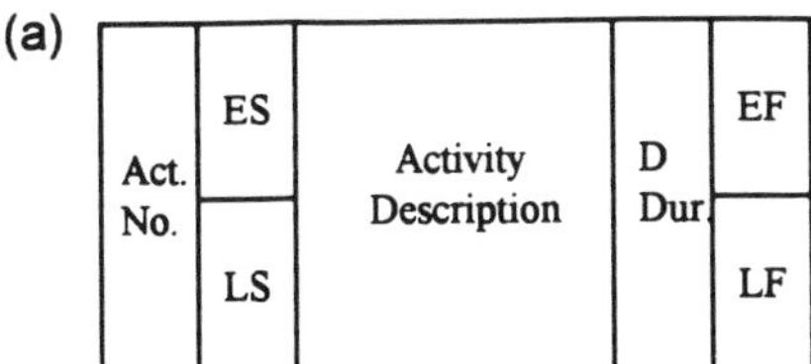

* Note: As used in this book

(b)

ES	Duration	EF
	Task No. Description	
LS	TF	LF

(c) Primavera Project Planner 5.0 (Sample Node)

Activity Number	TF
Activity Description	OD
	RD
ES	EF

* Note: OD = Original Duration
RD = Remaining Duration

FIGURE 7.3 Example of Nodes

FS_{ij} = lag time for a finish-to-start relationship. (The succeeding activity starts this amount of time after the completion of the preceding activity.

SF_{ij} = lead time for a start-to-finish relationship. (The preceding activity starts this much earlier than the completion of the succeeding activity).

7.3 PRECEDENCE LOGIC

Because there are more relationships to be considered than in AOA, more questions must be asked when drawing a PDM network. These are categorized into three groups.

1. Preceding activities.

   Which activities must be finished before this activity may begin? What is the time lag? (Finish to start.)

Which activities must be started before this activity may begin? What is the lead time? (Start to start.)

Which activities must be finished before this activity may be completed? What is the lag time? (Finish to finish.)

Which activities must be started before this activity is completed? What is the lead time? (Start to finish.)

2. Succeeding activities.

Which activities can begin after the finish of this activity? What is the time lag? (Finish to start.)

Which activities can begin after the start of this activity? What is the lead time? (Start to start.)

Which activities can be completed after the finish of this activity? What is the lag time? (Finish to finish.)

Which activities can finish after the start of this activity? What is the lead time? (Start to finish.)

3. Concurrent activities.

Which activities can be carried out at the same time? (Start to start equals zero, that is, SS = 0 in this case.)

7.4 FORWARD AND BACKWARD PASS

During the forward pass, when early starts and finishes are calculated, the predecessors of the activity being analyzed are used and have the subscript *i*. The activity being evaluated always has the subscript *j*. In the backward pass, which determines late starts and finishes, the successors have the subscript k, while the predecessor being evaluated has the subscript *j*. This is illustrated in Figures 7.4 and 7.5.

Similar to AOA, when more than one arrow is incident on an activity, the greatest value is taken on the forward pass and the least value on the backward pass. The latest finish time equals the earliest completion time for the entire project.

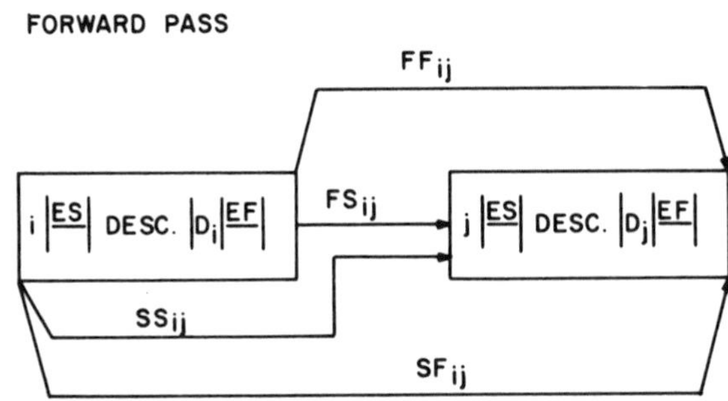

FIGURE 7.4 Lead/Lag Relationships for Forward Pass

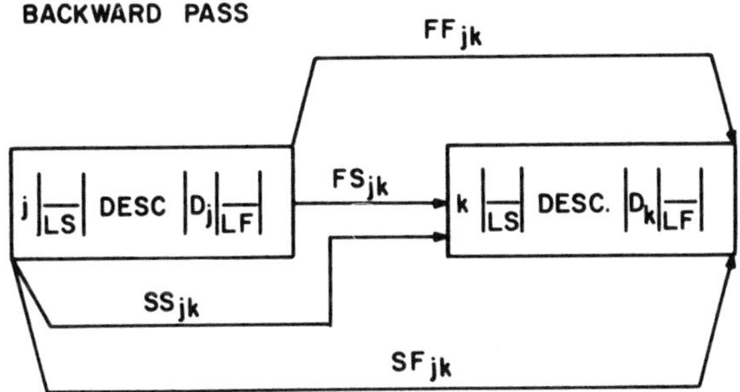

FIGURE 7.5 Lead/Lag Relationships for Backward Pass

The following equations show the necessary computations. (Refer to Figures 7.4 and 7.5.)

$$ES_j = \begin{cases} ES_i + SS_{ij} \\ \text{or} \\ EF_i + FS_{ij} \text{ (whichever is greater)} \end{cases}$$

$$EF_j = \begin{cases} EF_i + FF_{ij} \\ \text{or} \\ ES_i + SF_{ij} \\ \text{or} \\ ES_j + D_j \text{ (whichever is greater)} \end{cases}$$

If there is neither FS_{ij} nor SS_{ij}, and the activity cannot be split, then

$$ES_j = EF_j - D_j$$

and ES_j = project start time, if there is no FS_{ij} and if the activity can be split.

$$LF_j = \begin{cases} LF_k - FF_{jk} \\ \text{or} \\ LS_k - FS_{jk} \text{ (whichever is smaller)} \end{cases}$$

$$LS_j = \begin{cases} LS_k - SS_{jk} \\ \text{or} \\ LF_k - SF_{jk} \\ \text{or} \\ LF_j - D_j \text{ (whichever is smaller)} \end{cases}$$

If there is neither FF_{jk} nor FS_{jk}, and the activity cannot be split, then

$$LF_j = LS_j + D_j$$

and LF_j = project completion time, if activity can be split.

Generally the value of the lead/lag factor (SS_{ij} and FS_{ij}) is given in time units, but it can also be expressed as a percentage of the preceding activity's duration. For example $SS_{ij} = 10\%$ implies that activity *j* cannot start unless activity *i* is 10% completed. Similarly $FS_{ij} = 25\%$ implies that activity *j* can start after the completion of activity *i* and a lag time of 25% of the duration of activity *i*. This idea may be illustrated by Figure 7.4. In Figure 7.5 FF_{jk} may be expressed as a percentage of the succeeding activity's duration. These percentages must be converted to time units as only straight time units are used in all computations. Similarly, FF_{ij} can be expressed as a percentage of the following activity's duration. These are converted to time units as only straight time units are used in all computations.

An uninterruptable activity is an activity which cannot be split; it should be indicated before computations start. In such a case:

$ES_j = EF_j - D_j$ on the forward pass
$EF_j = ES_j + D_j$ on the forward pass
$EF_j - ES_j = D_j$ on the forward pass
$LS_j = LF_j - D_j$ on the backward pass
$LF_j = LS_j + D_j$ on the backward pass
$LF_j - LS_j = D_j$ on the backward pass

Since EF − ES and LF − LS do not necessarily equal the duration of the activity, an interruption is implied. This can occur in two ways: on the forward pass, as shown in Figure 7.6a, and on the backward pass, as shown in Figure 7.6b. The interruption concept is important in the comparison of AOA activity times with AON activity times.

In the case of interruptable activities these relationships will be modified as follows:

$ES_j = EF_j - D_j -$ interruption, on the forward pass
$EF_j = ES_j + D_j +$ interruption, on the forward pass
$EF_j - ES_j = D_j +$ interruption, on the forward pass
$LS_j = LF_j - D_j -$ interruption, on the backward pass
$LF_j = LS_j + D_j +$ interruption, on the backward pass
$LF_j - LS_j = D_j +$ interruption, on the backward pass

Example 7.1

Given the following information, construct the precedence network and using the durations given in the following table compute the early and late activity times.

a. *U* and *R* can be performed concurrently and both begin on the start of the project.

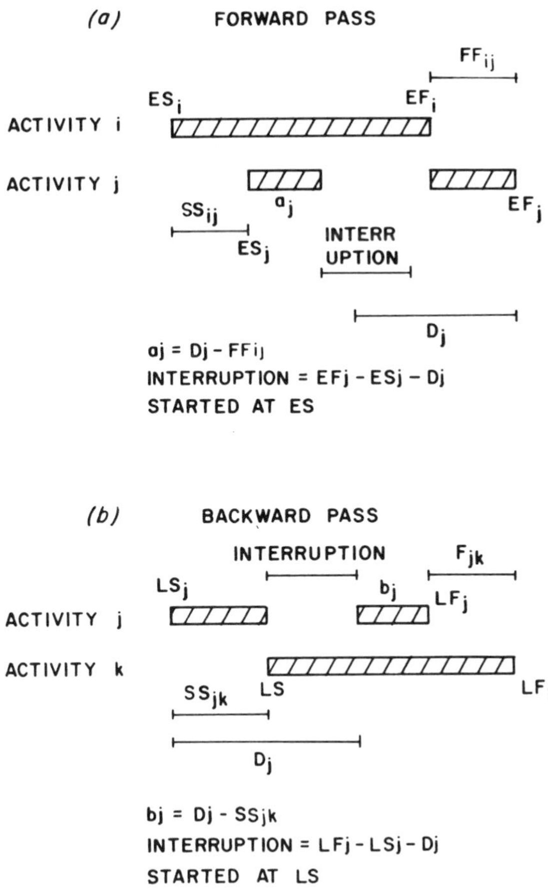

FIGURE 7.6 Interruptions

b. *E* must start before work on *K* can begin.
c. *X* is independent of both *Q* and *K*.
d. Neither *F* nor *G* can start before *R* is started, but *F* and *G* can be performed concurrently.
e. *U* must finish before the completion of *E* and *Q*.
f. *Q* must start before the completion of *J*.
g. *C* is independent of both *F* and *G* but follows *K*.
h. *E* and *Q* can be done concurrently.
i. *H* can begin only after *C*, *K*, and *J* are completed.
j. *H* is the last operation.
k. *X* is dependent on the start of *F* and *G*.
l. *G* cannot be completed until *F* has started.

Assume that activities Q, J, G, and R cannot be interrupted and that activities E and F can be interrupted. The lag times between U and Q, U and E, Q and J, and F and G are 2, 2, 1, and 1 respectively. The lead times from E to X, R to F, F to X, R to G, and G to X are 2, 3, 2, 2, and 5 respectively. There is no lag time between C and H, X and H, and J and H.

Activity	Number	Duration
U	1	1
R	2	3
F	5	5
G	6	4
E	3	2
Q	4	4
K	7	5
X	8	7
J	9	2
C	10	6
H	11	3

The first step in constructing the PDM network is to count the number of activities and draw a node for each activity. Then we join the nodes by arrows to represent the logic of the problem. Finally, we tidy up the network and fill in the descriptions, durations, activity identification numbers, and lead and lag times and compute activity times. Refer to Figures 7.7–7.9.

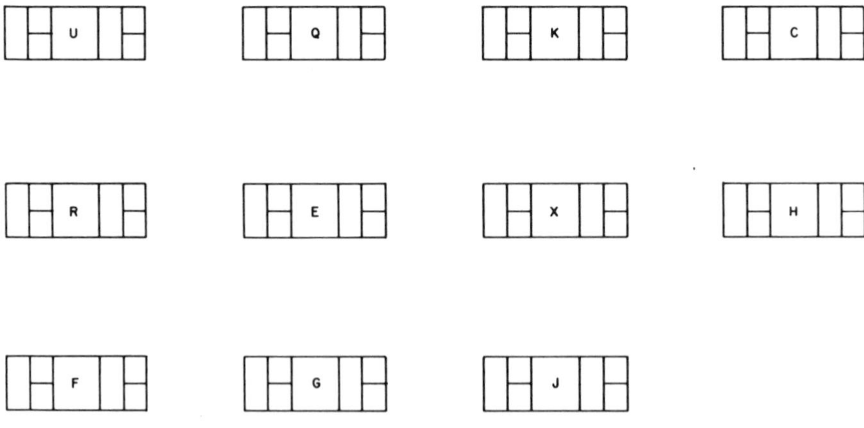

FIGURE 7.7 Starting with the Nodes

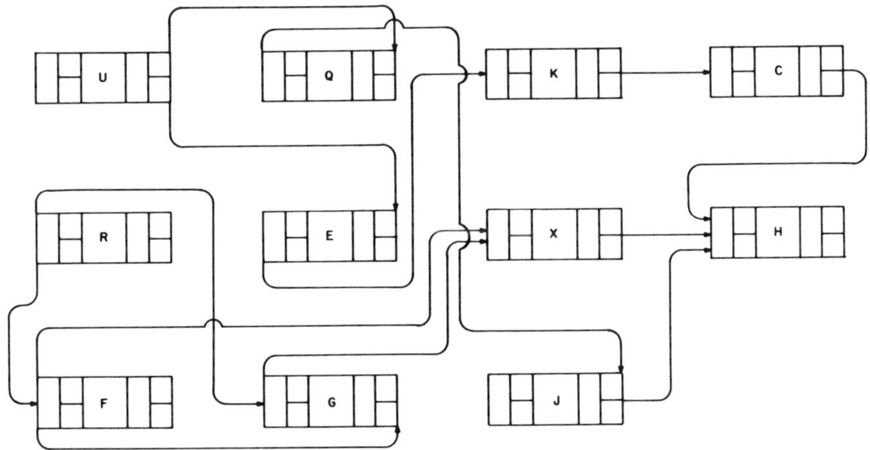

FIGURE 7.8 Connecting the Nodes

EXPLANATION OF ES AND EF TIMES FOR THE NETWORK (FORWARD PASS)

Activity *U* This is the start of the project, therefore the ES time = 0, and the EF time = $0 + D_1 = 0 + 1 = 1$, where D_i is the duration of activity *i*.

Activity *R* This is the start of the project, therefore the ES time = 0, and the EF time = $0 + D_2 = 0 + 3 = 3$.

Activity *E* It is preceded by activity *U* and has a finish-to-finish relationship. Since there is no restraint on the start time, the ES time = 0. On the forward pass we have two times: $0 + D_3 = 0 + 2 = 2$, and from the finish-to-finish relationship, $1 + FF_{1,3} = 1 + 2 = 3$. The greater of the two, which is 3, becomes the EF time for the activity.

Activity *Q* It is preceded by activity *U* and has a finish-to-finish relationship. Since there is no restraint on the start time, the ES time = 0. On the forward pass we have two times: $0 + D_4 = 0 + 4 = 4$ and from the finish-to-finish relationship, $1 + FF_{1,4} = 1 + 2 = 3$. The greater of the two, which is 4, becomes the EF time for the activity.

Activity *G* It has a start-to-start relationship with the preceding activity *R* and a start-to-finish relationship with the preceding activity *F*. The *ES* time = $0 + SS_{2,6} = 0 + 2 = 2$. It has two *EF* times, $2 + D_6 = 2 + 4 = 6$ and from start-to-finish relation $3 + SF_{5,6} = 3 + 1 = 4$. The greater of the two, which is 6, becomes the *EF* for the activity.

Activity *F* It is preceded by activity *R* and has a start-to-start relationship. The ES time = $0 + SS_{2,5} = 0 + 3 = 3$, and the EF

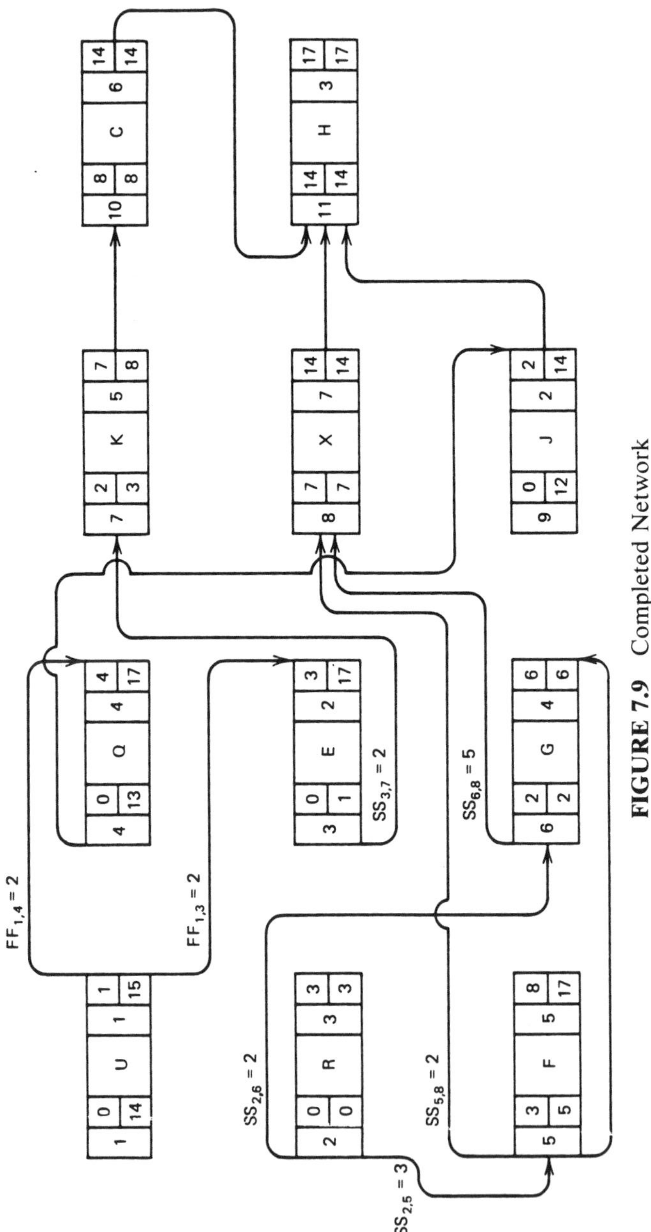

FIGURE 7.9 Completed Network

time = 3 + D_5 = 3 + 5 = 8 which is more than the 2 obtainable from the relationship with *G*.

Activity *K* It is preceded by activity *E* and has a start-to-start relationship. The ES time = 0 + $SS_{3,7}$ = 0 + 2 = 2, and the EF time = 2 + D_7 = 2 + 5 = 7.

The ES and EF of *J, X, C,* and *H* are determined in a similar manner and shown.

EXPLANATION OF LS AND LF TIMES FOR THE NETWORK (BACKWARD PASS)

Activity *H* The LF time = EF time of activity *H* = 17, and the LS time = 17 = D_{11} = 17 − 3 = 14.

Activity *C* The LF time = LS time of activity *H* = 14, and the LS time = 14 − D_{10} = 14 − 6 = 8.

Activity *X* The LF time = LS time of activity *H* = 14, and the LS time = 14 − D_8 = 14 − 7 = 7.

Activity *J* The LF time = LS time of activity *H* = 14, and since the activity cannot be interrupted, the LS time = 14 − D_9 = 14 − 2 = 12.

Activity *Q* There being no restraint on the LF time, the LF time can be the last day of the project, which is 17. The LS time = 17 − D_4 = 17 − 4 = 13, which is less than 14 obtainable from the SF relationship between *Q* and *J*. Also, since *Q* is uninterruptable, the difference between 17 and 13, which is 4, gives 4 days for uninterrupted work on the activity.

Activity *G* The LS time = 7 − $SS_{6,8}$ = 7 − 5 = 2. Since the activity cannot be interrupted, the LF time = 2 + D_6 = 2 + 4 = 6.

Activity *F* The LS time = 7 − $SS_{5,8}$ = 7 − 2 = 5. Since there is no restraint on the LF time, the LF time can be the last day of the project, which is 17.

Activity *R* The LS time = the smaller of 5 + $SS_{2,5}$ = 5 + 3 = 2, and 2 − $SS_{2,6}$ = 2 − 2 = 0. Therefore the LS time = 0. Because the activity cannot be interrupted, the LF time = 0 + D_2 = 0 + 3 = 3.

Activity *K* The LF time = LS time of activity *C*, which is 8, and the LS time = 8 − D_7 = 8 − 5 = 3.

Activity *E* The LS time = 3 − $SS_{3,7}$ = 3 − 2 = 1. Since activity *E* can be interrupted, and there is no restraint on the LF time, the LF time can be the last day of the project, which is 17.

Activity *U* The LF time = the smaller of 17 − $FF_{1,3}$ = 17 − 2 = 15, and 17 − $FF_{1,4}$ = 17 − 2 = 15. In this case, since they are both equal, the LF time = 15. The LS time = 15 − D_1 = 15 − 1 = 14.

Step 1. Eleven activities, therefore 11 nodes are drawn and the description printed in them.

Step 2. Connect the nodes to represent the logic.

Since U precedes Q and a finish-to-finish relationship exists, we draw an arrow from the end of node U to the end of node Q.

Since R precedes G, and a start-to-start relationship exists, we draw an arrow from the beginning of node R to the beginning of node G.

Since K precedes C, and a finish-to-start relationship exists, we draw an arrow from the end of node K to the beginning of node C.

Since Q precedes J, and a start-to-finish relationship exists, we draw an arrow from the beginning of node Q to the end of node J, and so on for the rest of the network.

Step 3. The calculations for the network are completed.

7.5 FLOAT: CRITICAL PATH ANALYSIS

Precedence networks are concerned with all types of floats as in the case of AOA in addition to two other types, namely, start float and finish float.

$$\text{SF} = \text{LS} - \text{ES}$$

$$\text{FF} = \text{LF} - \text{EF}$$

Start float is equal to finish float for an uninterrupted activity. This fact provides a check for the validity of computations.

In AON, criteria similar to those used with AOA are required to determine a critical activity:

1. The earliest start time and the latest start time must be equal.

$$\text{ES} = \text{LS (no start float)}$$

2. The earliest and latest finish times must also be equal.

$$\text{EF} = \text{LF (no finish float)}$$

3. The duration, D, for the activity is equal to the difference between the latest finish and earliest start.

$$\text{LF} - \text{ES} = \text{D}$$

To make the critical path itself more easily distinguishable, the precedence arrows, as well as the activities, are drawn with the use of double lines, as shown in Figure 7.10.

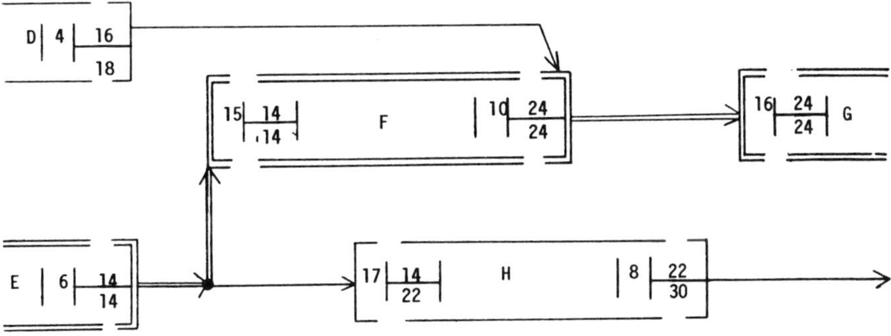

FIGURE 7.10 Highlighting the Critical Path.

When only a part of the work in an activity is critical, as occurs in an interrupted activity, the activity is considered critical. This is done to ensure that such an activity does not escape the additional care given to critical activities. An example of a partly critical activity is shown in Figure 7.11. Activity *D* is critical from time 13 to 18. This is so because six days of work must be done on *D* after activity *B* is complete at time 12; for one day's work to be done before day 13, day 10, day 11, or 12 could have been used. That this activity is critical is also indicated by the fact that it has no finish

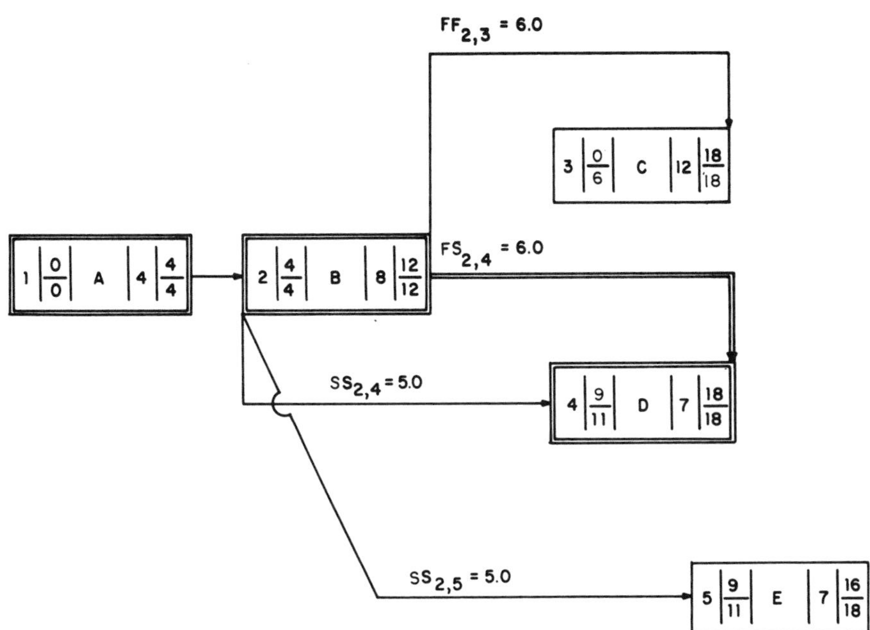

FIGURE 7.11 Illustration of Partly Critical Activity.

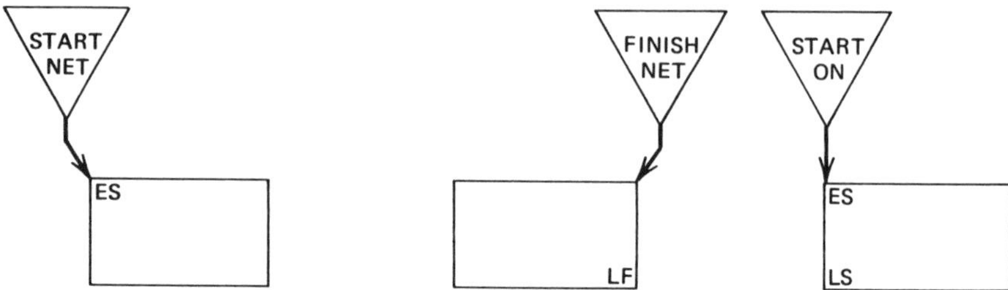

FIGURE 7.12 Scheduling Constraints

float. If this activity had been flagged as a nonsplittable activity, the earliest start of the activity would have been 18 − 7 = 11, and the whole activity would have been critical.

A scheduled event in AOA network represents either the start or finish of any activity. Scheduling in precedence, can be done in a similar manner by assigning zero duration to an activity which can then be treated as an event. For scheduling an activity the necessary time restraint can be placed on either the start or finish of the activity itself, rather than a specific node. Application of scheduling restraints is illustrated in Figure 7.12. Computations are similar to scheduling the start or finish of an activity in AOA.

7.6 ACTIVITY IDENTIFICATION CODE

Precedence diagraming method has another important advantage. Each activity has a unique activity number, rather than being identified by its preceding and succeeding nodes. In addition, a code can be used to identify such things as the organization division, project type, person responsible for the activity, type of resources required, or any other item that the management may wish to have identified. For example, the management of a particular company may decide to set up a code for the following three items:

1. The project involved.
2. The department responsible for the particular type of activity, for example, civil, electrical, or mechanical.
3. The individual responsible for the activity, either someone within the department or someone outside, such as a project manager, chief engineer, or resident engineer.

These items are tabulated, but the coding requirements may vary from company to company. The code can be alphanumeric, as the different elements that must be identified may use both numbers as well as letters:

	Code Number
Project	
Arterial road	1
Urban renewal	2
10-story office building	3
Department	
Civil engineering	C
Electrical engineering	E
Mechanical engineering	M
Structural engineering	S
Responsible individual	
Project manager	10
Chief engineer	20
Resident engineer	30

If the activity's identification number is eight digits long, the code might easily be set up as follows:

	Digit Meaning
Digit 1	Particular project
Digit 2	Overseeing department
Digits 3 and 4	Individual responsible for activity
Digits 5–8	Activity number

For example, if activity 53 refers to the urban renewal project and department E is installing some electrical facilities and is responsible to the chief engineer, identification for the activity would be 2E 10-0053. If a report is required by the project manager to the civil engineering department on the progress of a sewer installation which is activity 70 of the arterial road project, the activity identification would be 1C 10-0070. The activity number is set apart for clarity. When it is not necessary to code the project, department, and so on, the identification code is used only to identify the activity.

7.7 DRAWING PRECEDENCE NETWORKS

Scheduling software packages generally produce good graphic PDM network drawings in conjunction with a plotter or printer. Producing network drawings manually is not a serious production option in today's construction industry. However manual methods may be used in the preliminary stages of developing a network. If manual drawings are attempted there are drawing aids that will facilitate this arduous task. A time-saving technique is to use preprinted outlines of nodes which contain activity information. These nodes are moved about to suit the project logic and visual appearance of the network drawing. Connecting arrows are then drawn to join the nodes.

7.8 COMPARISON OF AOA AND AON COMPUTATIONS

Figure 6.5 shows an AOA network of 26 activities and 25 dummies. The same 51-arrow network is shown by an AON network of seven activities in Figure 7.13. Forward and backward pass computations are made in both cases.

A comparison between the two schedules may show that the ES and LS of activity 5 in the precedence network do not compare with the times computed for event 17 in the AOA network. Similarly, EF and LF times for activity 3 in precedence do not tally with times for event 22 in the AOA network. The reson for the discrepancy is that in AOA networks forward and backward passes generate event times and in AON the same computations result in activity times. Comparison of precedence times should be made with the activity times in AOA. The ES and LS of activity 17–23 in AOA are 10 and 13, as for activity 5 in Precedence. Similarly, EF and LF of activity 16–22 in AOA are 15 and 18, respectively, as for activity 3 in precedence.

Consider the case of activity 2 in precedence, which starts at time 3, has a duration of 8 and an LF of 17. This activity has a float of 3 days and an allowable interruption of up to three days. These interruptions are due to the noncompletion of AOA activities 2–3, 3–4, and 4–10, which causes delays in the start of AOA activities 6–7, 8–9, and 10–16, respectively. This information is not available from the AON network but can be found from AOA, where the activities are much smaller.

7.9 THE VALUE OF PRECEDENCE

It has been stated that the use of precedence diagraming method eliminates all dummy activities and more accurately reflects the sequence of engineering

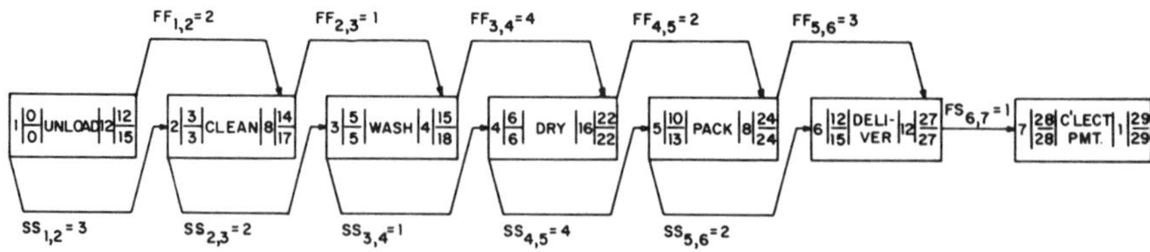

FIGURE 7.13 PDM Network for CPM Network shown in Figure 6.5.

TABLE 7.1 Tabulation for PDM Network—Figure 7.13

Description	Activity	Activities Preceding	Duration	ES	LS	EF	LF	TF
Unload	1	—	12	0	0	12	15	3
Clean	2	1	8	3	3	14	17	6
Wash	3	2	4	5	5	15	18	9
Dry	4	3	16	6	6	22	22	0
Pack	5	4	8	10	13	24	24	6
Deliver	6	5	12	12	15	27	27	3
Collect payment	7	6	1	28	28	29	29	0

operations as they occur in real life. The network is simpler because its size and complexity are greatly reduced by the removal of many unnecessary activities. Subsequently, less information needs to be reported with fewer chances for error. The staff and supervisory personnel also have an easier job, as there is less data to collect and report. It must be noted, however, that this is only true in situations where repetitive work is done and the operations overlap.

The three relationships in AON and the lead/lag factors aid the planner in many ways in the modeling of a project, but at the same time make the activity time computations more involved.

A standard coding system makes activity identification much more familiar across different projects. The coding system is flexible and is designed to meet the specific requirements of a company. Whenever a sequence of operations is rearranged, the activity identification does not require any change as in AOA. AON has the advantage of being able to refer to any aspect of the job, such as the project concerned, the financial section, particular people involved, and so on. It should be remembered that neither network system is inherently better than the other, but that each has certain advantages in particular situations or projects.

PROBLEMS

7.1 Draw an AON network for the land assembly project described in Problem 6.2. Draw up a schedule for the project, starting on December 1, 1994. Assume the crew of workmen observes the same holidays as are observed in your own organization.

7.2 Draw an AON network for the gas bar described in Problem 6.5. Tabulate earliest and latest start times and finish times, and floats. If you can overlap activities, where will you introduce lead and lag times?

Give reasons. Calculate and tabulate start and finish times and floats for the second case. Compare with the table from the first case.

7.3 The construction of a two-lane highway bridge is proposed to replace a single-lane bridge. This bridge will consist of eight pier-supported spans and two abutments, one at each bank (see the following illustration).

The abutments consist of reinforced concrete supported on a mat of 25-cm H-piles. Each of the seven piers located in the river consists of a reinforced concrete pier cap supported on five 600-mm diameter steel pipe piles, which will be driven to a depth of 25 m.

The deck of the bridge consists of a concrete slab supported on prestressed concrete beams which will be precast on shore by a subcontractor. Concrete will be supplied from a nearby town, and the reinforcing steel will be bent on site after it is delivered.

The bridge will be constructed in the following manner. First the excavation will be completed for abutment 1, the H-piles will be driven, and then formwork rebar, concreting, curing, and stripping will be completed. When the H-piles have been driven for this abutment, the crane and pile hammer will be transferred to a barge. During this period the 600-mm steel piles will be made ready for driving and brought to the site. The piles then will be driven in each pier, starting at pier 1 and ending at pier 4, successively further from the finished abutment 1.

As each set of five piles is finished for each pier, the pier cap will be formed, rebarred, poured, and stripped on top of the piles. When the crane and pile hammer have finished pier 4, they will be transferred to shore again, abutment 2 will be excavated, and the H-piles will be driven. Next abutment 2 will be formed, rebarred, poured, cured, and stripped. On completion of H-pile driving, the crane and pile hammer will be transferred to the barge again, and the piles in piers 5, 6, and 7 will be worked on with the caps poured in succession, as was done in piers 1 through 4.

While pile driving, concreting, and so on, take place, a subcontractor will make prestressed beams on shore. There will be 40 beams in all, five to each span. Twenty will be made on one shore, and 20 on the other shore.

On completion of pile driving, these beams will be placed in position using two cranes and barges, starting at one abutment and working toward the other abutment until the eight spans are finished. As the beams are placed on each span a concrete deck will be formed, rebarred, poured, cured, and stripped until the entire deck is completed. Then the curb and handrails will be poured and completed, and the bridge will be finished.

Draw an AON network for the bridge. Compute and tabulate activity times and the total float of each activity:

List of Activities	Weeks Required
1. Set up camp	1
2. Mobilize men and equipment	2
3. Build temporary dock	1
4. Excavation for abutment 1	2
5. Fabrication of 600-mm piles	4
6. Drive 25-cm H-piles	1
7. Fabricate test pile	1
8. Drive test pile	1
9. Form abutment 1	1
10. Rebar abutment 1	1
11. Pour and cure abutment 1	1
12. Strip abutment 1	2
13. Drive 600-mm piles piers 1–4	2/pier
14. Form cap piers 1–4	3/pier
15. Pour and cure caps 1–4	3/pier
16. Strip cap piers 1–4	1/pier
17. Fabricate beams (40)	10
18. Excavate abutment 2	2
19. Drive H-piles abutment 2	1
20. Form abutment 2	1
21. Rebar abutment 2	1
22. Pour and cure abutment 2	1
23. Strip abutment 2	2
24. Drive 600-mm piles piers 5–7	2/pier
25. Form pier caps 5–7	3/pier
26. Pour and cure caps 5–7	3/pier
27. Strip pier caps 5–7	1/pier
28. Place beams (40) starting spans 1–8	3/span
29. Form deck spans 1–8	1/span
30. Pour and cure deck spans 1–8	3/span
31. Strip beam spans 1–8	2/span
32. Form, pour, cure, strip handrail	3
33. Clean up and demobilize	1

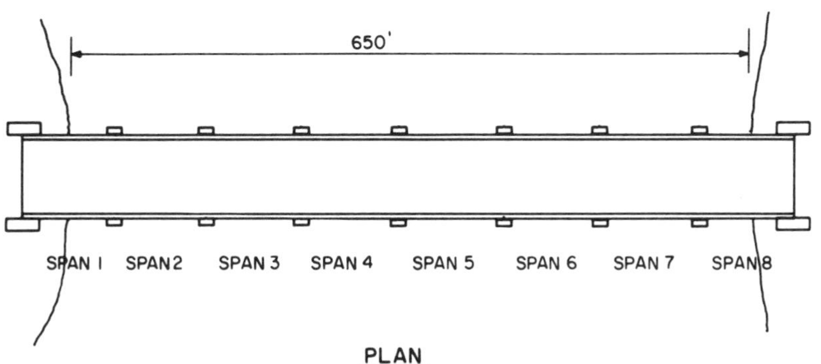

PLAN

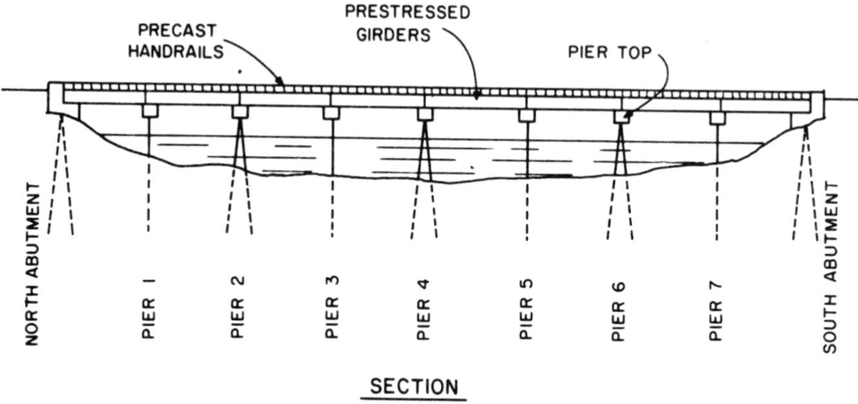

7.4 a. What questions should be asked for developing the logic of a precedence network?

b. What are lag and lead times?

c. Are there any event times in a precedence network?

d. Is it possible to be critical for a precedence activity which has the following times? Give reasons in support of your answer. Do not make any assumptions.

Duration	10 days
Earliest start	0
Earliest completion	10
Latest start	10
Latest completion	20

7.5 Foundations are to be excavated, and forms placed to receive poured concrete. The three activities take the following times:

Excavate foundations	24 days
Place forms	12 days
Pour concrete	18 days

If the three activities are carried out in sequence, each activity being completed before its successor is started, the total time for the project will be 54 days. To reduce this it is decided to start placing forms when only part of the excavating is complete, and pouring concrete when only part of the formwork is complete. Draw the AOA diagram for the situation when work on a successor starts after one third of a predecessor is complete. Assume that only one excavating gang, one formwork gang, and one pouring gang are available, and that these gangs must not be split. Calculate the total duration and show the critical path. Draw an AON network for this problem. Complete activity times, mark the critical path, and compare the start and finish times and float with those in the AOA network.

CHAPTER 8

PLANNING METHODS, TOOLS, AND PROCEDURES

In addition to the different techniques of network planning, there are methods and tools available to the planner that truly optimize the planning process. This process may involve the integration of many subplans, and the coordination of the various planning functions for any given project. In Chapter 2 the planning process was discussed and the process of project breakdown into manageable components was covered. Chapter 3 also covered the organization of projects while the other chapters covered scheduling mechanics. To put a professional project plan together, during the development process all of these plans must be coordinated and an integrated plan generated. The planning functions must also be modularized in the form of subnetworks so that more than one group can contribute to the overall plan. In addition, a hierarchy of plans is often useful as it facilitates different view levels of the project. This chapter covers many of these issues. It also considers planning methods and their interrelationship with network techniques to achieve the objective of project planning.

8.1 MILESTONE NETWORKS AND SUBNETS

Milestone networks are discussed in this book not as a recommended procedure but rather as a technique for breaking the project into subnets. Subnets are anchored at milestones, as illustrated in Figure 8.1.

Milestone schedules, whether they are bar charts or networks, are summary level schedules for project management. Summary level schedules are usually referred to as management level schedules or master plans.

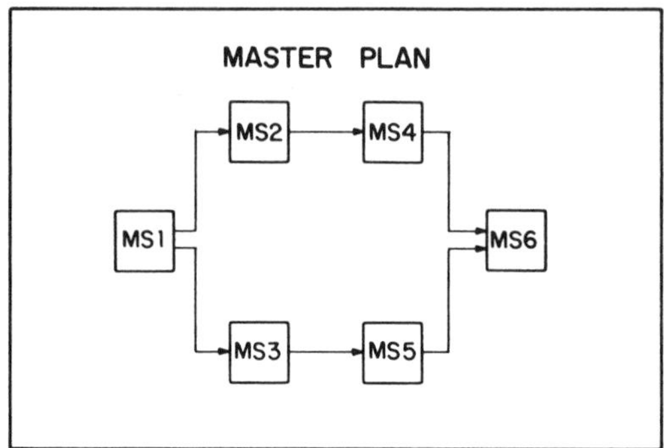

FIGURE 8.1 Milestone Network

Management is primarily interested in the attainment of project objectives rather than in the performance of individual tasks. The scheduling engineer obtains from management the necessary information during planning, and embeds these objectives as important milestones in the master plan. The development of a plan starts at the summary level, and details are filled in as the project proceeds and additional information becomes available. The summary level plan therefore is the master plan which, subject to periodical revisions, exists throughout the life of the project. Figure 8.2 further illustrates the use of milestone network as the basis for a master schedule or master plan.

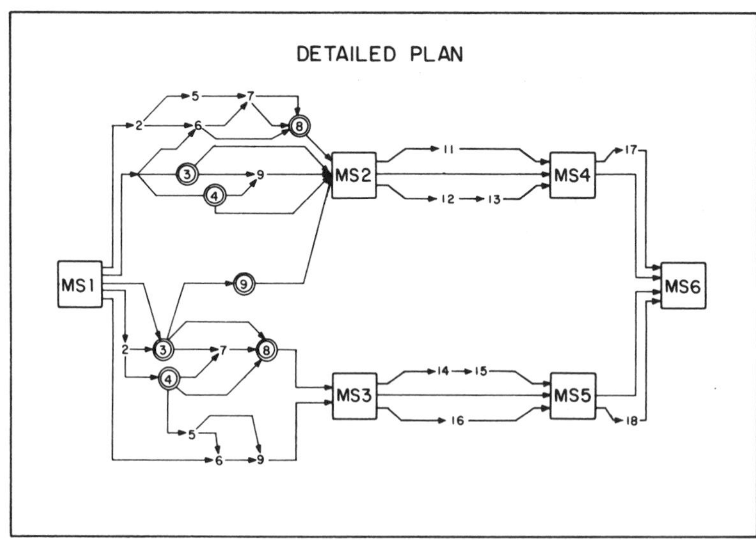

FIGURE 8.2 Master Plan

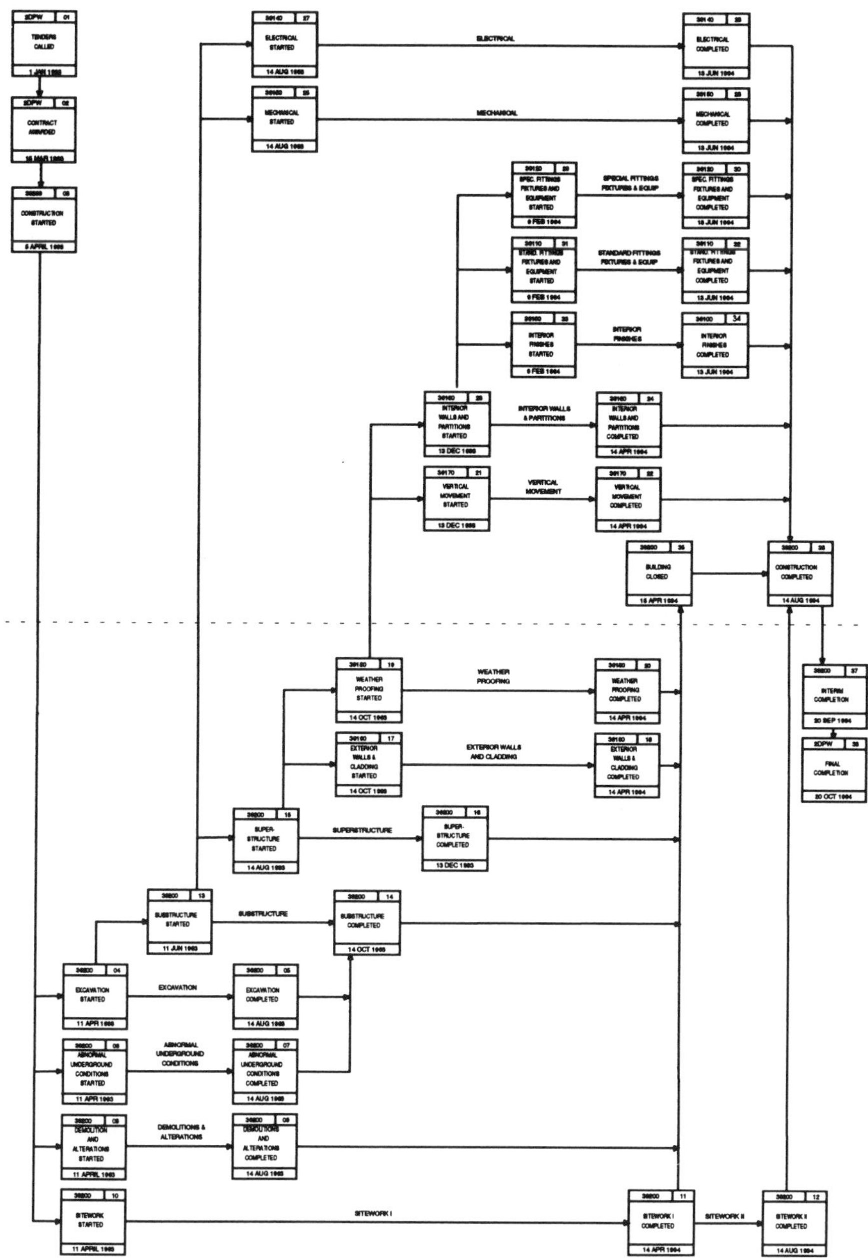

FIGURE 8.3 Milestone Network for a Building Project

Figure 8.3 shows a milestone network for a building project. It is at a summary level and can be used as a skeleton to which activities and milestones can be added. Each activity box includes the early start date. Note that the dates are arranged in chronological order from left to right. Activities at the front and finish ends are compressed together to suit the page size and format limitations of this book.

During the planning process the plan becomes more detailed as more information is received and added. In multiphase projects frequently the design is still being altered in later stages. By dividing the project into subprojects, a plan for each subproject can be developed by the scheduling engineer in conjunction with the team responsible for the actual work. The development of subnets is carried out within the framework developed through WBS, ORC, and milestone networks. The significant operation in the development of a subnet is planning the activities according to sets of milestones. Subprojects also enable changes to be made without upsetting the whole project. Remember that the project has already been broken down by means of the WBS, so each work package will be considered as a subproject. In addition the plans for the subprojects can be developed as subnetworks or subnets. This concept of a network composed of subnets being a part of the complete network is diagrammatically illustrated in Figures 8.4 and 8.5.

A subnet may belong to a network or stand alone. Breaking a network into subnets enables each manager to monitor his or her work independently, without having to wade through the whole network. For project management, division into subnets provides a means of monitoring and controlling the important or critical subnets rather than the entire network, thus saving time and effort.

The concept of subnets is used when several independent contractors are involved on a project. When a project is carried out by one general contractor, who is responsible for coordinating the work of several subcontractors,

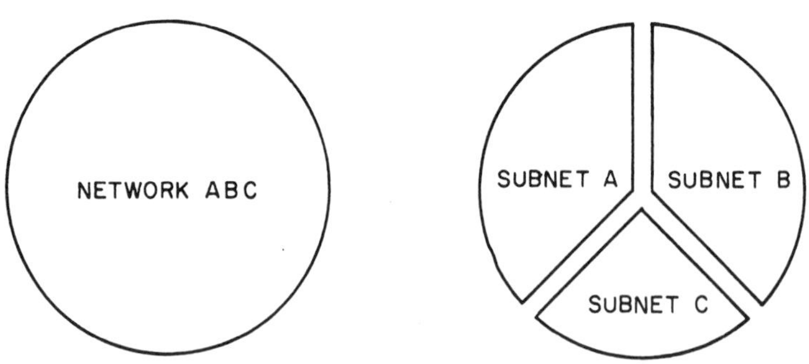

FIGURE 8.4 Subnet Concept

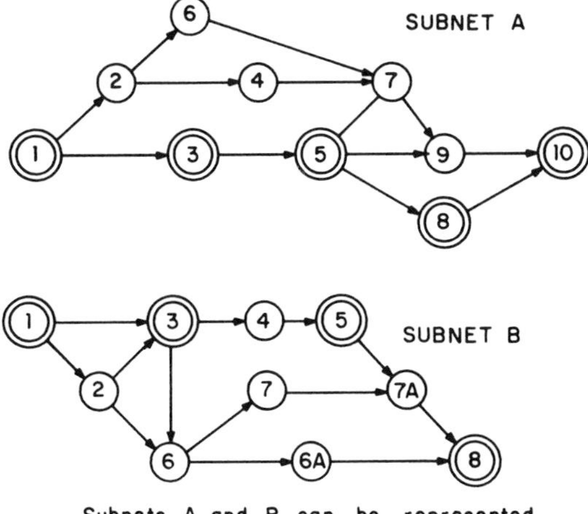

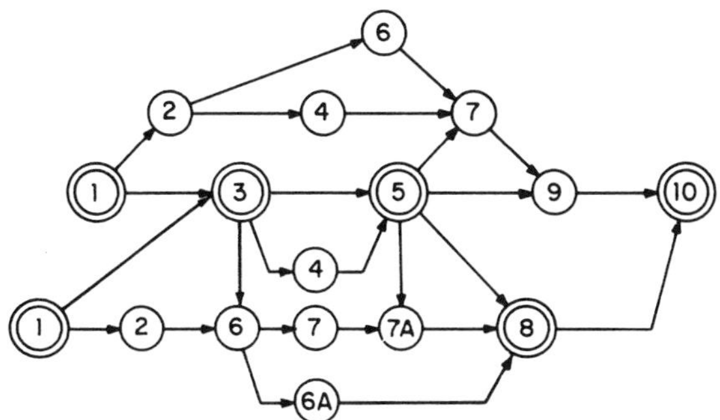

FIGURE 8.5 Example Subnet

it may not be necessary to divide the work into subnets. Separate schedules for the subcontractors may simply take the form of lists, stating specific activities for each, with these activities color-coded on the network plan.

A large construction company usually has more than one project in progress at a time. This may involve the continuous surveillance of several networks. By using the subnet approach, separate project networks can be treated as subnets and interfaced into a total operations network of the company. These projects can have different start and finish times. The interfaces may represent the times when resources from one project are transferred to another project or the target dates are imposed by the master schedule on the interfaced subnets.

8.2 SUBNET INTEGRATION

The detailed network received from the performing agencies form the subnets of the master network, where each subnet replaces a small group of activities. The number of activities in each subnet may run into several hundred or even a thousand or more, depending on the value of the contract and its complexity.

Subnets may have important events—other than start and milestones—which are points of contact with other subnets. Events at such points of contact are known as interfaces. The activities prior to the contact point, however, must be complete before activities leaving the point of contact may commence in both subnets.

In order to analyze subnets separately, each interface event is identified differently in each subnet. They are, however, labeled with the same event number and carry the subnet name as well as event number. Interfaces are indicated by a double circle on the network plan. Start and end events are also shown in double circles like other interfaces.

The interface between two subnets, X and Y, may be occasioned by inspection work, involving special testing equipment, to be performed by one agency upon the work of the two contractors involved separately in each subproject. If the interface for transfer of the equipment is at event 2 in Figure 8.6 (it was used in subnet X first), then activity 2–5 in subnet X will be delayed 5 days until activity 1–2 in subnet Y is completed. There is no necessity to delay the activity in subnet X, since the test equipment is needed for activity 2–4 in subnet Y. It can be seen that the circumstances of each contact point between subnet need to be considered carefully. The logic may be improved to indicate the sequential relationship at the point of contact shown in Figure 8.7.

In this case either subnet may be considered on its own, but the work to be performed in one subnet will affect the other subnet through the interfaces. These effects must be considered in the interface time calculations.

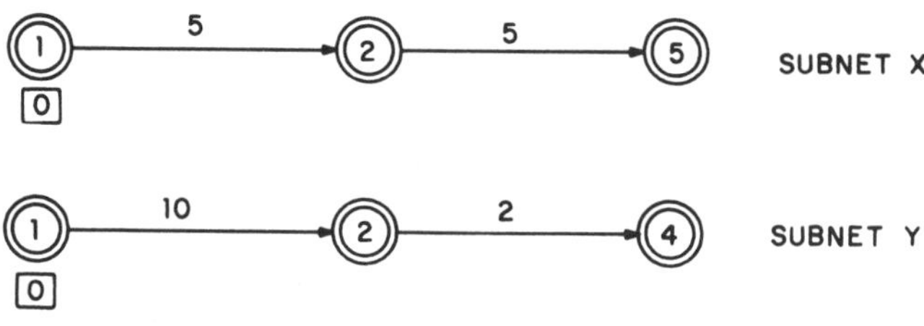

FIGURE 8.6 Subnet

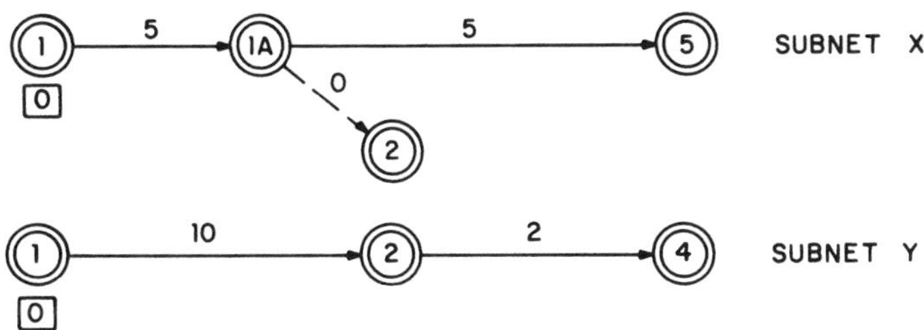

FIGURE 8.7 Subnet Interface

Most computerized scheduling programs perform subnet integration by a MERGE command. Except for very simple networks, a computer is used to facilitate the scheduling process.

8.3 INTERFACE TIME CALCULATIONS

The interrelationships of the two subnets act as mutual restraints, they must be considered when the subnet is analyzed. Computations for both subnets (using the forward and backward pass) should be carried out simultaneously. However, when an interface node is reached, the latest of the early event times in the forward pass calculations, and the earliest of the late event times in the backward pass at an interface node are used. Float can be calculated in the usual manner. The network from Figure 8.5 is reproduced in Figure 8.8 with activity durations and start times given. Subnet A and subnet B have different earliest start dates, as indicated in Figure 8.8. Forward and backward pass computations have been made to determine event times.

Real-time dummies may be inserted in the main network to replace a subnet. These represent the subnet and its effects on the main network. In Figure 8.9 the real-time dummies A, B, and C represent the effect of the subnet B, which they replace, on the given subnet A. Real-time dummies act as positioning restraints either to delay the interface event until the correct calendar time or to ensure that the particular interface events are achieved on time, and thus the scheduled completion date of the project is met.

After the event times are established, events other than start, end, interfaces, milestones, and those on the critical paths can be deleted so that just a skeleton of the original subnets remains, as shown in Figure 8.9. A skeleton network reduces the number of activities and events but maintains the relationships of the remaining activities and events. This skeletonization process enables computations to be performed on the master network independent of any subnet's details, and it ensures that a correspondence will be maintained with all the subnets comprising the network.

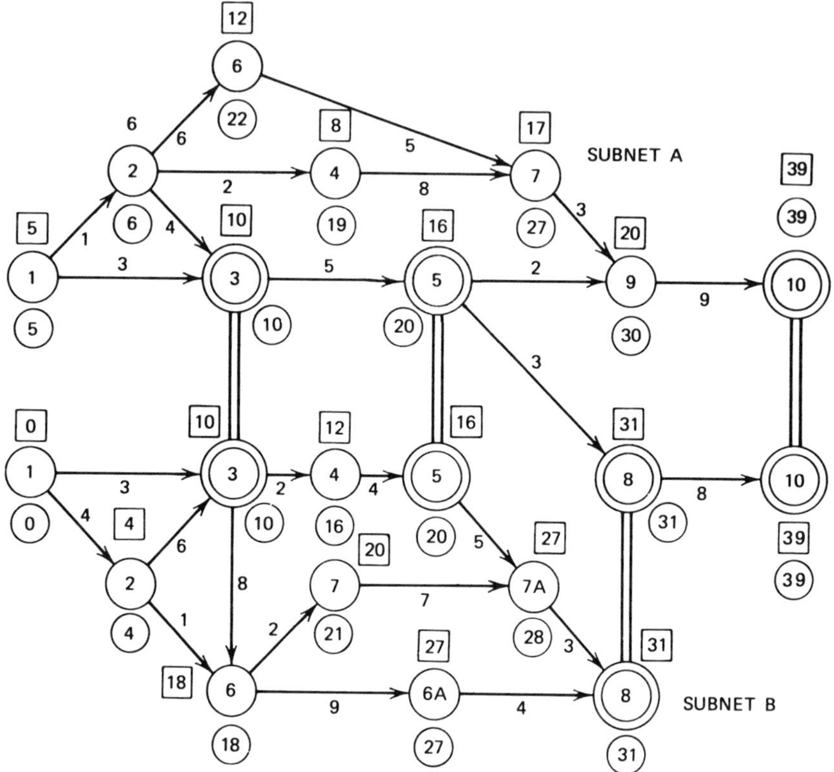

FIGURE 8.8 Interfaced Subnets

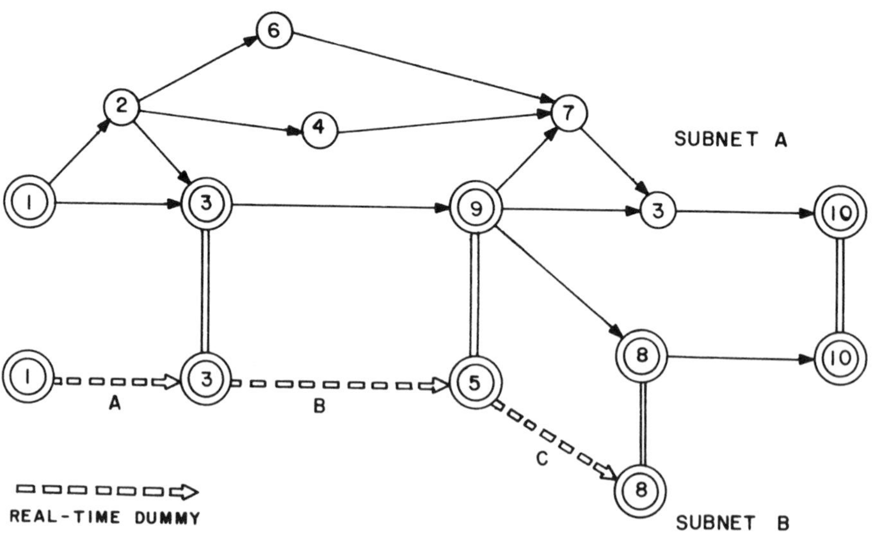

FIGURE 8.9 Skeleton Network

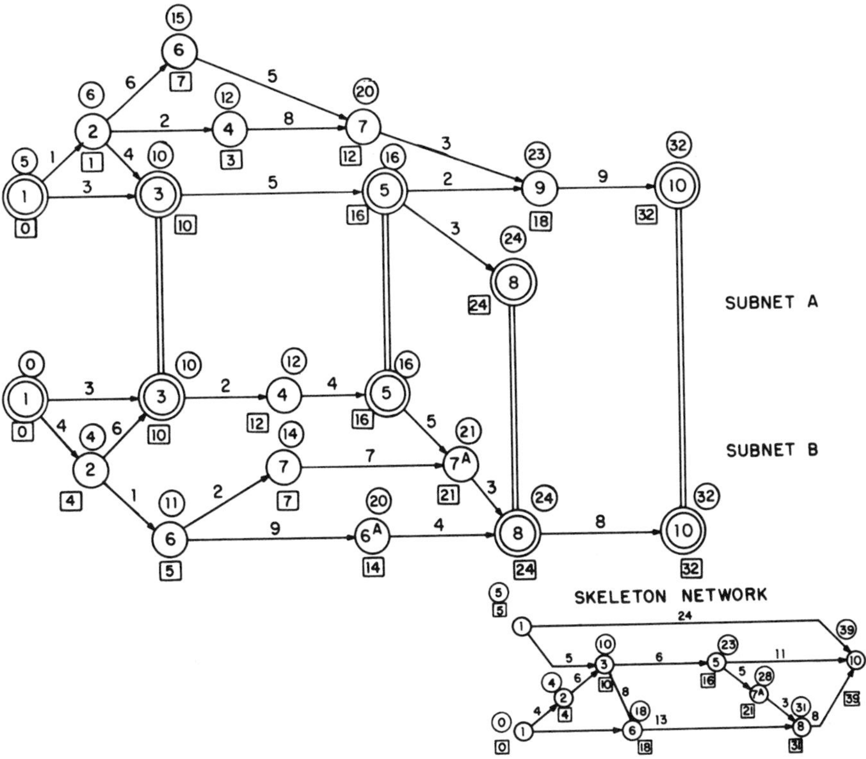

FIGURE 8.10 Subnet and Skeleton Networks

The critical path may not pass through all the subnets in the network. For example, in Figure 8.10 the critical path does not pass through subnet A. Alternatively, although a critical path passes through a certain subnet, it may enter and leave the subnet through interfaces other than the start and end, which are therefore not critical.

The contractor responsible for a subnet can delay the start of a subnet with float until it becomes critical. This often happens in practice. The contractor may mobilize at the earliest start time but suspend work until his subnet becomes critical, exhausting the float. If this subnet has interfaces with others, the float on the other subnets vanishes too. If the contractor runs into unexpected trouble, project duration may be affected, as there is no float left. This situation can be avoided by specifying definite target dates, including start, finish, and interface times, in the contract. The contractor treats these dates as the subnet's latest allowable times and bases the schedule on them. The contractor then plans the network to meet the target dates and can freely use any float without reducing the safety float and jeopardizing the project.

8.4 SKELETONIZATION

The preceding section mentioned the use of skeletonization in the computation of event times at interfaces; it has other uses as well. To derive maximum benefits, it is necessary to become familiar with milestones, subnetworks, subnet integration, and interface calculations.

If instead of the target dates for the milestones, the expected time between the milestones is obtainable, particularly if optimistic, pessimistic, and most likely times can be estimated, it is possible to make PERT-type* computations at this stage and to determine the probability of achieving the target dates. Probabilistic durations need not be carried out to the detail of activity networks, where deterministic estimates may be considered more practical. When used with discrimination, this method allows the management to estimate the probability of achieving target dates with the simplicity of CPM. Probability predictions are useful for projects with a very long durations, particularly where some technological development is being perfected along with the project's execution.

Another advantage of skeletonization is that it allows one to combine different networking techniques: CPM, PDM, PERT, and a simulation network for the various subnets. Using one type of networking technique may not be suitable for every subproject, for instance, PDM on a highly repetitive project. Through skeletonization the various different techniques for subnets may be integrated. The skeleton network itself may be drawn up using any of the networking techniques. This allows a great deal of flexibility in determining the most appropriate technique at the subproject level.

8.5 HIERARCHY OF SCHEDULES

The main output from the planning process is a series of schedules which are graphical representations of a total plan or parts thereof. Schedules are communication tools aimed at satisfying specific needs for the various levels of management or supervision. The differences in the schedules are the level of detail provided. An executive sponsor or owner needs certain information which is not at the level of detail required by a foreman. The higher up the organization structure, the broader is the view required, much like the different scales of maps.

The information needed should begin with as much detail as is available, summarizing this information for each level up into the organization.

Figure 8.11 shows this "roll-up" technique for summarizing information for the next higher level of supervision. Three levels of schedules are produced, that is, management, project, and control level schedules. We de-

* Program Evaluation and Review Technique—see Chapter 15

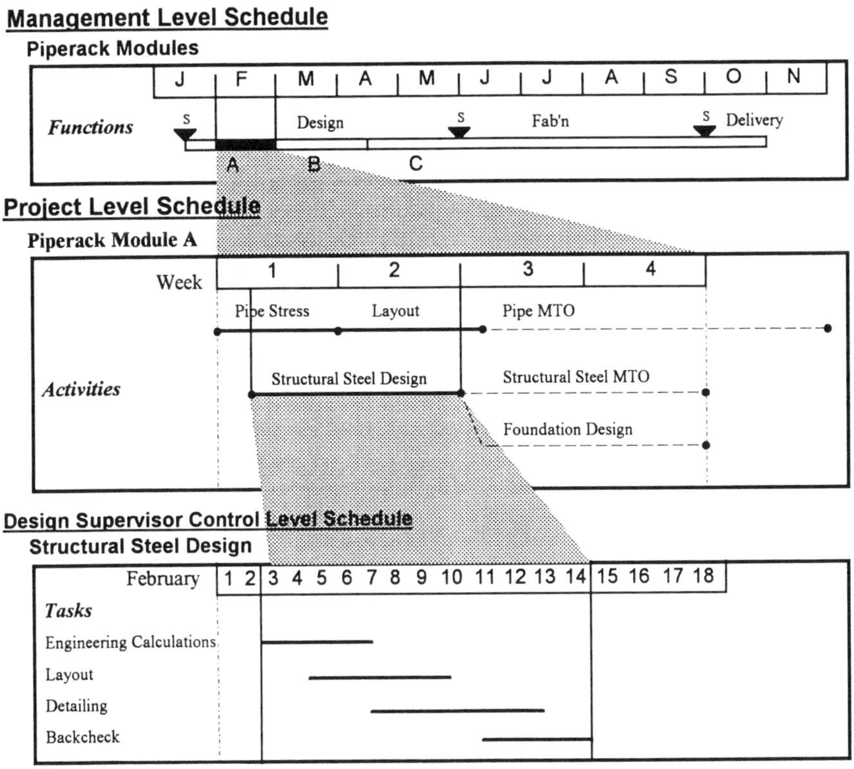

FIGURE 8.11 Roll-Up Technique

scribe these schedules from the top downward and it will become evident what is meant by the roll-up technique.

Management level schedules are usually in bar chart form and are the master level schedules which show the major milestones. The amount of detail is greatly reduced and what is shown are functions such as piperack modules. Figure 8.11 illustrates that this function consists of many components of piperack module A taken from the structure module of the project level schedule.

The project level schedule is the CPM network. Figure 8.11 shows activities in part of a CPM network which are all part of the design. The figure shows the detail that makes up some design activities.

The supervisor's control level schedule is made and used by those closest to the actual execution of the work, that is, by the front-line supervisors. This group of schedules includes a multitude of different schedules such as a 2-week or 3-month schedule for construction activities. These are the most accurate schedules. Figure 8.11 shows the design tasks for some structural steel design activities. This is the working schedule for designers and is produced by the supervisor. Scheduling engineers can assist at this level by gathering and presenting the information in a desired format. However, it is the supervisor closest to the work that contributes this input. In the field, the foremen and superintendents provide the knowledgeable input for construction control level schedules.

The scheduling process begins with some broad-brush schedules, which are developed and refined with increasing amounts of information. The first pass at a master plan begins at the higher management levels and as engineering and construction begin, more refined schedules are produced. These schedules in turn then provide better information so that the schedule can be updated. This is a feedback loop which is the roll-up technique.

PROBLEMS

8.1 A high-voltage transmission line was constructed. For simplicity, the erection of two towers in the 125-km line is discussed. Since it is an isolated view, the problem is modified; an activity such as conductor stringing, which normally takes place after 40 towers have been erected, is assumed to occur after the installation of two towers.

All activities were performed by one of three companies, A, B, or C. The activities are listed and coded by the name of the company responsible:

List of Activities

Activity	Company
1. Spot footing and anchor locations	A
2. Excavate for footing	A
3. Locate and lay out footing area	A
4. Check footing location	B
5. Place and compact footing	A
6. Place forms for footing base	A
7. Mix concrete for footing base	A
8. Pour footing base	A
9. Inspect placement and quality of concrete	B
10. Cur footing base	A
11. Remove forms from footing base	A
12. Install tower base	A
13. Check alignment of tower base	B
14. Place forms for footing top	A
15. Mix concrete for footing top	A
16. Pour footing top	A
17. Inspect placement and quality of concrete	B
18. Start cure footing top	A
19. Strip forms footing top	A

20. Finish cure footing top A
21. Backfill around footing A
22. Drill anchor holes A
23. Install anchors and ring bolts A
24. Grout anchor holes A
25. Inspect grouting B
26. Test anchors A
27. Inspect anchors' performance B
28. Package and deliver tower steel C
29. Assemble transmission tower C
30. Torque all tower bolts C
31. Attach guy wires to tower and to anchor bolts C
32. Erect transmission tower C
33. Tension and clamp guy wires C
34. Retorque all bolts and inspect tower for defects C
35. Inspect tower B
36. Install insulating plate and runners for conductor A
37. Install runners for sky wire A

Note that activities 1–37 are the same for both towers

75. String conductor A
76. Set up sag equipment A
77. Sag conductor A
78. Check conductor sag B
79. Remove runners and clip conductors A
80. Remove instruments and sag boards A
81. String sky wire A
82. Set up sag boards, instruments, and pulling equipment A
83. Sag sky wire A
84. Check sag on sky wire B
85. Remove runners and clip sky wire A
86. Remove sag boards and instruments A
87. Install separators on conductor A
88. Clean up site A

After the anchors are spotted for the first tower, the second tower can begin. All activities for both towers are identical. After footings and anchors are spotted, footing anchor and steel work can begin.

Anchor work is as follows: anchor holes are drilled and anchors are installed. Next anchor holes are grouted and simultaneously inspected. In both cases, when inspection is finished on one tower, the next tower must be ready for the same inspection. Next guy wires are attached to the tower and anchors. All tower bolts have to be torqued before this can occur.

The footing work is as follows: footings are excavated for, laid, and then checked, and the footing floor is placed. Next the formwork is installed for the footing base while concrete is mixed for it. This is followed by pour as well as inspection of the footing base. Similar work on the other tower is being made ready and is next in the sequence. The other tower must be ready for inspection. The footing base is then cured, and forms are stripped. Next, the tower base is installed, and alignment is checked simultaneously on one tower followed by a similar operation on the other tower. Next the forms are made for the footing top while concrete is mixed for it. Then the top is poured and inspected simultaneously. When the inspection for one tower is finished, the other top must be ready for inspection. Next the top is cured and the forms stripped. The footing is then backfilled. Next, the tower is erected, but guy wires must first be attached to the tower and anchors. When one tower is erected, the other tower is made ready for erection. After tower erection guy wires are clamped, the bolts are retorqued, and the tower is inspected. The inspection of the other tower follows. Runners for the conductor and then runners for the sky wire are installed. After sky-wire runners are installed on both towers, the conductor is strung between the towers, while sag equipment is set up. Then the conductor is sagged, and this sag is checked. Next the conductor is clipped, while the sag equipment is removed. Sky wire is then strung, while sag equipment is reset. Sky wire is sagged and then checked. Next sky wire is clipped, while sag equipment is removed. Finally separators are installed on the conductors, and then the site is cleaned up.

Draw three subnets, one for each construction company. Indicate the interface events. Integrate the three subnets to form one network. Make separate schedules for each construction company and a master schedule for the owner. Assume appropriate durations for the activities.

8.2 Write brief notes on the following:

a. Procedure for interfacing subnets.

b. Use of sorting parameters in reports.

c. Project management as management of interfaces.

d. Significance of a milestone to a project manager and to an owner.

CHAPTER 9

RESOURCE ALLOCATION

Network analysis arranges operations of a project in a sequence that, if followed, will bring the project to completion. If the estimated durations of the activities are not exceeded and time is not lost in delays, the project will be finished on time. Meeting the schedule is possible if the equipment and manpower on which the duration estimates are based are available on time. The analysis that compares these requirements against the available resources forms the physical feasibility check for a project plan and constitutes the subject matter of this chapter.

9.1 PHYSICAL FEASIBILITY OF A NEWORK PLAN

The readers are by now familiar with the critical path technique. To use this technique, the project is split into a set of interrelated activities, each with an estimated duration. Using these techniques, early start time, late start time, and floats are calculated. These calculations are based on the assumption that an activity can start as soon as its predecessors are completed—there are no constraints on the availability of resources. In practice, however, project completion requires the use of various resources, whose often limited availability directly influences the project duration.

After each stage of planning—initial, preliminary, control, and detail—it is necessary to ask the question, "Are all the resources available?" If so, the plan can be used; however, in most real-life projects resources are limited and hence impose constraints on the scheduling of activities.

If, in the initial planning stage, it is desired to ensure whether all the design engineers required for planning are available, a resource allocation exercise can be used to determine the number of different types of design engineers required and when they will be required. At the preliminary stage, it may be necessary to check on the number of draftspersons needed for steel fabrication drawings and welders needed for fabrication of steel tanks. A check for draftpersons and welders is required only if there is a scarce supply of people in these two trades.

At the control planning stage, it is useful to know whether the equipment required by the project is available locally, for hauling in a second pile-driving crane of the required capacity or a barge may cost up to $1 million. Later on, while accepting the network from contractors in the detail planning stage, it may be advantageous to check contractors' plans against their lists of resources. The question to be answered is whether the resources at their disposal are adequate for the work to be performed according to their plans.

Thus it can be seen that resource allocation is an essential part of planning. It answers the question whether adequate resources are available to implement the plan. Anyone can make a plan, but it will be useful only if it can be translated into actual work. It can be put into action only if the resources required are available.

The availability of resources is superimposed on the CPM project duration under two conditions:

1. Limited resources (variable project duration). It may be that the availability of resources is limited, and it is desired to evaluate the impact of such scarcity on project duration.
2. Unlimted resources (fixed project duration). When there are no constraints imposed on the availability of resources, the problem arises regarding what optimal level of resources is to be acquired to achieve the given target date.

Each of these cases, in the order presented here, is dealt with in detail in the following sections of this chapter.

9.2 ALLOCATING LIMITED RESOURCES

Allocation of limited resources to different activities is often known as "constrained resource scheduling." This type of technique is designed to produce schedules that will not require more resources than are available in any given period. The project will have to be completed using the given resources even if its duration has to be extended.

The limited resource problems are solved by two distinctly different approaches. The first category includes heuristic procedures which are designed to produce good resource feasibility schedules. The second category, in contrast, consists of procedures designed to produce the best (optimal) schedules.

9.3 THE HEURISTIC APPROACH

What is required in the heuristic approach is some basic criteria along with a procedure by which the resources may be allocated efficiently. The criteria comprise a set of predetermined priority rules. The combination of this procedure and the priority rules is known as the heuristic approach.

Various priority rules can be used to solve resource-constrained problems. To illustrate what the priority rules are, consider a network as shown in Figure 9.1. The number of resources required by each activity is shown in the square under the arrow depicting the activity. Assume that the resource availability is limited to a maximum of four. If it is desired to allocate resources to an eligible activity having the least slack, the activity 1–2 gets preference. The activity 1–4 will get preference if allocation is to be made to an activity having the largest number of resources or resource days. If shortest imminent activity is preferred, activity 1–3 gets first priority. If the minimum LFT (late finish time) heuristic is chosen, activity 1–2 will be allocated resources first. Some of the predominant priority rules commonly used follow. Allocate resources to an activity that:

1. Has the ealiest start time.
2. Has the minimum late start time.
3. Has the minimum early finish time.
4. Has the minimum late finish time.
5. Has the least float.
6. Has the largest duration.
7. Has the shortest duration.
8. Has the most immediate successors.

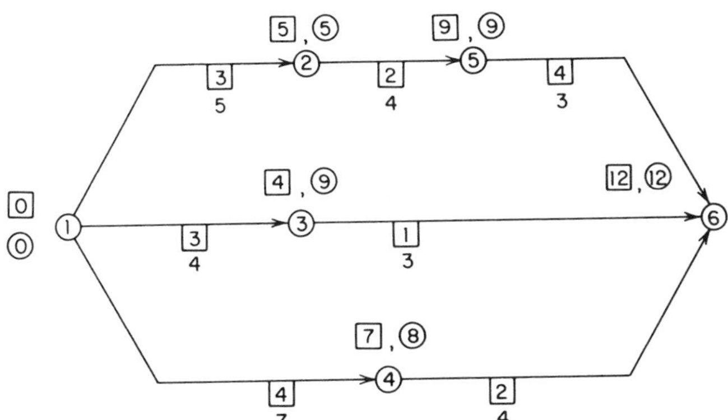

FIGURE 9.1 Example Network No. 1

9. Has most successors.
10. Has the least nonrelated activities.
11. Has the least nondependent jobs remaining.
12. Has the least immediate successors.
13. Has the least successors.
14. Can start first considering resources.
15. Has the least float per successor.
16. Has the longest path following.
17. Will finish first.
18. Has the largest resource requirement.
19. Has the largest resource days requirement.
20. Has the largest remaining resource days requirement.

In general, it is not possible to say which combination of priority rules will give the best results for a given project. A set of priority rules that performs poorly for a given problem may do well for others. However, the following combination of priority rules is used in this chapter:

1. Allocate resources to the activity having the least float.
2. Allocate to activity requiring the largest number of resource days.
3. Allocate to activity using largest number of resources (men or machines).
4. Allocate to an activity that precedes the largest remaining resource days requirement.
5. If a tie, allocate to the activity with the lowest sequence ($i - j$ value).

These priority rules may be employed throughout the two alternative methods. These are the series method and parallel method.

9.4 THE SERIES AND PARALLEL METHODS

The basic allocation procedure is a method of scheduling work by balancing need with availability of resources at a given time. The series method accomplishes this by allocating resources to activities in series (one activity at a time from start to finish). The parallel method allocates resources to activities one day at a time. An activity may be allocated resources one day but delayed the next day, while another activity using the resources is executed.Resource allocation for the two different conditions, using both methods, is discussed in separate sections; one covering the case when resources are limited and the other when they are not limited but the project duration is fixed.

9.5 LIMITED RESOURCES SOLUTION (SERIES METHOD)

To demonstrate how resources are allocated using the series method, consider Figure 9.2, which represents part of a project network.

The project manager has rented two cranes and wishes to perform the activities requiring cranes as soon as possible. The number of cranes required for each activity is shown in a square under the arrow depicting the activity. On day 35 he or she must select the activity to be started. His or her decision is based on total float. The project manager first considers the activities "pour concrete" and "waterproofing roof" as they each have a total float of zero. The activities cannot start simultaneously, as the cranes availabe do not meet resources required for both activities (i.e., three cranes). The project manager has a choice: start either one of the two activities or select the activity requiring maximum crane days. Again the project manager fails to make a decision, as both activities require the same number of crane days. He or she may pick the activity requiring the largest number or resources (cranes). "Pour concrete" is therefore given preference.

Consider the network shown in Figure 9.3. The data from this figure are entered in Table 9.1. Activities are listed in *i* major to *j* minor sorting sequence. Duration, resources needed, and earliest start time for each activity are read directly from the network. Total float is computed from the event times. Resource days, which are a product of resources required and the duration of an activity, are entered against each activity under the appropriate column. The remaining columns, which are numbered 0, 1, 2, 3, 4, and so

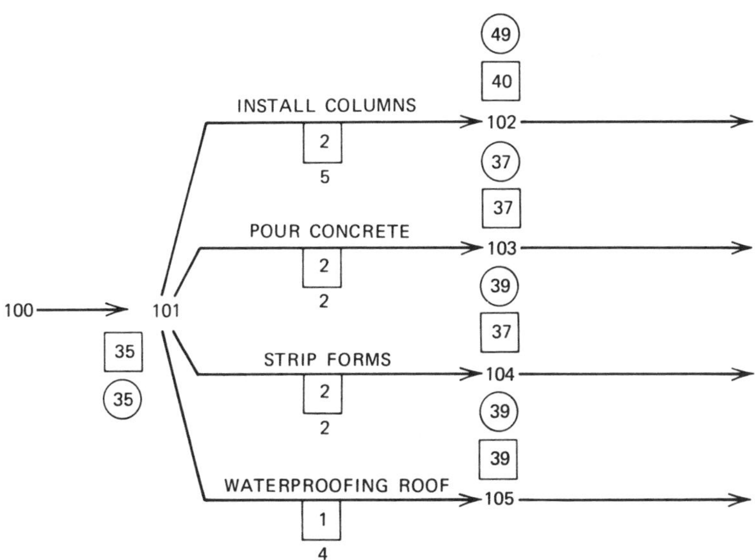

FIGURE 9.2 Part of Project Network

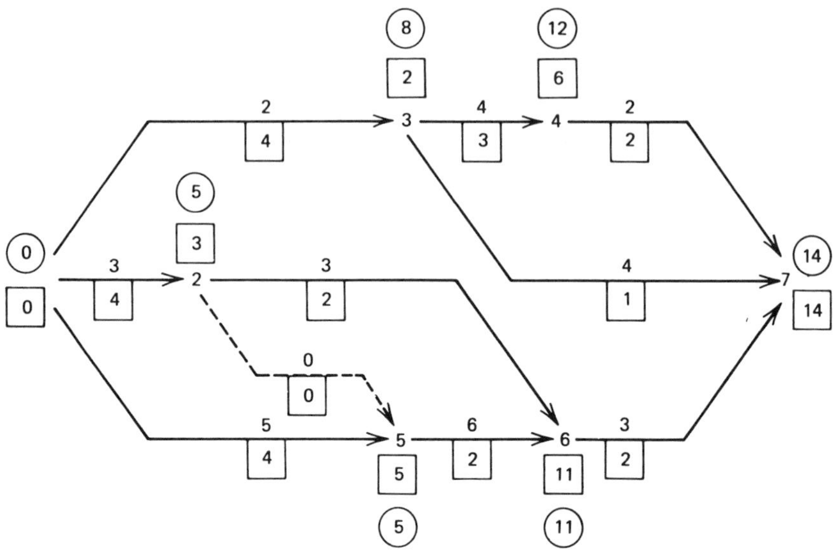

FIGURE 9.3 Example Network No. 2

on, denote the time units. These are referred to as project days, since the present example has its duration in days.

What is the level of available resources? Are there four workers or eight? If the entries in the column for resource days are summed, the total number of man-days necessary to complete this project is obtained. This sum is divided by the project duration to get the average number of workers required each day. This average number may not meet the requirements of all the activities, but it does provide a reasonable starting point. If resource needs for all the activities are met, this represents almost maximum utilization of available resources. There are two more things to remember. To begin with, this average number should not be less than the number of resources required by a single activity. Otherwise such an activity will never be assigned the required resources. Second, this number should always be an integer. If the quotient obtained by dividing resource days by project duration is not an integer, the next higher integer is taken. Resources, workers or machines, are available as whole units only.

For the present example the number of workers required to start the resource allocation exercise is obtained by dividing total resource days by the project duration. This resource limit is six (84/14 = 6).

For resource allocation, activities that can start on a particular day must be considered. To qualify for resource allocaton, these activities must follow completed activities; their start times should coincide with the present consideration of resource allocation. Resources are allocated to an activity for its entire duration.

TABLE 9.1 Single Resource Allocation Table: Limit of Resources Six per Day: Series Method

ACTIVITY	DURATION	RESOURCES	RESOURCE DAYS	EARLIEST START	TOTAL FLOAT	PROJECT CLOCK → PRIORITY	0		2		4		6		8		10		12		14		16		18		20
1-2	3	4	12	~~0~~ 5	~~2~~ -3	~~2~~ 1						4	4	4													
1-3	2	4	8	~~0~~ ~~5~~ 8	~~6~~ ~~1~~ -2	~~3~~ ~~2~~ 2									4	4											
1-5	5	4	20	0	0	1	4	4	4	4	4																
2-5	0	0	0	~~3~~ 8	~~2~~ -3	T																					
2-6	3	2	6	~~3~~ ~~8~~ ~~10~~ 14	~~5~~ ~~0~~ -~~2~~ -6	3 2 1															2	2	2				
3-4	4	3	12	~~2~~ 10	~~6~~ -2	1											3	3	3	3							
3-7	4	1	4	~~2~~ 10	~~8~~ 0	3											1	1	1	1							
4-7	2	2	4	~~6~~ 14	~~6~~ -2	2															2	2					
5-6	6	2	12	~~5~~ 8	~~0~~ -3	1									2	2	2	2	2	2							
6-7	3	2	6	~~17~~	0	1																			2	2	2

It is necessary to keep track of the time when resources are allocated, as this becomes the schedule time for the start of an activity. It is also necessary to maintain a count of the resources available for allocation, determining the start of an activity, as the decision depends on the availability of resources.

The first objective is achieved by maintaining a project clock. It is denoted by a vertical arrow (↓) placed over the day number lines. Since resource allocation starts at time zero, the clock is initially set at time zero. It is then reset to the time where it stops. It moves forward when either of the following two conditions prevails:

1. There are no resources left to be allocated.
2. There are no activities to which resources can be allocated.

The project clock stops when:

1. Resources are available for allocation.
2. Activities are available to which resources can be assigned.

When the project clock stops, start times and floats of all the eligible activities that have not been scheduled are updated. The start time of an unscheduled activity will never be less than the clock setting at the time. The symbol (↕) is used to denote the scheduling of dummies.

To account for resources, a resource pool is established with the given level of resources. The number of resources is decreased by the quantity allocated to activities. The resource pool is replenished by the resources returned from completed activities.

In the present case there are six resources in the pool. At every setting of the clock the resources are allocated from those available in this pool. If no resources are available, the clock moves to the next position. For the data given in Table 9.1, it can been seen that 1–2, 1–3, and 1–5 can initially start at time zero. The following is a step by step explanation of the solution.

Step 1. Since these activities are the start activities, there are no unfinished preceding activities. Initially these activities are waiting for resources. When there are several activities waiting for resources and insufficient resources to start them all simultaneously, priority rules as described earlier have to be applied. In the present case (Table 9.1) preference is given to activity 1–5 because it has the lowest total float. It is assigned priority 1; activity 1–2, priority 2; and 1–3, priority 3. Therefore resources are first assigned to activity 1–5. This activity requires four men for five days; resources are thus assigned from time zero to the end of day 5. Next consider priority 2: activity 1–2. It requires four men for three days, beginning at time zero. Because there is a limit of six resources and only two resources are available in the resource pool, it cannot start. Activity

1–3, which is priority 3, cannot be started for the same reason. Since there are inadequate resources in the resource pool to start any activity, the clock moves to the next position when more resources will be available. This position is day 5, when resources from activity 1–5 come back into the pool.

Step 2. On day 5 the early start column is updated to day 5. Activity 1–5 is not considered, as it is now complete. The earliest start date of the eligible activities 1–2 and 1–3 are increased by five days, and their total floats are reduced from 2 and 6 to −3 and 1, respectively. There is no other eligible activity for which updating is required.

Now that all eligible activities have been updated to the clock time and six resources are available in the resource pool, priorities are again assigned to the activities waiting for the resources. These are 1–2 and 1–3, the activities with earliest start time of 5. Activity 1–2 is given first priority and 1–3 second, based on total float. Activity 1–2 requires four men for three days, and these are assigned, beginning at day 5. The second priority activity cannot be assigned any resources, as there are only two resources available in the pool. The clock moves from day 5 to day 8 where resources from activity 1–2 come back to the pool, replenishing it to six.

Using the same procedure as discussed in Steps 1 and 2, the solution continues as outlined here. Whenever a dummy is encountered, it is assigned top priority denoted by "T."

Step 3.

Activities Considered	ES	Float	Priority
1–3	8	−2	2 assigned
2–5	8	−3	Top
2–6	8	0	3
5–6	8	−3	1 assigned

Step 4.

Activities Considered	ES	Float	Priority
2–6	10	−2	2
3–4	10	−2	1 assigned
3–7	10	0	3 assigned

Step 5.

Activities Considered	ES	Float	Priority
2–6	14	−6	1 assigned
4–7	14	−2	2 assigned

Step 6. The only remaining activity to be assigned resources if 6–7; it requires two men for three days beginning on day 17. The clock is updated to day 17, and resources are assigned. The complete solution is shown in Table 9.1. By using a limit of six resources, the project ends on day 20.

If the project duration so obtained is not suitable, further investigation will be necessary to arrive at a satisfactory solution.

9.6 FIXED PROJECT DURATION (SERIES METHOD)

The project manager, not being satisfied with the duration of 20 days, wishes to reduce to the minimum, using the minimum number of resources. To do this, the project manager may increase the resource level by one, allocate resources, and consider the resulting project duration. This can be done successively until an acceptable project duration is obtained. The solution using seven resources is also unacceptable. The solution using eight resources per day gives a project duration of 14 days.

9.7 LIMITED RESOURCES SOLUTION (PARALLEL METHOD)

Consider the example of the project manager and the number of available cranes. The project manager may decide to allocate resources on a daily basis. Each day the project manager decides which activity is more important that the others and accordingly allocates resources. The project manager thinks of the activities that can start on a certain day and the resources available from the resource pool and then allocates resources for this particular day only. On the next day he or she may not use the cranes on the same activities at all, even though these activities are not complete. They will be completed at a later date.

In order to decide which activities are more important, the project manager considers each day which activity should be given priority for timely completion of the project. He or she would also like to finish the activities that have been started. To decide which activities should be given preference, the project manager may consider the quantity of total resources required by an activity. This situation can be simulated by the parallel method. The rules for assigning priority to activities whose predecessors have been completed are as follows:

1. Allocate resources to the activity having the least total float.
2. If a tie, allocate to the activity that is in progress.
3. If a tie, allocate to the activity requiring the largest number of resource days.

4. If a tie, allocate to the activity requiring the largest number of resources per day.
5. If a tie, allocate to the activity that precedes minimum resource days requirement.
6. If a tie, consider sequence.

The given level of resources is available from the resource pool at the start of every day, and all resources are returned to the resource pool at the close of the day. On the starting day priority is assigned on the basis of minimum float. Allocation of resources is considered for every day when several operations are scheduled simultaneously for one day only. At the start of any day when resources are allocated, some jobs are completed, others are partly completed, and the remainder are not yet started. The previously stated priority rules are used to allocate resources among those operations that are in progress or can start as their predecessors are completed.

Example 9.1. Consider the data given in Table 9.1. The problem will be solved with the parallel method for a limit of six resources per day. The solution is shown in Table 9.2. A project clock is maintained in the same manner as for the series method, except that it moves only one day at a time. All unscheduled activities are updated when the clock stops.

Step 1. The clock is initially set at time zero. Activities 1–2, 1–3, and 1–5 have the earliest start time zero. Activity 1–5 is assigned priority 1, 1–2 priority 2, and 1–3 priority 3 based on total float. Activity 1–5 requires four men for five days. Because it has priority 1, it is assigned four resources on the first day. Its number of resource days is consequently reduced from 20 to 16, and its duration is reduced from 5 to 4. Priorities 2 and 3 cannot be assigned any resources because there are only two resources now available in the resource pool.

Step 2. The clock now moves to day 1. All eligible activities are updated. The duration of activities to which resources have been assigned is reduced, and the total float of the activities that could not be scheduled is changed. The total resource requirement of the activities that have been assigned resources is reduced by the number of resources used during the previous day. The earliest start of activity 1–5 now becomes 1, and its duration, 5, is reduced by one day to 4. The earliest start of activity 1–2 changes to 1. Its total float is reduced from 2 to 1. The same applies to activity 1–3. Now all eligible activities have been updated. At this stage all resources are in the resource pool.

Of the activities that can start on day 1, activity 1–5 is again given priority 1 (total float). Activity 1–2 is priority 2, and 1–3 priority 3. Again

TABLE 9.2 Single Resource Allocation Table: Limit of Six Resources per Day: Parallel Method

ACTIVITY	DURATION	RESOURCES	RESOURCE DAYS	EARLIEST START TIME	TOTAL FLOAT	PRIORITY	PROJECT CLOCK → 0 2 4 6 8 10 12 14 16 18
1-2	3 2 1 0	4	12 8 4 0	0 1 2 3 4 5 6 7 8	2 1 0 -1 -2 -3	2 2 2 1 2 1 2 1	4 4 4
1-3	2 1 0	4	8 4 0	0 1 2 3 4 5 6 7 8 9 10	6 5 4 3 2 1 0 -1 -2 -2	3 3 3 3 3 3 3 2 2 2	4 4
1-5	5 4 3 2 1 0	4	20 16 12 8 4 0	0 1 2 3 4 5 6 7	0 -1 -2	1 1 1 2 1 2 1	4 4 4 4 4
2-5	0	0	0	3	2		
2-6	3 2 1 0	2	6 4 2 0	3 8 9 10 11 12 13 14 15	5 0 -1 -2 -3 -4	3 3 3 2 2 3 1	2 2 2
3-4	4 3 2 1 0	3	12 9 6 3 0	2 10 11 12 13 14 15 16	6 -2 -3 -4	2 2 3 1 2 3 1	3 3 3 3
3-7	4 3 2 1 0	1	4 3 2 1 0	2 10 11 12 13 14	8 0	4 4 4 4	1 1 1 1
4-7	2 1 0	2	4 2 0	6 15 16 17 18	6 -3 -4	3 2 1	2 2
5-6	6 5 4 3 2 1 0	2	12 10 8 6 4 2 0	5 8 9 10 11 12 13 14 15	0 -3 -4	1 1 1 1 3 1 2	2 2 2 2 2 2
6-7	3 2 1 0	2	6 4 2 0	11 15 16 17 18	0 -4	2 1 2	2 2 2

1–5 is assigned four resources, and its required resource days are reduced to 12. This procedure continues as shown in Steps 3, 4, and 5.

Step 3.

Activities Considered	ES	Float	Priority
1–2	2	0	2
1–3	2	4	3
1–5	2	0	1 assigned

Step 4.

Activities Considered	ES	Float	Priority
1–2	3	−1	1 assigned
1–3	3	3	3
1–5	3	0	2

Step 5.

Activities Considered	ES	Float	Priority
1–2	4	−1	2
1–3	4	2	3
1–5	4	−1	1 assigned

By day 19 all the resources have been allocated. The complete solution is shown in Table 9.2.

Using six resources, the series method gives a project duration of 20 days, and the parallel method, a duration of 18. With seven resources it is 17 versus 17. With eight resources both methods give the minimum duration, 14 days.

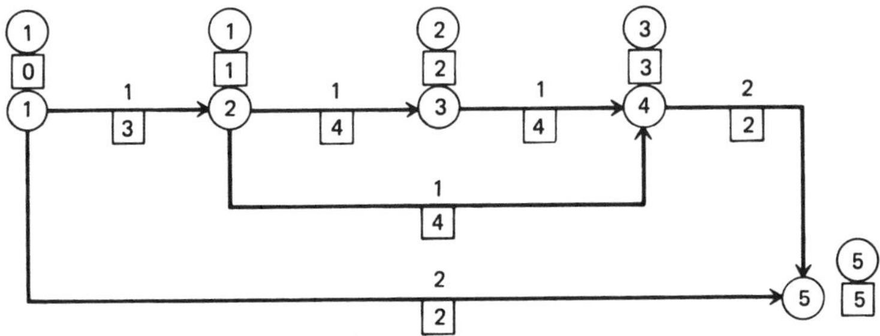

FIGURE 9.4 Example Network No. 3

Depending on whether the activities are splittable or not, any method—either the Series or Parallel Method—can be adopted for resource allocation. However, with the Series Method, despite an increase in the resource level, project duration may be increased. For example, while the duration for the project illustrated in Figure 9.4 is 6 days when 4 resources are used, it becomes 7 days when 5 resources are used instead. This is because the critical activity 2–3 cannot be started until resources are freed from the noncritical activity 1–5. The reader is forewarned so he or she can analyze the resource-adjusted schedule to avoid such discrepancies.

9.8 PARALLEL METHOD WITH LIMITED MULTIPLE RESOURCES

Although both series and parallel methods can be used for multiple-resource allocation, only the parallel method will be demonstrated in this and the next section. A study of these sections should enable the reader to adopt the single-resource series method to multiple-resource problems. The procedure for the parallel method is similar to that discussed in the previous section. The same priority rules apply.

For the example tabulated in the convenient form of Table 9.3, a resource level of each type of resource is determined and used in the first solution. The project duration is 14 days, as computed from the estimated durations. The total workdays for resources *A, B* and *C* are 73, 84, and 55, respectively. Dividing total workdays for each resource by the project duration gives the number of resources for the resource pool: six of *A*, six of *B*, and four of *C*. Resources are allocated one day at a time based on priorities. The project clock is initially set at time zero. For the first days activity 1–5 received priority 1; activity 1–2, priority 2; and activity 1–3, priority 3. Activity 1–5 is assigned the required resources. 1, 4, and 2 (corresponding to resources *A, B* and *C*) are placed in the three columns on the first day. Neither the second or third priority can be assigned any resources at this time, as there is a resource limit.

The clock moves to day 1. All eligible activities are updated. The duration of activity 1–5 is reduced by 1, resource days of *A* are reduced to 4, of *B* to 16, and of *C* to 8. Its earliest start now becomes 1, but its total float remains unchanged. Activities 1–2 and 1–3 are updated to the early start time 1, and their total float is reduced by one.

All eligible activities have now been updated and priorities can be assigned for day 1. Activity 1–5 is again priority 1, 1–2 priority 2, and 1–3 priority 3. Resources are assigned to activity 1–5, and again no other activity can begin, as they require more resources than what is available in the pools.

This procedure continues to day 23 when all the resources will have been assigned. The complete solution in shown in Table 9.3

TABLE 9.3 Multiple Resource Allocation, Parallel Method: Limit of Six of A, Six of B, and Four of C per Day

ACTIVITY	DURATION	RESOURCES			RESOURCE DAYS			EARLIEST START TIME	FLOAT	PROJECT CLOCK → PRIORITY
		A	B	C	A	B	C			
1-2	3 2 1 0	3	4	0	9 6 3 0	12 8 4 0	0 0 0 0	0 1 2 3 4 5 6 7 8	2 1 0 -1 -2 -2 -2	2 2 2 1 2 1 2 1
1-3	2 1 0	2	4	1	4 2 0	8 4 0	2 1 0	0 1 2 3 4 5 6 7 8 9 10	6 5 4 3 2 1 0 -1 -2 -2	3 3 3 3 3 3 3 2 2 2
1-5	5 4 3 2 1 0	1	4	2	5 4 3 2 1 0	20 16 12 8 4 0	10 8 6 4 2 0	0 1 2 3 4 5 6 7	0 0 0 0 -1 -2	1 1 1 2 1 2 1
2-5	0	0	0	0	0	0	0	3 8	2 -3	7
2-6	3 2 1 0	0	2	3	0 0 0 0	6 4 2 0	9 6 3 0	3 8 9 10 11 12 13 14 15 16 17 18	5 0 -1 -2 -3 -4 -4 -5 -5 -6 -7	3 3 3 2 1 2 1 2 2 1
3-4	4 3 2 1 0	2	3	0	8 6 4 2 0	12 9 6 3 0	0 0 0 0 0	2 10 11 12 13 14	5 -2 -2 -2 -2	2 3 3 4
3-7	4 3 2 1 0	2	1	3	8 6 4 2 0	4 3 2 1 0	12 8 6 3 0	2 10 11 12 13 14 15 16 17 18 19 20 21 22 23	8 0 -1 -2 -3 -4 -5 -6 -6 -7 -7 -7 -8 -9	4 4 4 3 2 3 1 2 2 2 2 1 1
4-7	2 1 0	3	2	4	6 3 0	4 2 0	8 4 0	6 14 15 16 17 18 19 20 21 22	6 -2 -3 -4 -5 -6 -7 -8	4 4 3 3 3 3 1 1
5-6	6 5 4 3 2 1 0	4	2	2	24 20 16 12 8 4 0	12 10 8 6 4 2 0	12 10 8 6 4 2 0	5 8 9 10 11 12 13 14 15 16	0 -3 -3 -3 -3 -3 -4 -4 -5	1 1 1 1 2 1 3 1
6-7	3 2 1 0	3	2	1	9 6 3 0	6 4 2 0	3 2 1 0	11 18 19 20 21 22 23	0 -7 -7 -7 -8 -9	1 1 3 3 2

TABLE 9.3 (*Continued*)

1	2	3	4	5	6	7	8	9	10	11	12	13	14	15	16	17	18	19	20	21	22	23
A B C	A B C	A B C	A B C	A B C	A B C	A B C	A B C	A B C	A B C	A B C	A B C	A B C	A B C	A B C	A B C	A B C	A B C	A B C	A B C	A B C	A B C	A B C
			3 4 0		3 4 0		3 4 0															
								2 4 1	2 4 1													
1 4 2	1 4 2	1 4 2		1 4 2		1 4 2																
												0 2 3		0 2 3			0 2 3					
										2 3 0	2 3 0	2 3 0	2 3 0									
																2 1 3		2 1 3	2 1 3			2 1 3
																				3 2 4	3 2 4	
								4 2 2	4 2 2	4 2 2	4 2 2		4 2 2		4 2 2							
																		3 2 1	3 2 1	3 2 1		3 2 1

9.9 FIXED PROJECT DURATION SOLUTION (MULTIPLE RESOURCES)

The project manager may not be satisfied with the duration of 23 days, as produced by using six resources of *A*, six of *B*, and four of *C*. In order to reduce this time, the project manager increases the number of each resource by one. The resources are allocated using the same procedure, and the duration now is 18 days. As this is still not the minimum; each resource is once again incremented by one so that there are eight of *A*, eight of *B*, and six of *C*. This level of resources gives a duration of 14 days.

Suppose the number of resources has to be incremented further. It could be possible that not all the resources of a certain type were used at any time. The increment of a certain type of resources can therefore be made dependent on the need for additional resources observed in the preceding resource allocation exercise. An additional resource will be necessary only if a delay occurs for want of a certain type of resource. Only the resource that has caused this delay need be increased.

An alternative approach is to increase each resource before starting on a subsequent resource allocation exercise and, after the resources have been allocated, to look for the resource type in which the additional resource has not been used at all. The number of this resource can be reduced by the extent of the unused quantity.

9.10 MULTIPROJECT RESOURCE ALLOCATION

So far the entire discussion has concentrated on the use of single and multiple resources needed on a single project. When an organization is responsible for several projects, its objective is to make the most efficient use possible of the resources required for these projects. Resources that become surplus on one project are transferred to others in the organization. The procedure that simulates such a situation is called multiproject resource allocation.

The approach to multiproject resource scheduling is fairly direct. All project networks are linked together, showing one initial event and one terminal event. Where necessary, key sequencing dummies are introduced to effect shipping of major crews or equipment from project to project. This approach is illustreated in Figure 9.5. Projects Alpha, Beta, and Gamma are linked to a common start S and common terminal T and are interfaced with key sequencing dummies 6*A*–3*B* and 4*B*–6*C*. The three projects share resources in such a manner that the resources used by 3*A*–6*A* in project Alpha are transferred to 3*B*–7*B* in project Beta as soon as they become available from 3*A*–6*A*. Similarly, resources from activity 1*B*–4*B* in project Beta are transferred to activity 6*C*–7*C* in project Gamma.

However, the insertion of scheduling interfaces such as 6*A*–3*B* and 4*B*–6*C* should be held to a practical minimum. In connecting the projects to be

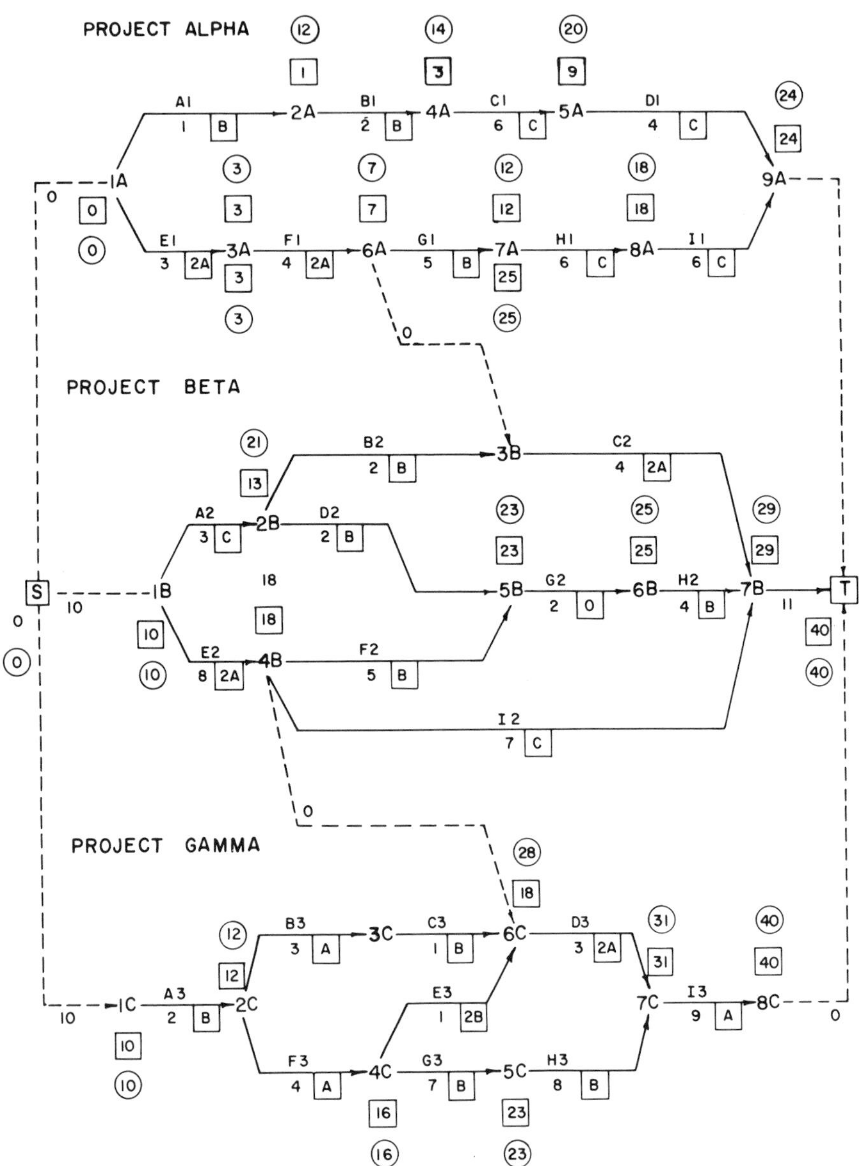

FIGURE 9.5 Multi-project Example

considered the start and finish dates must almost inevitably be adjusted. To do this, the lead arrow S–1*A*, S–1*B*, and S–1*C* and the lag arrows 9*A*–T, 7*B*–T, and 8*C*–T are given suitable durations. The start date of each of the projects Alpha, Beta, and Gamma is specified from event S which determines priority for each project's start. The three projects after linking are treated as one project starting at S and terminating at T. Resource allocation procedures followed by multiproject resource allocation are the same as delineated in previous sections. If the resource days for resources *A*, *B*, and *C* are totaled separately, and then each is divided by the project duration (40 days), an average resource level of two type *A*, two type *B*, and one type *C* is obtained. By using the parallel method and htese resource limits, the project will be completed in 46 days. Project Alpha can start at time zero; however, project

TABLE 9.4

Activity	Description	Duration	Resource	Scheduled Start	Scheduled Finish
1*A*–3*A*	*E*1	3	2*A*	0	3
3*A*–6*A*	*F*1	4	2*A*	3	7
1*A*–2*A*	*A*1	1	*B*	0	1
2*A*–4*A*	*B*1	2	*B*	1	3
6*A*–7*A*	*G*1	5	*B*	7	12
4*A*–5*A*	*C*1	6	*C*	3	9
5*A*–9*A*	*D*1	4	*C*	9	10
			C	13	16
			C	16	22
7*A*–8*A*	*H*1	4	*C*	16	22
8*A*–9*A*	*I*1	6	*C*	22	28
1*B*–4*B*	*E*2	8	2*A*	10	15
			2*A*	16	17
			2*A*	19	20
			2*A*	21	22
3*B*–7*B*	*C*2	4	2*A*	22	26
2*B*–3*B*	*B*2	2	*B*	13	15
2*B*–5*B*	*D*2	2	*B*	13	15
4*B*–5*B*	*F*2	5	*B*	22	27
6*B*–7*B*	*H*2	4	*B*	29	33
1*B*–2*B*	*A*2	3	*C*	10	13
4*B*–7*B*	*I*2	7	*C*	28	35
2*C*–3*C*	*B*3	3	*A*	15	16
			A	17	19
2*C*–4*C*	*F*3	4	*A*	15	16
			A	17	19
			A	20	21
6*C*–7*C*	*D*3	3	2*A*	26	29
7*C*–8*C*	*I*3	9	*A*	37	46
1*C*–2*C*	*A*3	2	*B*	10	12
3*C*–6*C*	*C*3	1	*B*	19	20
4*C*–6*C*	*E*3	1	2*B*	21	22
4*C*–5*C*	*G*3	7	*B*	22	29
5*C*–7*C*	*H*3	8	*B*	29	37

Beta and project Gamma cannot start until day 10. Project Alpha is complete on day 28, Beta on day 35, and Gamma on day 46. Schedules based on resource allocation using 2*A*, 2*B*, and 1*C* resources are generated for the integrated multiproject network. These are shown in Table 9.4. Resource profiles based on this schedule can be drawn separately for each resource type for each project.

In a multiproject environment project priorities may be superimposed over activities of a certain project because of its relative importance to the organization. At any clock setting such activities receive higher priority over all other activities competing for resources at the same time.

9.11 ALTERNATE HEURISTIC RULES

As discussed earlier, the resource-constrained problems are solved using minimum float priority (i.e., allocating resources to an eligible activity having the least float). In case of a tie the greatest resource days requirement and greatest resource requirement were given second and third priority, respectively. Now the question arises, "Is it the best heuristic?"

This question cannot be answered affirmatively because, as mentioned earlier, the heuristic rules that perform poorly for a given problem may do well on others. For example, if this resource allocation exercise is worked out using the series method with LFT (late finish time) heuristic, we get a duration of 19 days as against 20 days obtained by minimum flat heuristic. However, it has been concluded by several planners that, by and large, minimum float heuristic gives a better result (shortest duration) than the rest. The choice of the heuristic is less critical if the constraints on this project are severe.

The best result can be obtained by trial and error procedure, employing different heuristics for the same network and choosing the best. There is no limit to the number of trials that can be made to optimize the resource-time balance. However, usually less than six trials can generate a nearly optimum solution.

9.12 OTHER CRITERIA FOR PRIORITY RULES

The variations among the different solutions are due to the different construction priority rules adopted to simulate different construction environments. The rules listed earlier in this chapter are in no way exhaustive. There are other situations on construction projects that have not been covered by these priority rules.

For example, on a certain project a project manager wants to complete all the work that needs the use of a crane. The early return of this crane will mean savings in rent. The project manager therefore assigns a higher priority to the activities that need the use of this crane. Such a condition may also exist when a piece of equipment is required for some other project. These are jobs where temporary personnel are hired to perform certain operations, and in the interest of economy they are to be laid off at the earliest possible time.

In another situation a project manager may give a higher priority to those activities presenting the greatest potential difficulty. Resource allocation procedures would be required to assign resources to such activities in preference to others competing for resources at that time. Yet another project manager may prefer activities that require the least amount of work for their completion.

In another situation it may be necessary to assign priorities to activities whose completion will bring in a large amount of progress payment. The use of a certain resource may facilitate the use of other resources that are available in abundance. A case in point may be the use of a drilling machine, which facilitates blasting, followed by the hauling of muck and the laying of sewer and water mains in a land assembly project. These activities may optionally be given a priority number during a resource allocation exercise. If a higher priority number is assigned than the priority for total float, this activity is scheduled even if it has a float.

Uncertainty could be another criterion for scheduling activities. Many development projects have activities of both kinds—those for which time estimates can be accurately made and those that are highly uncertain. This suggests a possible modification of the priority list or at least an option to be used when desired. In place of or in addition to float as a criterion for a priority ordering of activities to be scheduled, activities could be characterized by the degree of uncertainty associated with their estimated duration. Those with the greatest uncertainty would receive highest priority for each scheduling; those with more certain time estimates would tend to be scheduled later or possibly to be postponed up to their latest start time.

If a piece of rented equipment is needed for four or five project activities, it may be appropriate to chain these activities so that the equipment can be returned as soon as possible. Chaining of activities implies that completion of an activity is immediately followed by the next activity in the chain. In starting this activity, any resource needs besides the one of prime interest are assigned to the activity on a preferential basis. Similarly there may be other situations where one activity must follow the other without loss of any time. For example, "dispatch of goods" is preceded by "processing of a product" in a factory, as the manufacturer does not want to lose any time after processing is complete. Similarly, if "commissioning" is preceded by "test assembly," the owner of a plant does not want to lose any time in between. Hence, after the first activity is scheduled by a resource-scheduling procedure, the immediately following activity in the chain assumes a higher priority for the assignment of resources.

On another project a project manager, short of a large bulldozer, may not stop work but use two available smaller bulldozers. Similarly, the project manager may use apprentices instead of journeymen to complete a certain operation. This situation can easily be simulated by resource allocation procedures that store all types of available resources, including their substitutes, in the resource pool and look for the substitutes if the requirement cannot be met by the regular resources.

If an activity does not require the same resource from beginning to end, the resource need for this activity is altered when the clock reaches a certain point in time.

So far the discussion has related to a fixed activity duration derived from a fixed resource need for each activity in a network. An assumption was made that work on a certain activity cannot start until the required number of workers or machines is available. This may not be true in real life. A project manager who does not have eight carpenters for formwork may go ahead with just two. However, because of space constraints the project manager may not be able to engage more than 16 carpenters on this activity. The normal duration is based on the level of resources the organization normally employs for such an activity. In the present case 2 and 16 could be the secondary level of resources. With the resource level of 8 carpenters this activity normally takes 8 days. With 2 carpenters the activity will take 32 days, and with 16 it will take just 4 days. A sound procedure can consider this need and allocate resources within the limits specified by the secondary levels, when the primary level is not available.

Some resources such as project equipment can be returned to the resource pool as soon as they are freed from an activity. Other resources, such as bricks or concrete, are consumed on an activity and are consequently not available for further use. If resource allocation is done simultaneously for both types of resources, discrimination is necessary. The availability of some resources varies regularly with time, and this variation is repeated in a cyclic pattern. One example of such a pattern is the weekly cycle of construction labor whose availability on a project is 80, 100, 100, 100, 90, and 80 for a six-day week. Resources are allocated to match this cyclic profile. Resource availability may vary with time because of the unavailability of equipment during preventative maintenance periods. For example, because of preventative maintenance on a certain project, tunneling equipment may not be available for the last two days of every quarter. On another project, because of worker absenteeism on Mondays and Fridays, it may be necessary to use overtime labor.

Resources, once used on an activity of a project, cannot be unscheduled and used on another operation instead. But, while allocating resources, an activity can be unscheduled without any economic loss to the project. If only a portion of an activity has been scheduled but not the remainder, it can be delayed up to the maximum split delay allowed for this activity. The remaining part of the activity is put in the queue of the next day for resource allocation. (This portion of the activity may be defined as the continuous operation, being the minimum time period for which the activity must continue before it can be started or restarted again.) If a decision is taken to assign priority to another activity competing for the same resources, then any portion of this activity that is already scheduled has to be unscheduled, and the entire activity put into the queue for the next day. Take the example of a bridge pier, where not more than a one-day split is allowed between pours for two lifts; if resources are not available, the part of the activity that is already scheduled will be unscheduled because the remaining part cannot be scheduled within one day of the first part. The resource clock will move back by the number of days unscheduled and start allocating once again, this time after moving "pouring concrete for bridge pier" down the priority list. If this activity has no float left and resources are not available, the project duration is extended. If this activity must be done by a certain date, either the resource allocation exercise is abandoned or a negative float is generated that is indicative of a need for additional resources.

A project manager usually looks ahead by allocating resources to certain sets of activities in alternative ways, comparing the effects on the project duration. He does this by studying the time at which a certain common target at the end of two sets of activities is reached. If this target lies on the critical path, it is obvious that the alternative that achieves the target without any slippage or with minimum slippage will be used. The alternative is selected, and the clock is moved back to the common starting point. Resources are allocated to activities in the selected manner. By using this feature of the resource allocation procedure, it is not necessary to allocate resources right up to the end of a project in order to pick the best alternative.

The preceding cases give an idea of the many situations that can be simulated by resource allocation procedures. The selection of priority rules and the establishment of the procedure to follow depend on the needs of project management and the size, type, and environment of a project.

However, because of the complexity involved in realistically simulating a project for resource allocation for maximum practical benefit, the resource allocation exercise should be kept simple by applying it to a summary level network such as the master network. If it must be used for a detailed network, it should be limited to the part of the project duration that will be worked on immediately, so guidance is obtained for the deployment of resources. Resource allocation, besides being unwieldy, is meaningless for a detailed network of the entire project duration, since there are bound to be so many changes.

9.13 OPTIMAL PROCEDURES

The readers may by now be convinced that the heuristic approach is an approximate procedure, or it is just a combination of priority rules and procedure that guides the allocation exercise and hence cannot give an optimal solution.

The optimal procedures that have been developed can be divided into two groups:

1. Procedure based on Linear Programming (LP).
2. Procedure based on enumerative and other mathematical techniques.

These procedures attempt to get the optimal solution by going through all possible solutions. Since many variables are involved in resource allocation problems, these procedures cannot be adopted for large networks or for projects where the number of resources required are many. Hence heuristic procedures are widely used and are the only available means of solving the complex problems that occur in practice. However, the increased availability of more powerful computer systems may improve the usage of optimal procedures.

9.14 UNLIMITED RESOURCE ALLOCATION

The discussions in the preceding sections were exclusively concerned with allocation of limited resources. If ample quantities of all required resources are available, it may have to be seen how best they can be used. Consider,

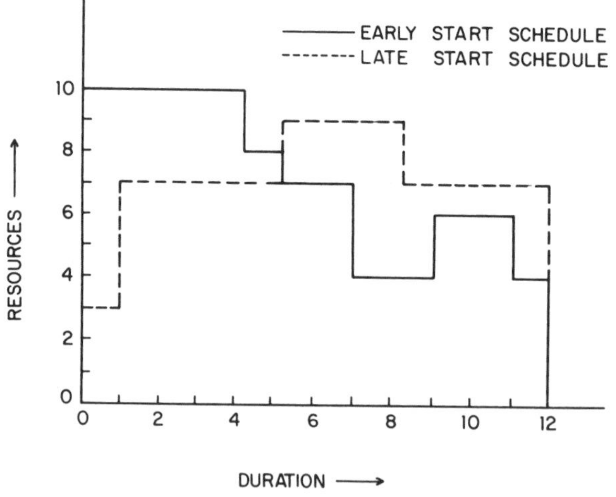

FIGURE 9.6 Resource Profile

for example, the network shown in Figure 9.1, and assume that the resource availability is unlimited. Now we have three options for scheduling the activities. They can be scheduled at their earliest start time, their latest start time, or at any time between these two. The resource profile for the first two cases is shown in Figure 9.6. The objective in unlimited resource scheduling is to obtain the least costly profile, that is, to lower the project cost as much as possible. This may be achieved by resource leveling.

9.15 RESOURCE LEVELING

In the resource-leveling procedure the resources are allocated in such a way that the resource profile is gradually built up to a peak and slowly brought down to the end without another rise, as shown in Figure 9.7*a* and *b*. The hiring and dismissal of resources as well as resource idleness are concerned with special costs. Generally speaking, the smoother the resource profile, the lower is the overall cost.

As in the case of constrained resource allocation, the leveling problems can be solved by heuristic as well as optimal procedures. The optimal procedures are normally not adopted because they are too expensive for even small networks, and medium- and large-sized problems cannot be solved by them. Hence only the heuristic approach is further discussed in this chapter.

The essential idea of a heuristic approach centers around rescheduling the activities within the limits of available float in order to achieve better distribution of resource usage. This can be done by rescheduling all the critical jobs first followed by selective scheduling of the noncritical jobs to obtain the leveled profile. The heuristic rules normally adopted may be summed up as follows:

1. Schedule all the critical jobs first.
2. Start the noncritical jobs whenever there is a rise up to the point where the peak is reached.

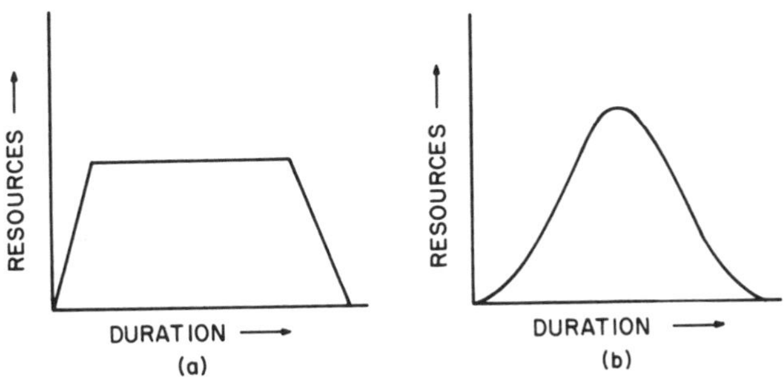

FIGURE 9.7 General Resource Profiles

3. Start the noncritical jobs whenever there is a drop so that no ups and downs occur in the resource profile.

There are many heuristic algorithms available to level the resource profile. The rules mentioned here are just some of them. The best leveling may be obtained by a trial and error procedure, applying different heuristics for the same problem and choosing the best.

Example 9.2. Heuristic rules may be employed using either of the parallel or series methods. However, only the parallel method is discussed here. Consider the network shown in Figure 9.3. The project clock is initially set at zero. Activity 1–5 is critical at this time, as it has a total float of zero. There are no other critical activities at this time. Hence the clock moves to day one. The duration of activity 1–5 is reduced to four, and its resource days become 16. The earliest start of activity 1–2 is updated to one, and its

TABLE 9.5 Resource Leveling Table

ACTIVITY	DURATION	RESOURCES	RESOURCE DAYS	EARLIEST START	PROJ. CLOCK / TOTAL FLOAT	DAYS 0	1	2	3	4	5	6	7	8	9	10	11	12	13	14
1-2	~~3~~ ~~2~~ ~~1~~ 0	4	~~12~~ ~~8~~ ~~4~~ 0	~~0~~ ~~1~~ ~~2~~ ~~3~~ 4	~~2~~ ~~1~~ 0			4	4	4										
1-3	~~2~~ ~~1~~ 0	4	~~8~~ ~~4~~ 0	~~0~~ ~~1~~ ~~2~~ ~~3~~ ~~4~~ ~~5~~ 6	~~6~~ ~~5~~ ~~4~~ ~~3~~ ~~2~~ 1						4	4								
1-5	~~5~~ ~~4~~ ~~3~~ ~~2~~ ~~1~~ 0	4	~~20~~ ~~16~~ ~~12~~ ~~8~~ ~~4~~ 0	~~0~~ ~~1~~ ~~2~~ ~~3~~ 4	0	4	4	4	4	4										
2-5	0	0	0	~~3~~ ~~4~~ 5	~~2~~ ~~1~~ 0															
2-6	~~3~~ ~~2~~ 1	2	~~6~~ ~~4~~ ~~2~~	~~3~~ ~~4~~ ~~5~~ ~~6~~ 7	~~5~~ ~~4~~ 3						2	2	2							
3-4	~~4~~ ~~3~~ ~~2~~ 1	3	~~12~~ ~~9~~ ~~6~~ ~~4~~ 0	~~2~~ ~~3~~ ~~4~~ ~~5~~ ~~6~~ ~~7~~ ~~8~~ ~~9~~ ~~10~~	~~6~~ ~~5~~ ~~4~~ ~~3~~ ~~2~~ 1								3	3	3	3				
3-7	~~4~~ ~~3~~ ~~2~~ 1	1	~~4~~ ~~3~~ ~~2~~ ~~1~~ 0	~~2~~ ~~3~~ ~~4~~ ~~5~~ ~~6~~ ~~7~~ ~~8~~ ~~9~~ 10	~~8~~ ~~7~~ ~~6~~ ~~5~~ ~~4~~ 3								1	1	1	1				
4-7	~~2~~ ~~1~~ 0	2	~~4~~ ~~2~~ 0	~~6~~ ~~7~~ ~~8~~ ~~9~~ ~~10~~ ~~11~~ ~~12~~ 13	~~6~~ ~~5~~ ~~4~~ ~~3~~ ~~2~~ 1												2	2	2	
5-6	~~6~~ ~~5~~ ~~4~~ ~~3~~ ~~2~~ 1	2	~~12~~ ~~10~~ ~~8~~ ~~6~~ ~~4~~ ~~2~~ 0	~~5~~ ~~6~~ ~~7~~ ~~8~~ ~~9~~ 10	0						2	2	2	2	2	2				
6-7	~~3~~ ~~2~~ ~~1~~ 0	2	~~6~~ ~~4~~ ~~2~~ 0	~~11~~ ~~12~~ ~~13~~ 14	0												2	2	2	

Resource profile

total float is reduced by one. The earliest start of activity 1–3 is updated to one, and its total float becomes five. Now all the eligible activities have been updated; activity 1–5 is the only critical activity for day one. It is again assigned the resources.

The clock moves to day two. After all the necessary updating is carried out as described, both activities 1–2 and 1–5 are critical. They are assigned the resources. This procedure is repeated until the fifth day is reached, when there are two critical activities, namely, 2–5 and 5–6. Activity 2–5 is a dummy and 5–6 requires only two resources. On the preceding day eight resources were allocated to critical activities, and now only two resources can be allocated. The peak has been reached. At this point noncritical jobs are scheduled; therefore activities 1–3 and 2–6 are assigned resources. The clock moves to day six, and the preceding allocation is repeated. From this time to the end of the project all noncritical activities are scheduled at their earliest start time.

The complete solution and the profile obtained are shown in Table 9.5. The significant factor is that the starting times for only noncritical jobs are varied to produce a leveled schedule. The project duration is never extended.

It should be noted that leveling can produce several alternative solutions. These solutions are acceptable if one peak is maintained, and buildup to the peak and the subsequent decline are gradual.

If it is desired to determine which alternative schedule presents the most leveled solution, a comparison can be made of the sum of squares of resource requirements within each time unit. The lowest sum of squares indicates the most leveled solution.

9.16 MULTIRESOURCE LEVELING

Multiresource leveling problems are clearly more complicated than single-resource ones. Since each activity may require different quantities of several resources for its execution, attempts at balancing one resource type may spoil the balance of the others. If manual solutions are desired, this problem can be solved by applying a weighting factor to each type of resource and computing the optimum solution. However, if even one of the resources' availability is limited, then the problem has to be treated as one of resource-constrained allocation.

To minimize the cost of site facilities it is necessary to keep the peak manpower to a minimum level. The scheduling engineer can accomplish this by plotting manpower required for each one of the major activities on a project and by adjusting the start and finish dates of such major activities. Figure 9.8 shows a manpower profile obtained by adding up the manpower requirements of several major activities on a project.

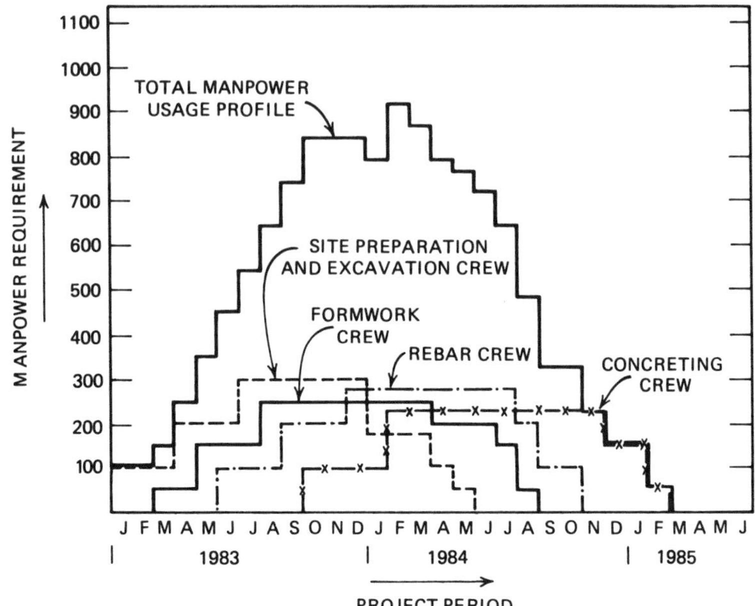

FIGURE 9.8 Manpower Resource Profiles

9.17 COMPUTER-AIDED RESOURCE ALLOCATION

The methods of resource allocation discussed in this chapter involve simple arithmetic and data manipulation. It is obvious, however, that to perform complete resource allocation manually even for a summary level network for an average-sized project would be impractical and almost impossible for a large project. Fortunately, there are many computer programs available to do this type of network analysis.

When using resource allocation programs, it is first necessary to perform a time analysis on the network. The schedule is then combined with additional data on resource requirements and limitations to produce a daily resource requirement schedule and a modified project schedule.

Program packages of this type possess considerable flexibility. The user may perform allocation by the series method, the parallel method, a combination of the two, or a variation of either method. For instance, some activities may be split (stopped while in progress and restarted later, as occurs in the parallel method), whereas others are not permitted to be split. The priority rules can be specified by the user to follow those outlined in either method or to meet other requirements. In addition the programs can perform both fixed-resource and fixed-duration scheduling.

Resource leveling also can be carried out by using any of the various resource allocation computer programs available today. The program should

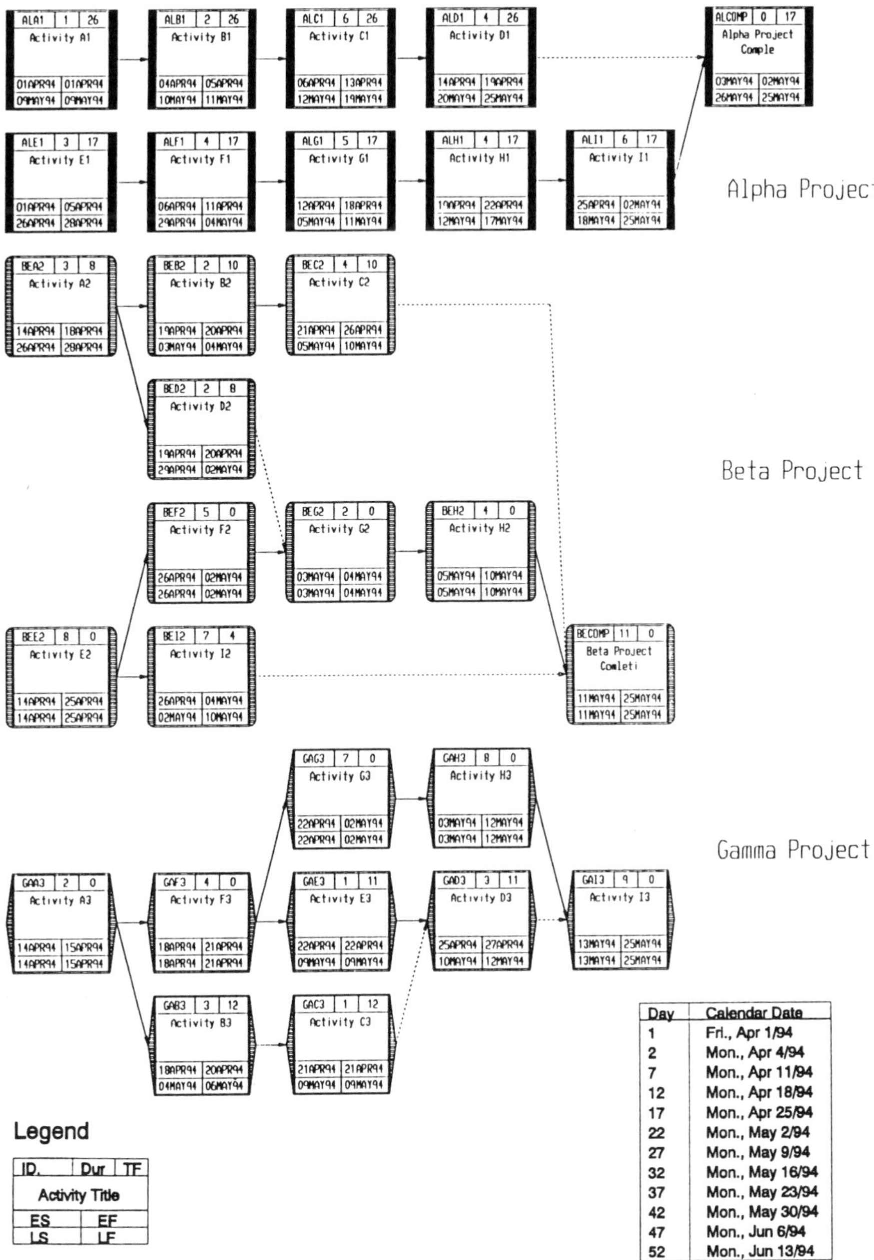

Day	Calendar Date
1	Fri., Apr 1/94
2	Mon., Apr 4/94
7	Mon., Apr 11/94
12	Mon., Apr 18/94
17	Mon., Apr 25/94
22	Mon., May 2/94
27	Mon., May 9/94
32	Mon., May 16/94
37	Mon., May 23/94
42	Mon., May 30/94
47	Mon., Jun 6/94
52	Mon., Jun 13/94

FIGURE 9.9 Multiproject Example—No Constraints on Resources

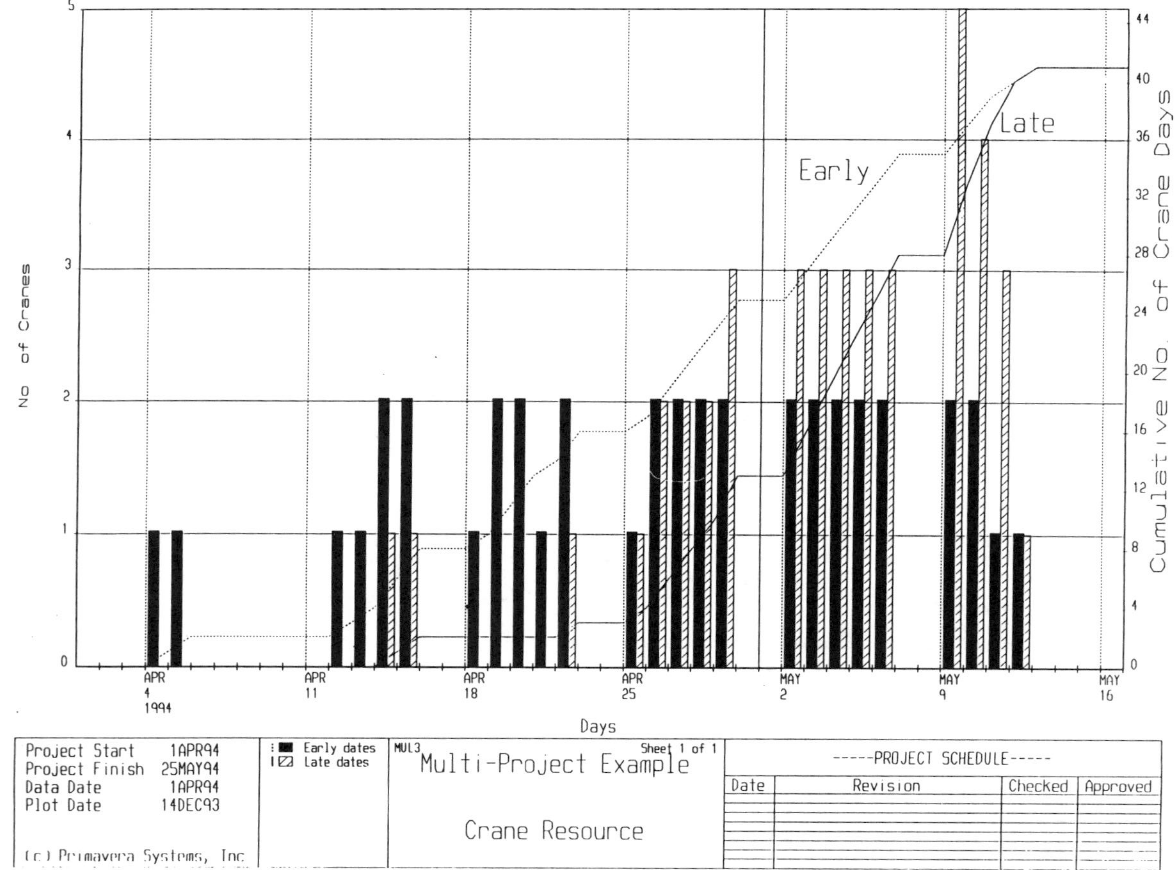

FIGURE 9.10 Crane Resource for Unconstrained Schedule

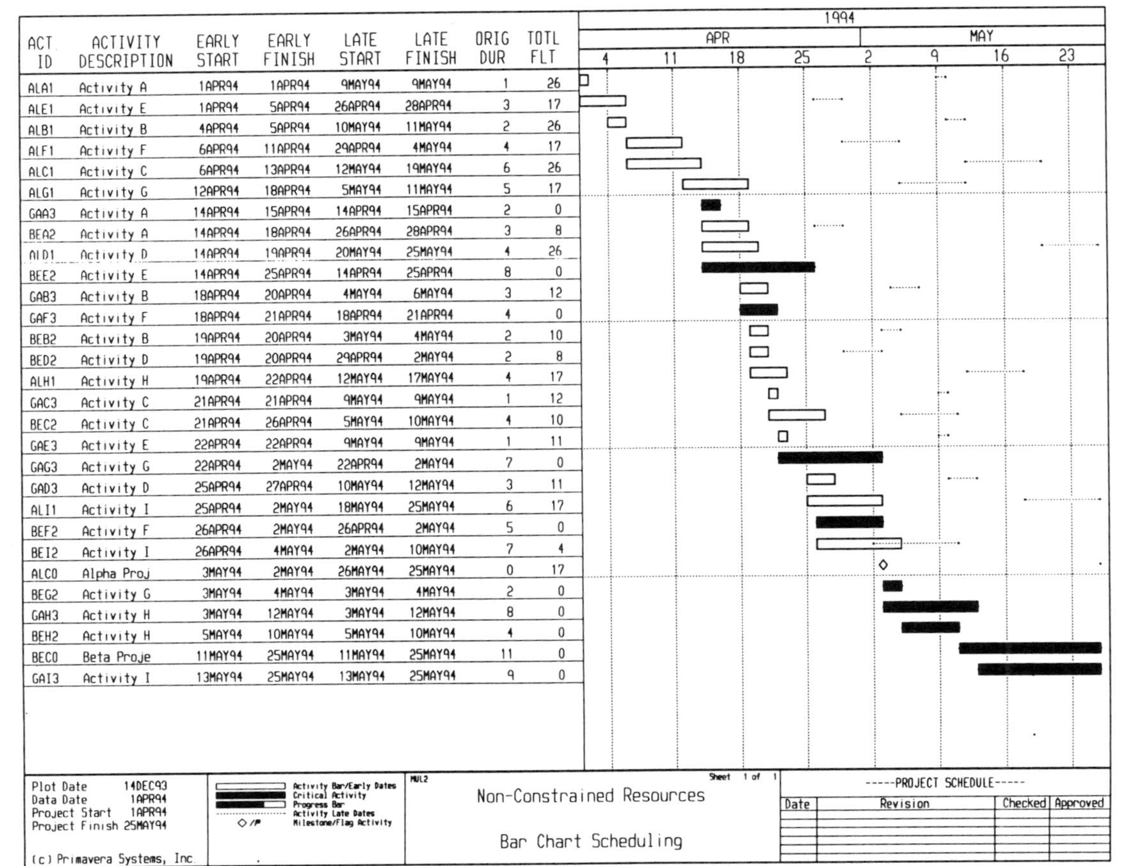

ACT ID	ACTIVITY DESCRIPTION	EARLY START	EARLY FINISH	LATE START	LATE FINISH	ORIG DUR	TOTL FLT
ALA1	Activity A	1APR94	1APR94	9MAY94	9MAY94	1	26
ALE1	Activity E	1APR94	5APR94	26APR94	28APR94	3	17
ALB1	Activity B	4APR94	5APR94	10MAY94	11MAY94	2	26
ALF1	Activity F	6APR94	11APR94	29APR94	4MAY94	4	17
ALC1	Activity C	6APR94	13APR94	12MAY94	19MAY94	6	26
ALG1	Activity G	12APR94	18APR94	5MAY94	11MAY94	5	17
GAA3	Activity A	14APR94	15APR94	14APR94	15APR94	2	0
BEA2	Activity A	14APR94	18APR94	26APR94	28APR94	3	8
ALD1	Activity D	14APR94	19APR94	20MAY94	25MAY94	4	26
BEE2	Activity E	14APR94	25APR94	14APR94	25APR94	8	0
GAB3	Activity B	18APR94	20APR94	4MAY94	6MAY94	3	12
GAF3	Activity F	18APR94	21APR94	18APR94	21APR94	4	0
BEB2	Activity B	19APR94	20APR94	3MAY94	4MAY94	2	10
BED2	Activity D	19APR94	20APR94	29APR94	2MAY94	2	8
ALH1	Activity H	19APR94	22APR94	12MAY94	17MAY94	4	17
GAC3	Activity C	21APR94	21APR94	9MAY94	9MAY94	1	12
BEC2	Activity C	21APR94	26APR94	5MAY94	10MAY94	4	10
GAE3	Activity E	22APR94	22APR94	9MAY94	9MAY94	1	11
GAG3	Activity G	22APR94	2MAY94	22APR94	2MAY94	7	0
GAD3	Activity D	25APR94	27APR94	10MAY94	12MAY94	3	11
ALI1	Activity I	25APR94	2MAY94	18MAY94	25MAY94	6	17
BEF2	Activity F	26APR94	2MAY94	26APR94	2MAY94	5	0
BEI2	Activity I	26APR94	4MAY94	2MAY94	10MAY94	7	4
ALC0	Alpha Proj	3MAY94	2MAY94	26MAY94	25MAY94	0	17
BEG2	Activity G	3MAY94	4MAY94	3MAY94	4MAY94	2	0
GAH3	Activity H	3MAY94	12MAY94	3MAY94	12MAY94	8	0
BEH2	Activity H	5MAY94	10MAY94	5MAY94	10MAY94	4	0
BEC0	Beta Proje	11MAY94	25MAY94	11MAY94	25MAY94	11	0
GAI3	Activity I	13MAY94	25MAY94	13MAY94	25MAY94	9	0

FIGURE 9.11 Schedule with no Resource Constraints

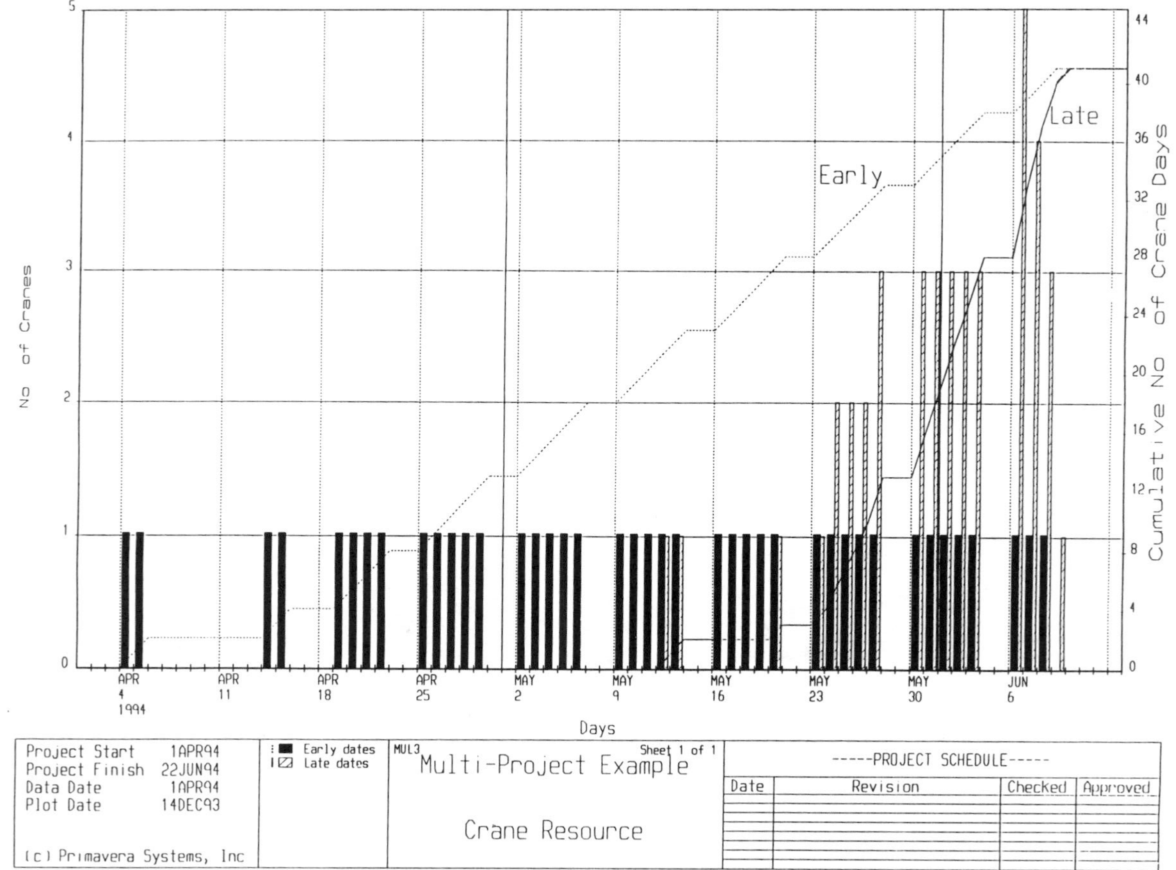

FIGURE 9.12 Levelled Crane Resource

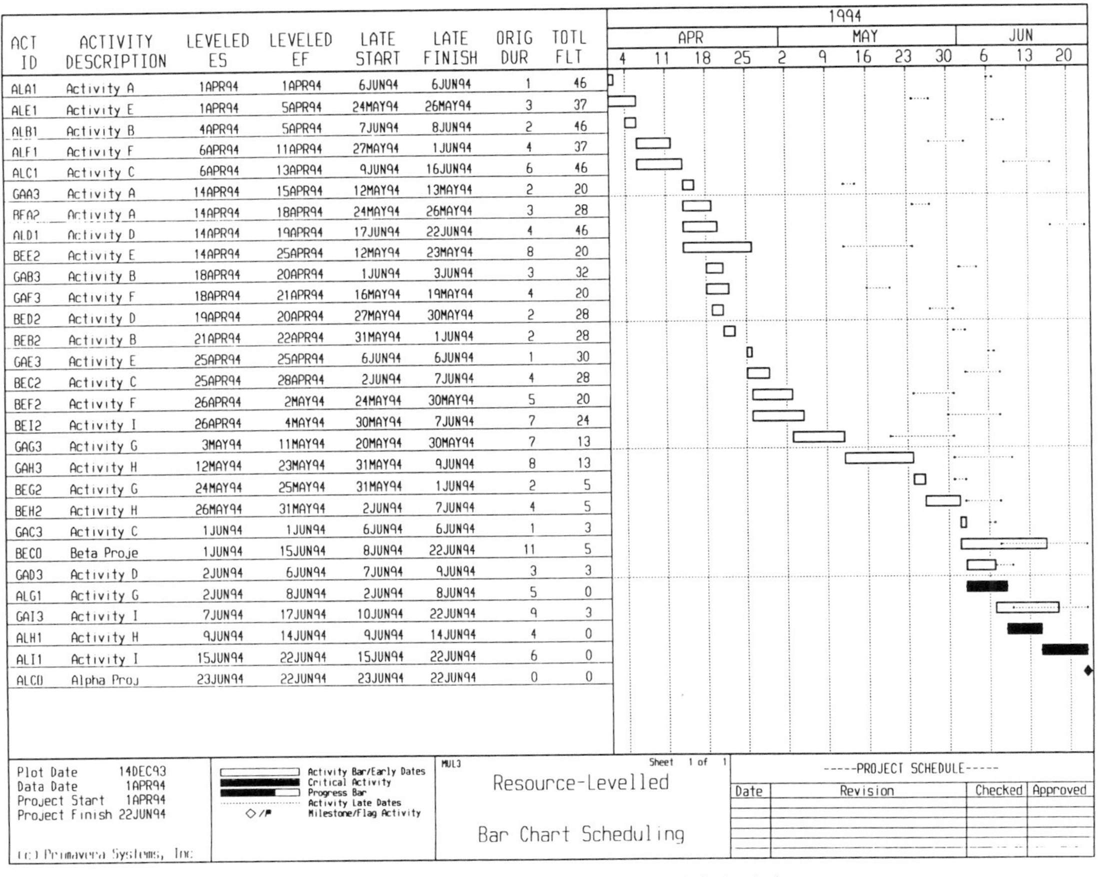

ACT ID	ACTIVITY DESCRIPTION	LEVELED ES	LEVELED EF	LATE START	LATE FINISH	ORIG DUR	TOTL FLT
ALA1	Activity A	1APR94	1APR94	6JUN94	6JUN94	1	46
ALE1	Activity E	1APR94	5APR94	24MAY94	26MAY94	3	37
ALB1	Activity B	4APR94	5APR94	7JUN94	8JUN94	2	46
ALF1	Activity F	6APR94	11APR94	27MAY94	1JUN94	4	37
ALC1	Activity C	6APR94	13APR94	9JUN94	16JUN94	6	46
GAA3	Activity A	14APR94	15APR94	12MAY94	13MAY94	2	20
BEA2	Activity A	14APR94	18APR94	24MAY94	26MAY94	3	28
ALD1	Activity D	14APR94	19APR94	17JUN94	22JUN94	4	46
BEE2	Activity E	14APR94	25APR94	12MAY94	23MAY94	8	20
GAB3	Activity B	18APR94	20APR94	1JUN94	3JUN94	3	32
GAF3	Activity F	18APR94	21APR94	16MAY94	19MAY94	4	20
BED2	Activity D	19APR94	20APR94	27MAY94	30MAY94	2	28
BEB2	Activity B	21APR94	22APR94	31MAY94	1JUN94	2	28
GAE3	Activity E	25APR94	25APR94	6JUN94	6JUN94	1	30
BEC2	Activity C	25APR94	28APR94	2JUN94	7JUN94	4	28
BEF2	Activity F	26APR94	2MAY94	24MAY94	30MAY94	5	20
BEI2	Activity I	26APR94	4MAY94	30MAY94	7JUN94	7	24
GAG3	Activity G	3MAY94	11MAY94	20MAY94	30MAY94	7	13
GAH3	Activity H	12MAY94	23MAY94	31MAY94	9JUN94	8	13
BEG2	Activity G	24MAY94	25MAY94	31MAY94	1JUN94	2	5
BEH2	Activity H	26MAY94	31MAY94	2JUN94	7JUN94	4	5
GAC3	Activity C	1JUN94	1JUN94	6JUN94	6JUN94	1	3
BEC0	Beta Proje	1JUN94	15JUN94	8JUN94	22JUN94	11	5
GAD3	Activity D	2JUN94	6JUN94	7JUN94	9JUN94	3	3
ALG1	Activity G	2JUN94	8JUN94	2JUN94	8JUN94	5	0
GAI3	Activity I	7JUN94	17JUN94	10JUN94	22JUN94	9	3
ALH1	Activity H	9JUN94	14JUN94	9JUN94	14JUN94	4	0
ALI1	Activity I	15JUN94	22JUN94	15JUN94	22JUN94	6	0
ALC0	Alpha Proj	23JUN94	22JUN94	23JUN94	22JUN94	0	0

MOC

FIGURE 9.13 Resource Levelled Schedule

allow the user to select the priority rules. By assigning first priority to float and second priority to late start, a profile close to a leveled profile can be obtained. Thorough manual intervention adjustments can be made by profiling the start of certain activities. A leveled resource profile and schedule is obtained in a few runs.

The network was created using Primavera Project Planner (P3) version 5.0 in the precedence format as shown in Fig. 9.9 with three sub-projects to enable allocating resources at the company level and not specifically for each project. The network analysis of the 'mobile crane' resource for an unconstrained schedule yields the resource loading diagram given in Fig. 9.10. Should there be only one crane available to serve all three sub-projects, the schedule would not be feasible. A demand of 1, 2, 3, 4 and 5 crane days is reflected in Fig. 9.10 at specific points of time, thus making it impossible to maintain the unconstrained schedule shown in Fig. 9.11 with only one crane (project completion on May 12, 1994). Resource levelling was performed in P3 specifying that only one crane day may be assigned at any point of time. The new resource loading diagram is shown in Figure 9.12. The new constrained project schedule is shown in Figure 9.13 with a project completion time of June 3, 1994 compared to the original finish of May 12. The extension of completion time was due to the limited availability of the crane resource.

OTHER AREAS OF USE

Resource allocation usually includes manpower, machines, and materials. Sometimes its use is extended to other areas, such as the availability of space. If space is limited, it must be carefully utilized.

Besides being useful for projects, resource scheduling is also carried on in factory or manufacturing planning, to optimize the investment in capital equipment based on sales forecasts and projections. For already established plants the technique can generate an optimal schedule for the fixed number of machines available.

Another use of the resource allocation exercise is to simulate generation of resources and to assign them according to availability. Since assignment of resources to an activity carries a positive sign, their generation by an activity is given a negative sign. The daily production of concrete from a batching and mixing plant is a negative resource. Availability of this resource results in scheduling, and its unavailability means postponement of a "pour concrete" activity on a large project.

PROBLEMS

9.1 Differentiate between "Constrained Resource Scheduling" and "Resource Leveling."

9.2 For the given project eight skilled workers are available.

Activity i–j	Description	Duration	Resources Required (skilled workers)
1–2	*A*	1	4
1–6	*B*	4	6
2–3	*C*	4	5
2–4	*D*	2	2
3–9	*F*	2	4
4–5	*F*	5	3
5–9	*G*	6	1
6–7	*H*	5	5
6–8	*I*	4	7
7–8	Dummy	0	0
8–9	*J*	7	2
9–10	*K*	3	2

Determine the minimum project completion time using both the Series and Parallel Methods.

9.3 In the preceding problem, if the project completion time is to be restricted to 16 days, determine the minimum level of resources to be considered to achieve that time. Solve the problem using both the Series and Parallel Methods.

9.4 Explain with an example the resource leveling technique.

9.5 The following project consisting of 10 activities is considered for multiresource scheduling. Develop a resource schedule using the Series and Parallel Methods assuming that the availability of resources is limited as below:

Resource *A*	7
Resource *B*	7
Resource *C*	6

If the project completion time is limited to 14 days, estimate the minimum resource level required for each craft.

Activity			Resources Required		
i–j	Description	Duration	*A*	*B*	*C*
1–2	*A*	3	4	4	2
1–3	*B*	4	3	4	1
1–5	*C*	5	1	3	2

Activity i–j	Description	Duration	Resources Required		
			A	B	C
2–4	D	2	1	0	0
2–6	E	3	2	1	0
3–4	F	4	2	2	1
4–7	G	3	3	1	2
5–6	H	6	4	4	4
5–7	I	4	3	2	1
6–7	J	3	1	4	5

9.6 Explain the basic difference between the Series and Parallel Methods of resource allocation.

9.7 Suggest a priority list for allocating single limited resource on a large hydropower project.

CHAPTER 10

DETERMINATION OF MOST ECONOMICAL PROJECT DURATION

In designing a system, components may be put together in several different ways to form alternative systems so long as the resulting alternatives can accomplish the mission for the system. A network plan is evolved through alternative arrangements of activities even though the normal duration alloted to the activities remains unchanged. This is followed by resource allocation for the most suitable alternative, which is similar to the physical feasibility check in systems design.

After a physical feasibility check the network plan has to pass through the economic feasibility sieve. Its cost has to be minimized. Since the total cost includes overhead costs, which are dependent on the project duration, as well as direct costs, a delay in the completion of a project will usually increase its cost. Also, if the project duration is specified, the owner may award a bonus for early completion or ask for liquidated damages for late completion. The project manager may investigate different project durations to determine which will provide the minimum cost. Economic feasibility analysis means more than finding the minimum cost solution that justifies the investment. The fact that a project is being undertaken indicates that the project has an adequate rate of return for the owner. The contractor submits a bid because he or she finds this project more attractive, economically, than other available alternatives. The economic feasibility analysis of a network plan can improve the rate of return or enhance the attractiveness of the project by further reducing project costs through the minimum cost solution. This is achieved by evaluating the effects of decreases or increases in project duration (compression or decompression of the network). The achievement of a minimum cost duration plan is discussed in this chapter.

10.1 TIME VERSUS COST

In the economic feasibility analysis of a system, the cost of each component is minimized, and the effect of this reduction on the total cost of the system is observed. The activities in a network system are its components. The time and cost of each one is altered so as to minimize the total cost. In the process of reducing the cost to the minimum, several alternative solutions are generated, each with a different project duration and its associated cost. The most suitable solution is selected. The criteria for selection of a solution depend on the balance between time and cost parameters that the planner wants to achieve.

On a construction project a project manager may be paying interest on an investment in an apartment building. For every day that this building is not in use, he may be losing rental income as well. In order to expedite the project without unduly raising capital costs, the project manager wants to determine the project duration that will yield the maximum payoff.

In the military establishment there is usually some desired cutoff date for a project, reflecting the need to commission a system by a given date. It may be required to complete the project as early as possible before a given deadline while keeping the increase in cost to a minimum.

In the business world it is imperative, or at least very desirable, to have a product on the market before the competitor does. As a result the criterion may be to maximize the probability of marketing it as early as possible while minimizing the increase in cost.

10.2 COST DURATION CURVES

To achieve the objectives discussed in the previous section, some additional information is required. In network planning the normal duration for each activity is specified; it is the time required to complete an activity with the resources normally available in the organization and with no extra inputs into the project. Besides normal duration, normal cost is also specified. It is the estimated cost associated with the normal duration.

Scheduling is the determination of the timing of the activities of a project and their coordination to give the overall project duration. The start time of any given activity depends on factors such as resource availability, appropriate manpower requirements, float time, management decisions, and the work pattern of the organization performing the activity. For reliable planning, the manpower and equipment needed for each activity to be completed in normal duration at a normal cost is derived from data collected on similar projects completed in the past. It is assumed that the same type of technology will be used as in the past and the competence level of manpower remains the same.

Normal cost may also be considered as simply the cost to the owner. The bid value given in the form of a tender of the contractor who is awarded the contract, and who subsequently carries out the work, may also be considered the normal cost. It is possible to complete the activity in less time than scheduled. For example, by employing additional crews and incurring extra cost, the time required for formwork can be reduced. This approach is known as "crashing."

The crash duration and crash cost, however, must be known. The crash duration is the time, which is less than the normal duration time, required to complete an activity, possibly with extra funds or resources. The crash cost of an activity is the cost of completing an activity by its crash duration. Crash project duration is the time, which is less than the normal duration time, computed from a combination of crash and normal durations of activities to complete the project, and crash cost for the project is the total cost associated with the crash project duration.

Consider any activity that has a normal duration and normal cost and a crash duration and crash cost. The normal duration and normal cost may be 10 days and $1000; the crash duration and crash cost may be 5 days and $2000. Besides these two durations and their associated costs, it may be useful to know the costs of intermediate durations. This depends on the relationship between normal and crash durations and costs.

There are four types of relationships between the time and cost of an operation. The procedure necessary for deriving the relationships defined by the cost-duration curves in Figures 10.1–10.4 is as follows:

1. Select several methods by which an activity can be performed.
2. Determine the duration that would result using each method as well as the direct cost of this method.

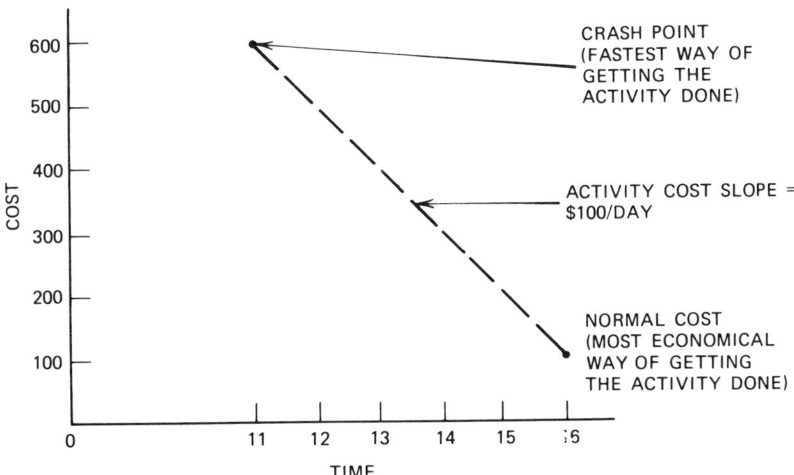

FIGURE 10.1 Linear Relationship between Time and Cost.

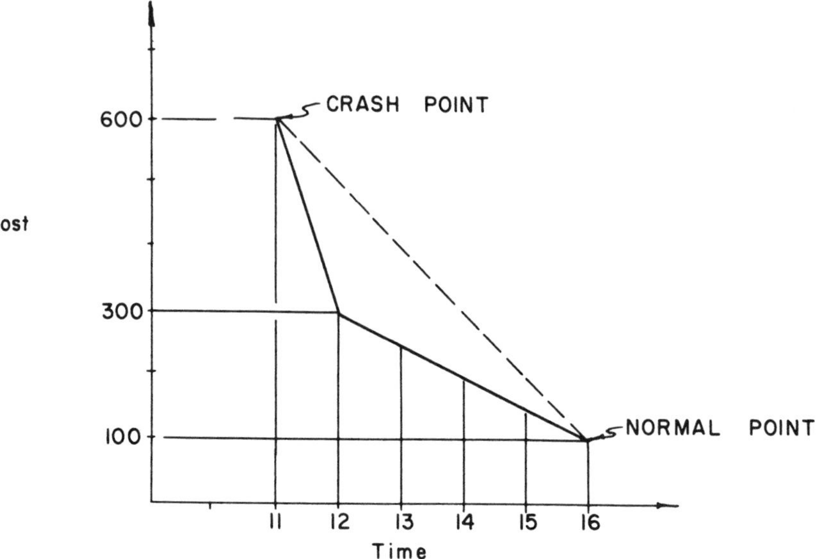

FIGURE 10.2 Multilinear Cost-Time Relationship.

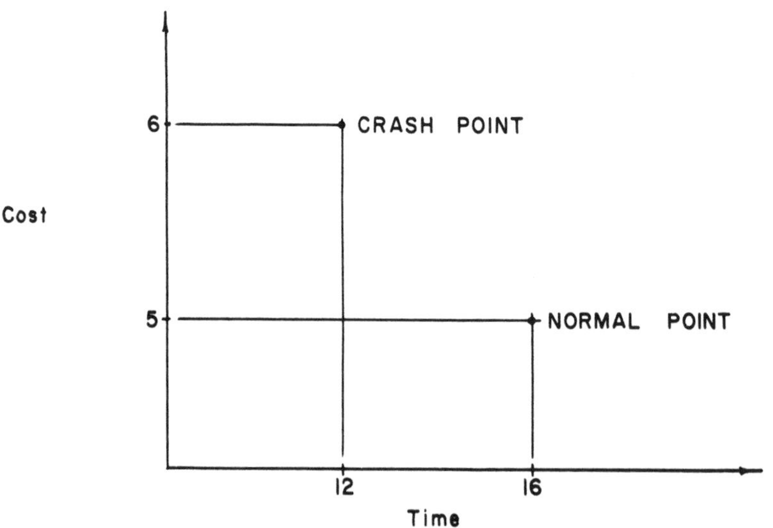

FIGURE 10.3 Discrete Cost-Time Relationship

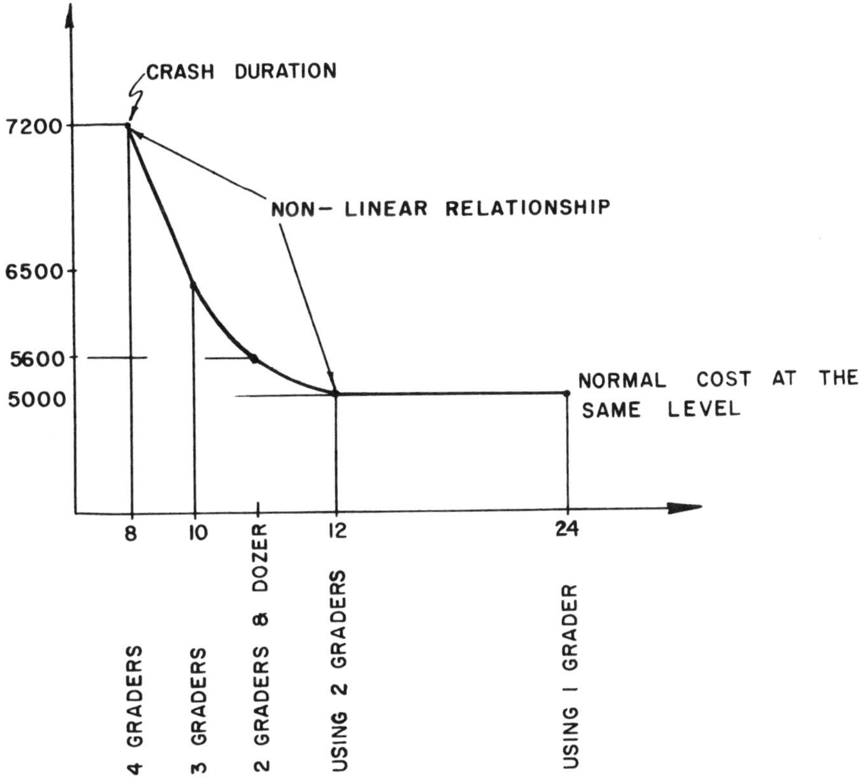

FIGURE 10.4 Curvilinear Continuous Cost-Time Relationship.

3. Plot the results of Step 2 on a graph of duration versus direct cost.
4. Connect the points with straight lines sloping upward to the left as follows: from the lowest direct cost point draw a straight line to the next point, which represents the cost slope between these two points. From this second point draw a straight line to the next point, which denotes the cost slope from the second to the third point, and so on, until the final higher direct cost point is connected.

From the plot, the cost slope is determined. It is the amount of funds required to reduce the duration of an activity by one day:

$$\text{Cost slope} = \frac{\text{crash cost} - \text{normal cost}}{\text{normal duration} - \text{crash duration}}$$

Case 1: Linear Relationship Between Time and Cost

Figure 10.1 shows the straight-line relationship. This kind of linear relationship represents the case where overtime work can result in savings in direct cost. An additional expenditure of \$500 is necessary to save five days. Each day saved cost \$100. Thus the slope for this case is

$$\frac{600 - 100}{16 - 11} = \frac{500}{5} = 100$$

Case 2: Multilinear Relationship Associated with Different Time Intervals

Figure 10.2 is an example of a linear relationship for different time intervals. The additional cost per day of time saved is not uniform over the entire period but varies. To vary the time from 16 to 21 days, which may be done by using a loader of different capacity on an earthwork job, the cost slope is uniform for four days. Beyond day 12 the job may require two loaders instead of one. The mobilization cost of the second loader is not covered by the same slope. However, the cost for saving one day beyond day 12 to \$300. In this case we have two slopes:

$$\text{From day 16 to day 12} = \frac{300 - 100}{4} = 50$$

$$\text{From day 12 to day 11} = \frac{600 - 300}{1} = 300$$

Case 3: Discrete Function

Figure 10.3 shows only two costs. The activity can be performed in normal time for normal cost or it can be performed in crash time at crash cost. There is no relationship between the normal and crash costs in this case. This may be a tunneling project where, by using a drill jumbo, the project duration is 16 months and the cost is \$5 million, but by using the mole, the cost jumps to \$6 million and the project duration is reduced to 12 months. This could also happen on a pile-driving job where the additional cost would represent the mobilization and demobilization of an additional pile-driving rig. There is no slope in such a case. Either the normal cost or the crash cost is used for compression and decompression of networks.

Case 4: Curvilinear Continuous Relationship

In the fourth case, shown in Figure 10.4, there is no straight-line relationship between normal and crash costs but a continuous curve that represents the relationship between different plots of crash costs. A nonlinear relationship between normal and crash costs may occur in certain cases such as a dike construction project. Using one grader, the operation may take 24 days, whereas with two graders it will take just half the time. The operation can be further expedited by using two graders and one dozer, resulting in a saving of one day but a cost of $5600. Three graders will bump the cost to $6500, but the operation will be completed in 10 days. With four graders it will be finished in eight days for $7200. For nine days the cost may be measured from the curve; it is $6800 and represents the use of graders and dozers of certain capacities.

There is no reason to expect that every contractor will have the same normal and crash costs for a selected operation. The normal cost involves the contractor's methods. Experience of workers, availability of equipment, and so on, differ from one organization to another. Thus the normal point reflects the capability of a specific construction organization. It is important to keep in mind that the crashing does not involve changing material from one alternative to the other. The time compression is due only to increased labor and/or machinery. However, crash cost may include additional cost for using alternative transportation to expedite delivery of materials.

The analogous argument can be applied to the crash cost. This point is associated with the method the contractor believes work forces can apply successfully to complete the operation in the shortest possible time.

It is not always true that crashed activities require more man and equipment hours. Even though more workers and equipment are needed, they are required for shorter periods; so in effect the total man and equipment hours of a crashed activity could be the same or only slightly higher than the normal duration. The additional cost could result from mobilization of additional equipment, overtime work, or reduced productivity resulting from nightshift work or congested work space.

Given dozens, hundreds, or even thousands of activities, it is possible to generate more than one schedule that will get the job done within the limits of the network diagram. Each of these alternatives has associated with it a unique rate of resource application which will effectively change the duration of activities from alternative to alternative. The next section illustrates a simple example.

Example 10.1. This example is based on a straight-line relationship between normal and crash costs. Consideration will be given to the other three types of relationships at the end of the chapter.

Consider a project comprised of four activities, *A*, *B*, *C*, and *D*, as shown in Figure 10.5. For each activity, both normal and crash duration is given. Associated with these durations are the normal and crash costs of the activity, which are listed in Table 10.1. The normal project duration is 12 days.

If the project manager decides that the project must be completed in the shortest time possible, additional funds are necessary. The additional funds permit the rental of additional equipment and use of overtime labor. The project manager decides which activity should be expedited and how much

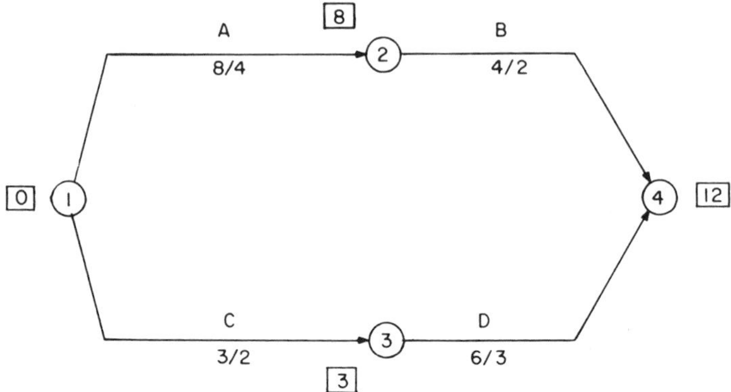

FIGURE 10.5 Example 10.1 Network

its duration should be shortened in order to finish the project in the minimum possible project duration.

The normal cost of the project for the project duration of 12 days is $1900, as indicated in Table 10.1. If all the activities are crashed, the project duration is reduced to six days, and the project cost becomes $2700. The network with all crash durations has a project duration of six days. The total costs, both normal and crash, are totaled in Table 10.1.

Is it necessary to crash all the activities? For example, if activities *A* and *B* have been reduced to their minimum crash duration and we know that the project duration cannot be reduced below six days, there is no point in trying to reduce the duration of activities *C* and *D* to less than six days. On the other hand, is it more economical to crash *C* or *D*? This can be answered by examining the slope of each activity. The slopes of the activities of the project in question are listed in Table 10.2.

Naturally, the activity that has a lower cost for saving a day is crashed. Activity *D* has a slope of 66.67 compared with activity *C*, which has a slope of 200. It is therefore economical to crash activity *D*. The duration of *D* is reduced to three days, and the duration of activity *C* is not changed. The

TABLE 10.1 Data for Example 10.1

Activity	Normal Duration	Crash Duration		Normal Cost	Crash Cost
A	8	4		$ 500	$ 800
B	4	2		300	400
C	3	2		300	500
D	6	3		800	1000
			Total	$1900	$2700

TABLE 10.2 Example 10.1-Cost Slopes

Activity	Normal Duration	Crash Duration	Slope
A	8	4	75
B	4	2	50
C	3	2	200
D	6	3	66.67

total cost resulting from selectively crashing activities for this project is listed in Table 10.3. The reader will agree that there is no point in spending $2700 by crashing activity *C* to save one day when the project can be completed in six days for $2500.

10.3 RELAXATION

The opposite of crashing is relaxation. Sometimes it may be necessary to relax activities, particularly noncritical activities, to make a project more economical. It may be possible for an activity that can be completed in six days for $2100 by using overtime labor to be completed in nine days for $1800 without overtime labor. If the increase of three days in the duration of this activity does not increase the project duration, no overtime and therefore no resulting productivity loss will be involved. Hence the relaxation can result in a saving of $300. The investigation of such activities is important in minimizing the cost of a project.

The usefulness of exercising an option in crashing activities was demonstrated for a small project in the previous example. It is not as easy to select the activities for crashing at the lowest cost when we have a large network. The theorem discussed in the following section is helpful in making such a selection.

TABLE 10.3 Example 10.1 Selective Crashing

Activity	Reduced by	Additional Cost	Total Cost
A	4	$300	$ 800
B	2	100	400
C	0	0	300
D	3	200	1000
Total		$600	$2500

10.4 THE CRITICALITY THEOREM

C_1 and C_2 are two paths with common end nodes. The duration of the paths is γ_1 and γ_2, respectively, such that $\gamma_1 \geqslant \gamma_2$ where the duration of all the activities is normal. It is desired to determine whether it is necessary to consider both paths in crashing their activities to arrive at the minimum cost solution. Let the path whose activities must be crashed be called the path of maximum duration. C_1 is to be compared with other paths to determine if it has the maximum duration. Path duration of path C_1 is maximum if and only if

$$\gamma_1 - \gamma_2 \geqslant \sum_{(i,j)\in C_1} b_{(i,j)} - \sum_{(i,j)\in C_1} a_{(i,j)} \tag{10.1}$$

where $b_{(i,j)}$ is the normal duration and $a_{(i,j)}$ the crash duration, and their difference is zero or positive for any path. $(i,j)\in C_1$ means (i,j) belongs to C_1. The normal path duration of C_2 is $\sum_{(i,j)\in C_2} b_{(i,j)}$. Substituting into Equation 10.1 yields

$$\sum_{(i,j)\in C_1} b_{(i,j)} - \sum_{(i,j)\in C_2} b_{(i,j)} \geqslant \sum_{(i,j)\in C_1} b_{(i,j)} - \sum_{(i,j)\in C_1} a_{(i,j)}$$

or

$$- \sum_{(i,j)\in C_2} b_{(i,j)} \geqslant - \sum_{(i,j)\in C_1} a_{(i,j)}$$

or

$$\sum_{(i,j)\in C_1} a_{(i,j)} \geqslant \sum_{(i,j)\in C_2} b_{(i,j)} \tag{10.2}$$

C_1 is greater than C_2 if and only if the crash path duration of C_1 is greater than the normal path duration of C_2.

This theorem is applied to the network shown in Figure 10.6. The numbers above the arrows show normal durations, and numbers beneath the arrows denote crash durations of the activities. C_1 represents the path of 1–4, 4–5, and 5–6, and C_2 represents the path of 1–2, 2–3, and 3–6; C_1 is greater than C_2 only if the crash path duration of C_1 is greater than the normal path duration of C_2:

$$\begin{aligned} 30 - 18 &\geqslant 30 - 21 \\ 12 &> 9 \text{ (using Equation 10.1)} \end{aligned}$$

which is the case here. Alternatively,

$$6 + 5 + 10 \geqslant 4 + 6 + 8 \text{ (using Equation 10.2)}$$

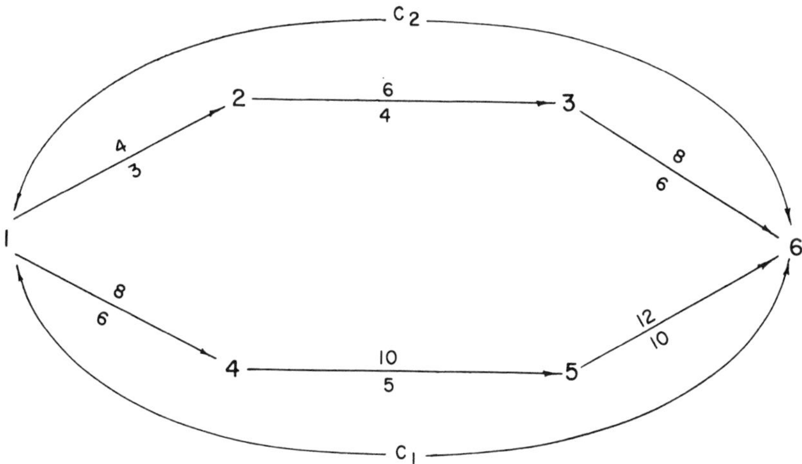

FIGURE 10.6 Crashing Example

In simpler terms, since even the crash duration of path C_1 is greater than the normal duration of path C_2, path C_1 is for all purposes the greater path. Therefore path C_2 need not be considered for crashing.

Figure 10.7 shows another network with path C_1 and C_2. Here $C_1 > C_2$ if and only if $6 + 5 + 10 > 7 + 9 + 8$, $21 > 24$, which is not the case. Hence path C_2 must be considered for crashing activities in combination with activities on path C_1.

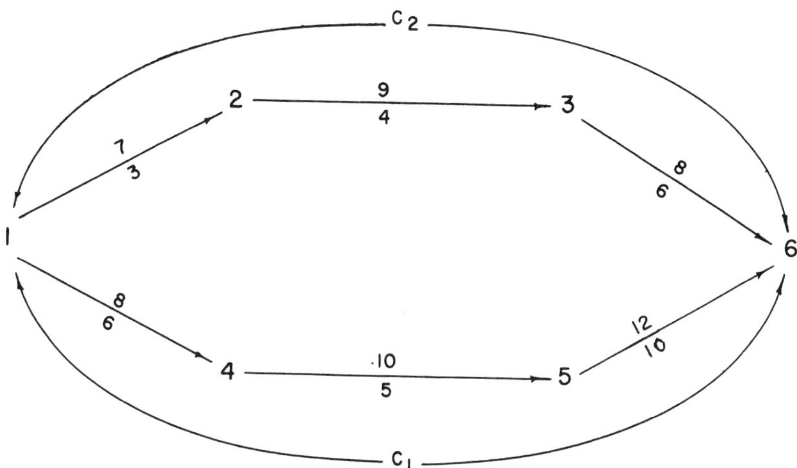

FIGURE 10.7 Another Crashing Example

Having solved a simple problem to demonstrate the advantages of selective crashing and having described the criticality theorem, a procedure for determining the minimum cost solution for a project must be outlined. The next section describes the steps of this procedure and the following section demonstrates its use.

10.5 PROCEDURE FOR COMPRESSION OF PROJECT DURATION

1. Determine normal project duration and normal project cost.
2. Identify normal duration critical path.
3. Eliminate all noncritical activities that don't need to be crashed. This is done by successively comparing normal duration of paths parallel to the critical path to the latter path's crash duration, using the criticality theorem.
4. Tabulate normal and crash durations and normal and crash cost for all the activities.
5. Compute and tabulate the cost slope of each activity from the following formula:

$$\text{Cost slope} = \frac{\text{crash cost} - \text{normal cost}}{\text{normal duration} - \text{crash duration}}$$

6. Proceed to determine the project time cost curve by shortening the critical activities beginning with the activity having the lowest cost slope. Each activity is shortened until (a) its crash time is reached or (b) a new critical path is formed.
7. When a new critical path is formed, shorten the combination of activities having the lowest combined slope. Where several parallel paths exist, it is necessary to shorten each of them simultaneously if the overall project time is to be reduced.
8. At each step check to see whether float time has been introduced in any of the activities. If so, perhaps these activities can be expanded to reduce cost.
9. At each shortening cycle compute the new project cost and duration. Tabulate and plot these points on a time–cost graph.
10. Continue until no further shortening is possible. This is the crash point.
11. Plot the indirect project costs on the same time–cost graph.
12. Add direct and indirect costs to find the total project cost at each duration.
13. Use the total project cost curve to find the optimum time (completion at lowest cost) or the cost of any other desired schedule.

Example 10.2. This example illustrates the use of the criticality theorem. Consider the network shown in Figure 10.8. The normal project duration is 70 days, and the normal direct cost is $6600, as shown in Table 10.4. Critical path is indicated by double strokes on the arrows. The project duration is to be minimized, using the time compression procedure outlined here.

Using the criticality theorem and considering all the alternative paths that join 3 and 8, activity 3–8 is eliminated. Similarly, activities 2–4, 3–5, and 6–8 are dropped. Having eliminated all the noncritical activities that need not be crashed, a skeleton network is developed, as shown in Figure 10.9. Next the crash duration, crash cost, and slope for each activity are tabulated in Table 10.4. The slope is determined assuming a linear relationship between normal and crash costs.

The crash duration and crash cost of the project are determined by shortening the critical activities, beginning with the activity having the lowest cost slope. Each activity is shortened until (a) its crash time is reached or (b) a new critical path is generated. Figure 10.10 illustrates the crashing of activities for which a step-by-step description follows.

Activity 7–8 which has the least slope though equal to that of 5–6 is crashed from 10 to 9 days. The project duration is reduced from 70 to 69 days, and the cost of the project is increased to $6650. This information in entered in Table 10.4.

Activity 5–6 also has a slope of 50. It is reduced by one day, changing the duration from 10 to 9. It cannot be reduced any further without reducing

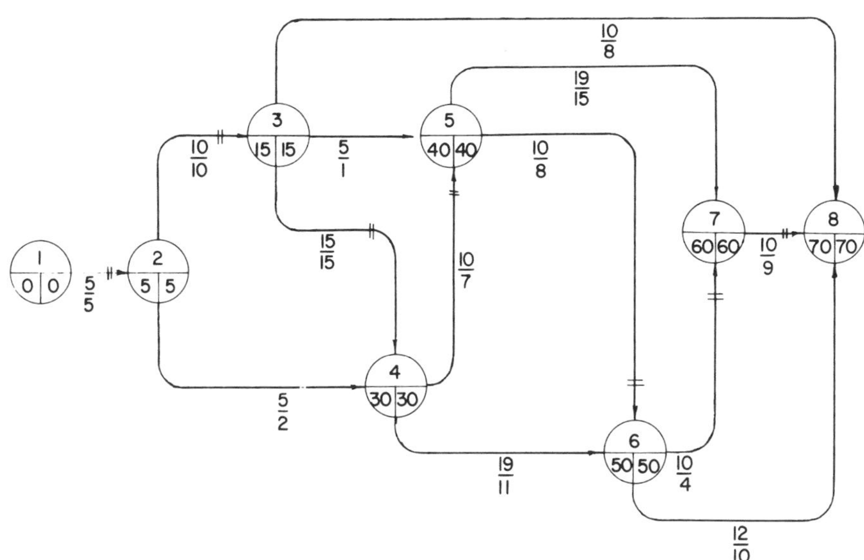

FIGURE 10.8 Network for Example 10.2

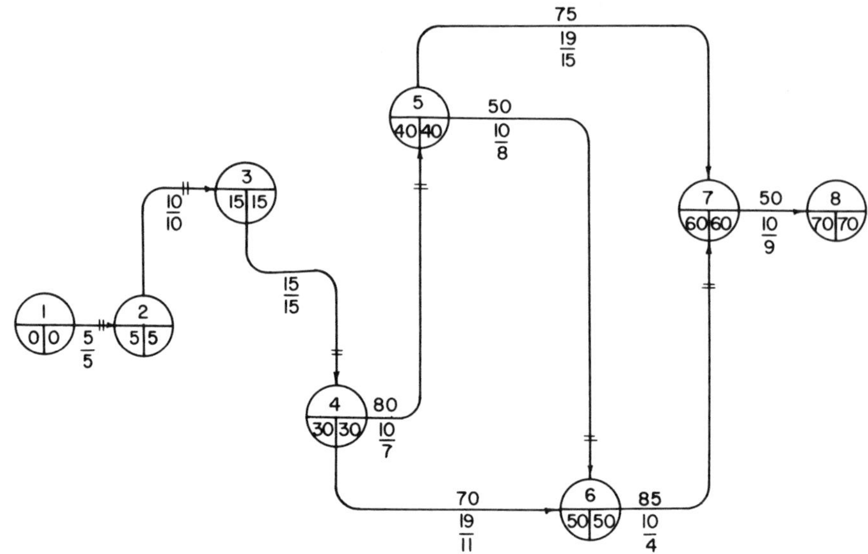

FIGURE 10.9 Skelton Network for Example 10.2

4–6 and 5–7 along with it. The early event time of event 6 is updated to 49, of event 7 to 59, and of event 8 to 68. Project cost amounts to \$6700. Crashing this activity by one day has generated three parallel critical paths between events 4 and 7. These are: (1) activities 4–5 and 5–7; (2) activities 4–5, 5–6, and 6–7; and (3) activities 4–6 and 6–7. It is now necessary to shorten the combination of activities having the lowest combined slope. The slopes of the parallel critical paths (1), (2), and (3) are respectively \$150, \$195, and \$160.

Activities 4–5 and 4–6 form the combination with the least slope. This combination costs \$150 (\$80 for activity 4–5, and \$70 for activity 4–6) per

TABLE 10.4 Example 10.2 Data

Project Duration	Activity	Normal Duration	Crash Duration	Number of Days Reduced	Slope	Project Cost Before Crashing	Project Cost After Crashing
70						6600	
69	7–8	10	9	1	50		6650
68	5–6	10	8	1	50		6700
65	4–5	10	7	3	80		7150
	4–6	19	11	3	70		
61	5–7	19	15	4	75		7790
	6–7	10	4	4	85		

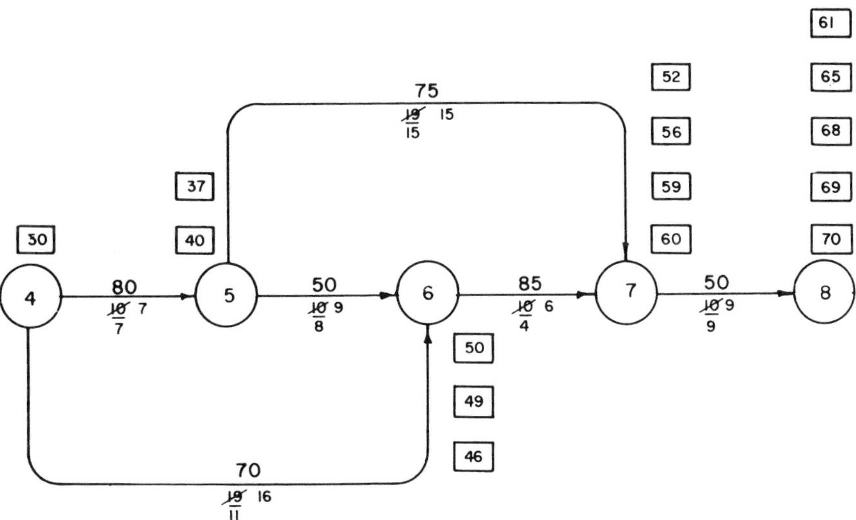

FIGURE 10.10 Example 10.2 Step by Step Crashing

day in comparison to activities 5–7 and 6–7, whose combined costs are $160 ($75 for activity 5–7 and $85 for activity 6–7). Thus activities 4–5 and 4–6 are crashed by three days, and the event times are updated. This combination cannot be crashed any more because activity 4–5 has been crashed to the limit. Activities 5–7 and 6–7 form the next combination to be crashed at a cost of $160 per day for four days. The project duration is now reduced to 61 days. The last combination, 4–6, 5–6, 5–7 cannot be crashed any further because activity 5–7 has reached its limit at 15.

None of the activities has generated float. Hence there is no opportunity to relax any activity to save cost. Because no further crashing is possible, the minimum project duration has been obtained.

The network in Figure 10.11 shows all the activities of the project along with the new critical paths that have been generated. The crash durations of the activities achieved through the foregoing computations are shown in the figure, giving a project duration of 61 days.

Assuming a linear relationship between time and cost, the slope, which is the cost of crashing an activity for any day between the limits of normal and crash durations, is found. In the other three cases of time and cost distribution—multilinear, discrete, and continuous—the slope cannot be computed for the entire period between the normal and crash cost thresholds. If there is a linear relationship associated with different time intervals, different slopes may be computed for each interval. For discrete functions, expediting to the crash duration can be done at the given cost, but there is no slope to determine costs at intermediate points. Thus, if the normal cost of an activity is $200 and its crash cost is $300, for a saving of four days the crash

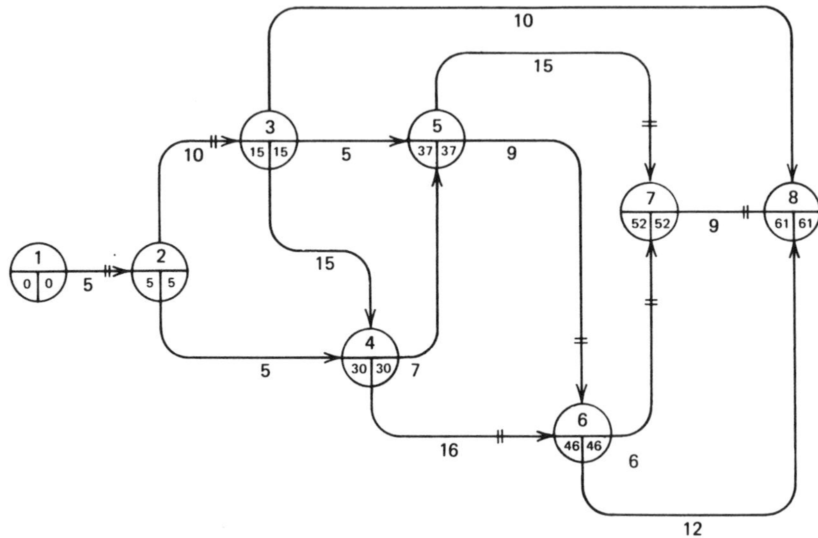

FIGURE 10.11 Example 10.2 Crashed Network

cost will be \$300 whether one, two, three, or all four days are saved in the performance of the operation. In a nonlinear continuous relationship the cost curve is divided into many small, straight lines which approximate it. The nonlinear relationship is thus reduced to a linear relationship associated with different time intervals.

The information listed in Table 10.4 can be plotted. (It should be remembered that the cost listed in the last column is the direct cost of the project;

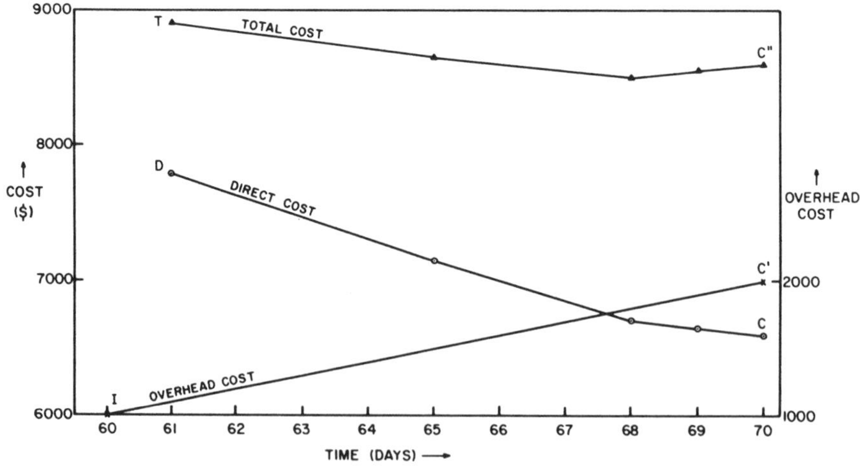

FIGURE 10.12 Time Cost Graph

the cost of labor, materials, and equipment.) This information is plotted on a time-cost graph, shown in Figure 10.12 and a curve DC is drawn through these plots. Every project has an additional cost for supervision, design, engineering, and administrative staff. This cost, which is usually called overhead or indirect cost, can be added to the direct cost of the project to obtain the total cost. Assuming that there is a fixed cost of $1000 for 60 days and a cost of $100 per day for overhead expenses for the duration of the project, cumulative overhead cost is plotted on the time–cost graph as curve IC'. It is added to the direct cost to obtain the total cost curve TC''.

From the total cost curve TC'' in Figure 10.12 the lowest total cost of the project is $8500 for 68 days. Now 68 days is the minimum cost project duration that was desired. The total costs for different project durations ranging from 61 days to 70 days are listed as follows:

Project Duration	Total Project Cost
61 days	$8890
65 days	8650
68 days	8500
69 days	8550
70 days	8600

Notice that there are many alternative ways of doing this project; however, the most economical alternative is 68 days and $8500.

The exercise of determining the most economical project duration proves to be most economical on repetitive type of projects such as construction of single-span bridges, multistory buildings, franchise business premises, prefabricated buildings, or installation of equipment by manufacturer's representatives in process plants. Here, once the most economical project duration is determined, it can result in accumulative savings from similar projects to be done in the future. Many computer packages are available to solve such time–cost trade-off problems.

10.6 LINEAR PROGRAMMING FOR DETERMINATION OF MINIMUM COST DURATION

Besides the method already discussed in this chapter of determining the most economical project duration there is also the more analytical technique of linear programming. Due to space limitations and the nature of this book it has not been feasible to examine linear programming in general. Instead the technique will only be discussed here as a tool for determining the most economical project duration.

Before the reader unfamiliar with linear programming techniques begins this section, the author recommends that he/she first consult a text on the

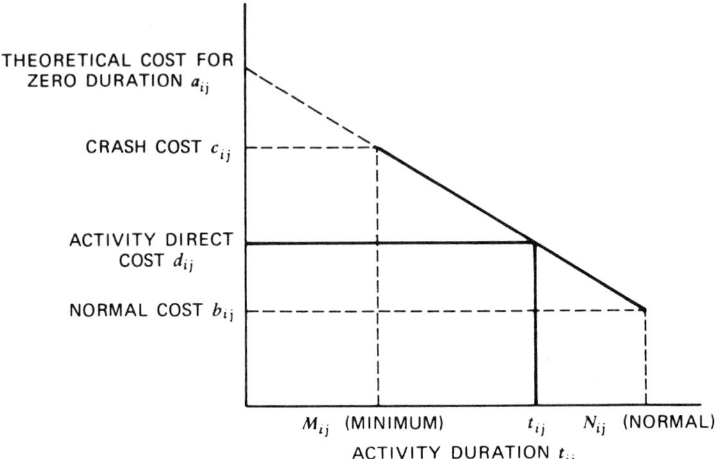

FIGURE 10.13 General Linear Time-Cost Model.

subject and become familiar with the basics. There are several authoritative sources listed in the bibliography. Those already acquainted with the technique should encounter no problems in this section, which shows how a network-based project plan is modeled.

10.6.1 Approach Used for the Model

To determine the minimum cost project duration, time, and cost information relating to a network has to be formulated into a linear programming model. Linear relationships between activity duration and estimated cost are assumed. Where nonlinear relations exist, the nonlinear curve is replaced by straight lines that represent this curve for all practical purposes. Figure 10.13 shows durations and costs. For an activity *ij*, normal duration is designated as N_{ij} (duration with zero crash cost) and crash duration as M_{ij}. Similarly, normal cost is designated as b_{ij} and crash cost as c_{ij}. The cost curve is denoted by a solid line that, when extended, intersects the vertical axis of cost at a_{ij}, which is the theoretical cost of activity if performed in zero duration.

Since the cost curve has negative slope, the equation for theoretical cost may be written as follows:

$$a_{ij} = b_{ij} + S_{ij}N_{ij}$$

The direct cost equation is

$$d_{ij} = a_{ij} - S_{ij}t_{ij}$$

where S_{ij} is $(a_{ij} - b_{ij})/N_{ij}$. The total direct cost for all activities with duration $t_{ij}(M_{ij} \leq t_{ij} \leq N_{ij})$ for appropriate i, j is

$$\sum_i \sum_j d_{ij} = \sum_i \sum_j a_{ij} - \sum_i \sum_j S_{ij} t_{ij}$$

The first part of the expression for total direct cost is a nonvariable cost, once the slope for each activity is decided. Total direct cost can therefore be minimized by maximizing the expression $\sum_i \sum_j S_{ij} t_{ij}$. Hence the objective function is

$$\text{Maximize} \quad 0 \sum_i \sum_j S_{ij} t_{ij}$$

subject to the constraints $M_{ij} \leq t_{ij} \leq N_{ij}$.

This ensures that the selected duration t of an activity lies between N and M and that the objective function is positive. Another constraint imposed indirectly on t_{ij} and related to event times is

$$E_i + t_{ij} - E_j \leq 0$$

The less than or equal to sign allows for float time.

The E_i variables serve the bookkeeping function and the E variable has an impact on the objective function. The time of the final event E_L constitutes the completion date. If project duration must be less than a specified length of time T, another constraint would be introduced:

$$E_L \leq T$$

If the objective in a project is to minimize total cost and O is the daily overhead cost and OE_L the total overhead cost, then the objective function is

$$\text{Maximize} \quad \sum_i \sum_j S_{ij} t_{ij} - OE_L$$

Since the objective function is used to minimize project cost, negative and not positive values of overhead cost are maximized.

Depending on the conditions of a particular project, certain other constraints may be considered. If a project is subejct to a daily penalty of p dollars for delay beyond the fixed project duration, and the project is delayed for q days, the management would desire to keep the cost to the minimum. Here the objective function can be restated as

$$\text{Maximize} \quad \sum_i \sum_j S_{ij} t_{ij} - OE_L - pq$$

The variable q is subject to the constraint: $E_L - q \leq t_p$ where t_p is the time when the daily penalty would begin to be charged.

The preceding objective function is correct if overhead is incurred at a uniform rate. If the rate is variable, a different rate may be chargeable to the project between the time of events E_x and E_z; then the objective function would become:

$$\text{Maximize} \quad \sum_i \sum_j S_{ij} t_{ij}$$

with the constraint $E_z - E_x - R \leq 0$, where O_A is the daily additional overhead and R is the number of days this additional overhead is charged.

When not one but n cost slopes per activity exist, as shown in Figure 10.14, the problem of minimizing project cost is treated by starting with all

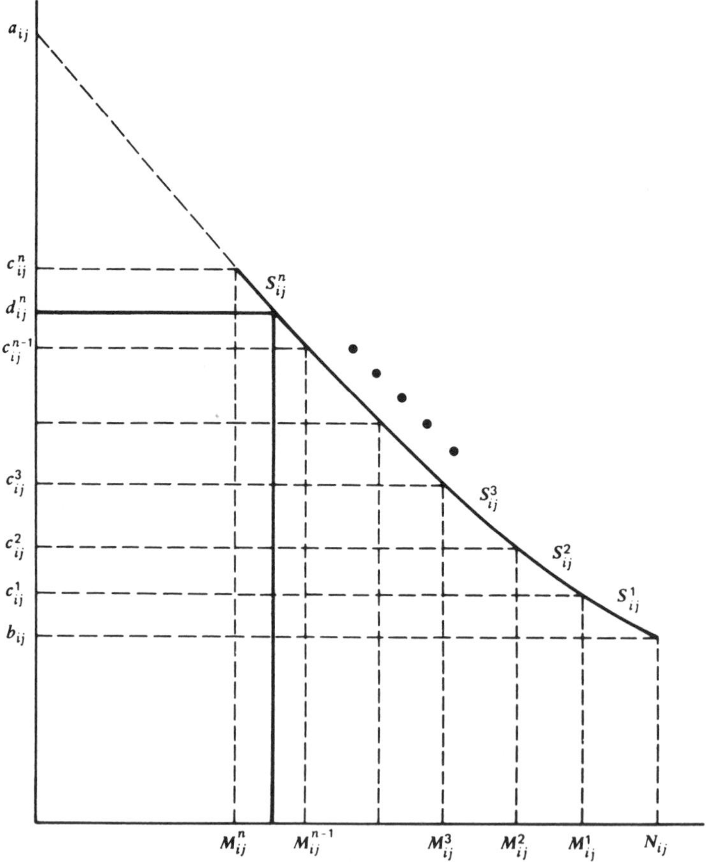

FIGURE 10.14 Activity with **n** Cost Slopes

activities at crash level and uncrashing them. The direct costs associated with the n slopes of activity ij are

$$S^{ij}(\text{slope}) = \frac{C^{n}_{ij} - C^{n-1}_{ij}}{M^{n}_{ij} - M^{n-1}_{ij}}$$

The direct costs associated with the n slopes of activity ij are

$$d^{n}_{ij} = a_{ij} - S^{n}_{ij}t^{n}_{ij} \qquad M^{n}_{ij} \leq t^{n}_{ij} \leq M^{n-1}_{ij}$$

$$d^{n-1}_{ij} = (c^{n-1}_{ij} - c^{n-2}_{ij}) - S^{n-1}_{ij}t^{n-1}_{ij} \qquad t^{n-1}_{ij} \leq M^{n-2}_{ij} - M^{n-1}_{ij}$$

$$\vdots$$

$$d^{2}_{ij} = (c^{2}_{ij} - c^{1}_{ij}) - S^{2}_{ij}t^{2}_{ij} \qquad t^{2}_{ij} \leq M^{1}_{ij} - M^{2}_{ij}$$

$$d^{1}_{ij} = (c^{1}_{ij} - b_{ij}) - S^{1}_{ij}t^{1}_{ij} \qquad t^{1}_{ij} \leq N_{ij} - M^{1}_{ij}$$

The objective function can be written as

$$\text{Maximize} \quad \sum_{i}\sum_{j} S^{1}_{ij}t^{1}_{ij} + \sum_{i}\sum_{j} S^{2}_{ij}t^{2}_{ij} + \cdots$$

$$\sum_{i}\sum_{j} S^{n}_{ij}t^{n}_{ij} - OE_{L}$$

where $E_i + t^{1}_{ij} + t^{2}_{ij} + \cdots t^{n}_{ij} - E_j \leq 0$.
Also

$$t^{1}_{ij} \leq N_{ij} - M^{1}_{ij}$$

and

$$t^{1}_{ij} \leq M^{1}_{ij} + M^{2}{}_{ij}$$

$$t^{3}_{ij} \geq M^{2}_{ij} - M^{3}_{ij}$$

$$t^{n}_{ij} \leq M^{n-1}{}_{ij}$$

$$t^{n}_{ij} \geq M^{n}_{ij}$$

where N_{ij}, M^{1}_{ij}, M^{2}_{ij} . . . , M^{n}_{ij} and t^{n}_{ij} are all distances from the origin on the x axis.

These constraints will hold only if the S^{n} is greater than S^{n-1} because it is necessary to uncrash the nth crash before uncrashing the $(n - 1)$th crash. In a maximization problem in linear programming, the variables with the largest coefficient in the objective is brought into the solution first. The nth slope, being the maximum, will be uncrashed first, followed by the $n - 1$, and so on.

10.6.2 Example 10.3

An example will explain the use of the equations formulated so far. Consider a small network with six activities, as shown in Figure 10.15. It is desired to determine the minimum cost project duration. The time–cost data for activities A–F are listed in Table 10.5. The absolute cost slopes for the activities are listed in Table 10.6. Assume an overhead expenditure of \$160 per day; that is, $O = 160$. The objective function can now be written as:

$$\begin{aligned}\text{Maximize} \quad & 100t_{12}^1 + 120t_{12}^2 + 130t_{13}^1 + 250t_{13}^2 + 80t_{24}^1 \\ & + 100t_{24}^2 + 100t_{35}^1 + 140t_{45}^1 - 160E_5\end{aligned}$$

subject to the following bounds and constraints:

$$t_{12}^1 \leq 5 - 4$$
$$t_{12}^2 \leq 4$$
$$t_{12}^2 \geq 3$$
$$t_{13}^1 \leq 6 - 5$$
$$t_{13}^2 \leq 5$$
$$t_{13}^2 \geq 4$$
$$t_{24}^1 \leq 4 - 3$$
$$t_{24}^2 \leq 3$$
$$t_{24}^2 \geq 2$$
$$t_{35}^1 \leq 3$$
$$t_{35}^1 \geq 2$$
$$t_{45}^1 \leq 2$$
$$t_{45}^1 \geq 1$$
$$E_1 + t_{12}^1 + t_{12}^2 - E_2 \leq 0$$
$$E_2 + t_{24}^1 + t_{24}^2 - E_4 \leq 0$$

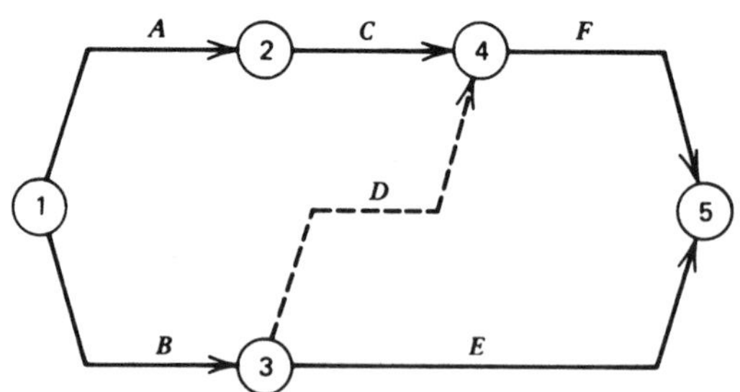

FIGURE 10.15 Example 10.3 Network

TABLE 10.5 Example 10.3 Time-Cost Data

Activity Identification	Description	Normal		First Crash		Second Crash	
		Time	Cost	Time	Cost	Time	Cost
1–2	*A*	5	4000	4	4100	3	4220
1–3	*B*	6	4200	5	4330	4	4580
2–4	*C*	4	3200	3	3280	2	3380
3–4	*D*	0	0	0	0	0	0
3–5	*E*	3	1000	2	1100	—	—
4–5	*F*	2	2600	1	2740	—	—

$$
\begin{aligned}
&E_3 - E_4 && \leq 0 \\
&E_4 + t_{45}^1 - E_5 && \leq 0 \\
&E_1 + t_{13}^1 + t_{13}^2 - E_3 && \leq 0 \\
&E_3 + t_{35}^1 - E_5 && \leq 0
\end{aligned}
$$

where E_i, E_j are the start and end events for an activity ij.

The constraints can be converted into equalities by the addition of slack variables. Then the objective function and the constraints can be arranged into a simplex table to determine the values of t_{12}^1, t_{12}^2, t_{13}^1, t_{13}^2, t_{24}^1, t_{24}^2, t_{35}^1, t_{45}^1, and E_5. To find the durations for the activities, it is necessary to combine the durations of different crash levels.

Total cost of the project can be determined either by computing the sums of the individual theoretical cost levels a_{ij} and subtracting the value of the objective function or by summing the direct costs of the activities for the durations determined from the linear programming solution and adding the total overhead cost.

TABLE 10.6 Example 10.3 Absolute Cost Slopes

Activity Identification	Description	First Crash Cost Slope	Second Crash Cost Slope
1–2	*A*	100	120
1–3	*B*	130	250
2–4	*C*	80	100
3–4	*D*	—	—
3–5	*E*	100	—
4–5	*F*	140	—

It is difficult to find a manual solution to this problem because of the large number of variables involved. A computer is almost essential. However, the important thing is to be able to formulate the model in the linear programming framework to be able to feed it into a computer.

10.6.3 Example 10.4—Computer Solution

The example given in the previous section is solved using QS (Chang and Sullivan, 1991). It was required to maximize $100t_{12}^1 + 120t_{12}^2 + 130t_{13}^1 + 250t_{13}^2 + 80t_{24}^1 + 100t_{24}^2 + 100t_{35} + 140t_{45} - 160E_5$. This is equivalent to minimizing: $-100t_{12}^1 - 120t_{12}^2 + 130t_{13}^1 - 250t_{13}^2 - 80t_{24}^1 - 100t_{24}^2 - 100t_{35} - 140t_{45} + 160E_5$. The objective function and constraints for the problem are shown in Figure 10.16. The list of variables and their column numbers for this example are shown in Table 10.7.

The output from this program is given in Fig. 10.17.

The values of E_5, E_4, E_3, E_2, and E_1 may be read in the solution column as:

$E_1 = 0.0$
$E_2 = 4.0$
$E_3 = 6.0$
$E_4 = 7.0$
$E_5 = 9.0$

```
Free Format Model for Example 10.4

>> Min -1001T12-1202T12-1301T13-2502T13-801T24-1002T24-100T35-140T45+160E5
>> Subject to
>> (1)    11T12 <= 1
>> (2)    12T12 <= 4
>> (3)    12T12 >= 3
>> (4)    11T13 <= 1
>> (5)    12T13 <= 5
>> (6)    12T13 >= 4
>> (7)    11T24 <= 1
>> (8)    12T24 <= 3
>> (9)    12T24 >= 2
>> (10)  1T35 <= 3
>> (11)  1T35 >= 2
>> (12)  1T45 <= 2
>> (13)  1T45 >= 1
>> (14)  11T12 + 12T12 + 1E1 - 1E2 <= 0
>> (15)  11T24 + 12T24 + 1E2 - 1E4 <= 0
>> (16)  1E3 - 1E4 <= 0
>> (17)  1T45 + 1E4 -1E5 <= 0
>> (18)  11T13 + 12T13 1E1 - 1E3 <= 0
>> (19)  1T35 + 1E3 - 1E5 <= 0
```

FIGURE 10.16 Free Format Model for Example 10.4

Final Solution for 10.4 Page : 1

Variable No.	Variable Names	Solution	Opportunity Cost	Variable No.	Variable Names	Solution	Opportunity Cost
1	1T12	0	0	16	S3	+1.0000000	0
2	2T12	+4.0000000	0	17	A3	0	0
3	1T13	+1.0000000	0	18	S4	0	+70.000000
4	2T13	+5.0000000	0	19	S5	0	+190.00000
5	1T24	0	+19.999998	20	S6	+1.0000000	0
6	2T24	+3.0000000	0	21	A6	0	0
7	T35	+3.0000000	0	22	S7	+1.0000000	0
8	T45	+2.0000000	0	23	S8	0	0
9	E1	0	+160.00000	24	S9	+1.0000000	0
10	E2	+4.0000000	0	25	A9	0	0
11	E3	+6.0000000	0	26	S10	0	+40.000000
12	E4	+7.0000000	0	27	S11	+1.0000000	0
13	E5	+9.0000000	0	28	A11	0	0
14	S1	+1.0000000	0	29	S12	0	+40.000000
15	S2	0	+20.000000	30	S13	+1.0000000	0
31	A13	0	0	35	S17	0	+100.00000
32	S14	0	+100.00000	36	S18	0	+60.000000
33	S15	0	+100.00000	37	S19	0	+60.000000
34	S16	+1.0000000	0				

Minimized OBJ. = -1300 Iteration = 20 Elapsed CPU second = 2.089844

FIGURE 10.17 Computer Solution for Example 10.4

Thus the project duration (E_5) is nine days.

The solution column in Figure 10.17 gives the following values:

$$
\begin{aligned}
t_{12}^1 &= 0.0 \qquad & t_{12}^2 &= 4.0 \\
t_{13}^1 &= 1.0 \qquad & t_{13}^2 &= 5.0 \\
t_{24}^1 &= 0.0 \qquad & t_{24}^2 &= 3.0 \\
t_{35} &= 3.0 \qquad & t_{45} &= 2.0
\end{aligned}
$$

This shows that activity 1–2 is compressed to four days and activity 1–3 has a duration of six (5 + 1 = 6) days. Activity 2–4 is reduced to three (2 + 1) days. The durations of activities 3–5 and 4–5 remain unchanged.

The optimum valve for the objective function is given in the bottom of the solution table of Fig. 10.17 as $1,300.

Project cost can be determined by adding the theoretical cost at zero duration for each activity and deducting from it the value of the objective function, as shown in Table 10.8. The maximum cost for the project to be completed in nine days is $17,920 − $1,300 = $16,620. This can be verified as follows:

Direct cost of the project	=	$15,000
Add crash cost	=	180
Total for nine days	=	$15,180
Add overhead for nine days	=	1,440
Total cost for nine days	=	$16,620

TABLE 10.7 Example 10.4 List of Variables and Their Column Numbers

Col. 1	Col. 2	Col. 3	Col. 4	Col. 5
t^1_{12}	t^2_{12}	t^1_{13}	t^2_{13}	t^1_{24}
Col. 6	Col. 7	Col. 8	Col. 9	Col. 10
t^2_{24}	t_{35}	t_{45}	E_5	E_4
Col. 11	Col. 12	Col. 13		
E_3	E_2	E_1		

PROBLEMS

10.1 A researcher in a manufacturing plant has prepared a network for a research proposal. The following activities can be crashed at the costs shown against each:

		Normal Duration	Crash Duration	Normal Cost	Crash Cost
2–5	Design equipment	3 wks	1 wk	$ 5000	$ 7000
5–7	Order equipment	6 wks	2 wks	25,000	30,000
5–9	Write computer programs	8 wks	4 wks	3000	5000

TABLE 10.8 Example 10.4 Theoretical Cost

Activity	Theoretical Cost ($)
1–2	4,580
1–3	5,580
2–4	3,580
3–5	1,300
4–5	2,880
Total for the project	17,920

		Normal Duration	Crash Duration	Normal Cost	Crash Cost
6–7	Build room	6 wks	3 wks	4000	5500
7–8	Install equipment	2 wks	1 wk	4000	6000
10–13	Analyze data	4 wks	2 wks	1600	2000

On completion of the experiment the company expects to save $500 per day. What is the most economical duration for this research project?

10.2 For the network discussed in Problem 6.5 the normal and crash costs of the activities are listed here. There is a penalty of $5000 for finishing the project later than the planned project duration and a bonus of $200 per day for time saved. Overhead cost to the contractor is $40 a day. Compute the project duration that will provide maximum profit to the contractor assuming that the only crashable activities are the ones listed below.

Activities	Normal Duration	Crash Duration	Normal Cost	Crash Cost
Excavate for sales island base	1 wk	1 wk	$ 150	$ 150
Construct sales island base	1 wk	0.8 wk	200	225
Construct cash office	2 wks	1.6 wk	1000	1200
Obtain pumps	16 wks	10 wks	1500	1500
Erect pumps	1 wk	1 wk	300	300
Connect pumps	2 wks	0.4 wk	100	120
Inspector approves pump installation	2 wks	2 wks	0	0
Obtain office furnishings	8 wks	1 wk	1000	1100
Paint and furnish office and toilets	2 wks	1 wk	200	250
Connect office and toilet lighting	1 wk	0.4 wk	0	0
Excavate for office island	1 wk	1 wk	300	300
Construct office island base	1 wk	1 wk	150	150

Activities	Normal Duration	Crash Duration	Normal Cost	Crash Cost
Build offices and toilets including all services	2 wks	1.4 wks	9000	10,000
Install burglar alarm	1 wk	0.6 wk	200	250
Connect up burglar alarm	2 wks	1 wk	100	120

CHAPTER 11

PROJECT ESTIMATES

Estimating is a fundamental part of the construction industry. It is a business skill of utmost importance. The success or failure of a project is dependent on the accuracy of several estimates throughout the course of the project, that is, from conceptual and feasibility estimates through to the detailed or bid estimates.

The estimate at best is an approximation of the expected cost of the project. At each stage estimates are required and financial commitments are made based on these estimates. Initially an owner commits to the project based on feasibility estimates. Later designers develop the scope of a project within target cost restraints, with estimates providing the necessary information for making decisions. Finally the contractors will commit their resources to the project, with profit being the motive. The financial profit or loss of each of the participants depends on the accuracy of anticipated cost targets established based upon the estimates. The other component for success is an execution plan which is compatible with the forecast of events that constitute the estimate.

Accurate estimates optimize good contracting. As a corollary, inaccurate estimates provide improper guidelines for the project management. Unrealistic targets produce unrealistic expectations.

Construction is a unique industry which is inherently risky because most projects must be priced before they are constructed, whereas in other industries the selling price is based on known manufacturing costs. A serious industry problem is inaccurate estimating. Numerous failures of construction companies can be attributed to faulty or inaccurate estimating. Inaccurate estimates contribute to the squandering of valuable resources.

Estimating is partly a science and partly an art and relies heavily on the experience of the estimator and his "feel" and perception of the project. Anyone can produce an estimate. However, the challenge is to produce a reliable estimate.

Estimates are an aid to decision making. Project management must know what cost information is needed and how to use this information in making decisions. Different types of estimates are made at various stages of a project and each type has a different basis and therefore an understanding of the estimating basis and process is necessary for good decision making. To properly use an estimate as a tool, the decision maker must also understand its limitations.

The overall project economics are determined by the project's ability to attract, service, and repay capital. These financial restraints limit the capital cost of most projects in the private sector of the economy. Capital cost is only one of the several variables that affect the economic feasibility of a project. These variables are capital, operating costs, and the combination of inventory and work-in-progress costs.

This chapter deals with the estimating of the capital costs of a project. Capital costs can be grouped into three categories—feasibility, implementation, and commissioning or start-up costs. To determine the feasibility, expenditure would be undertaken in the areas of administration, research, surveys, studies, and in some instances, pilot plants. Implementation costs would include expenditures to engineer, procure and construct the facilities, licenses, land, and equipment needed at start-up and to operate the facility, including the initial inventory of spare parts.

Commissioning costs can easily be overlooked and considered as operating costs, but are really part of the capital costs. Included in this category are items such as start-up, training, debugging, debottlenecking, and initial inventory costs. The scope of an estimate must be well defined to include all or some of the capital cost categories.

The foregoing provides a basis for understanding the estimating environment, but we must also understand the type of estimates that are required during the various stages of a project.

11.1 ESTIMATE CATEGORIES

There are numerous categories and names of estimates that originate in the various sectors of the engineering and construction industries. Table 11.1 shows the American Association of Cost Engineers (AACE) classifications and accuracy range. Because of the wide variety of estimate types and methods, the authors have added other nomenclature, approximate engineering progress, and estimating methods which apply to these AACE classifications.

Order of magnitude estimates are used for feasibility studies, choosing between alternatives and thus determining the early economics of a project. These estimates are based on very preliminary information and have a wide

TABLE 11.1 AACE Classification of Estimates and Methods

AACE Classification	AACE Accuracy Range (%)	Other Nomenclature	Approximate Engineering Progress	Estimating Method
Order of Magnitude	−30 to +50	Conceptual Screening	0–5%	Indices Capacity Curves Capacity Ratios
Budget	−15 to +30	Preliminary Appropriation	5–20%	Component Ratio, Equipment Factored
		Semidetailed	30–50%	Square Foot, Parameter or Elemental
Definitive	−5 to +15	Engineer's, Bid, Detailed	60%+	Detailed pricing and takeoff

range of accuracy. The point is that the accuracy of this category is not necessarily −30 to +50%, but that potentially it could be, and this potential spread should be considered in the early economic evaluation of projects. The accuracy of any estimate is directly related to the amount of information available. Budget estimates are used for budgeting and authorization of funds, and provide an early basis for cost control.

Definitive estimates are the most accurate category because the available information consists of working drawings, detailed specifications, and subcontractor and supplier price quotations. These estimates include direct and indirect cost estimates of materials, labor, equipment, engineering, support staff, insurance, bonds, taxes, allowances, contingencies, and profit.

11.2 ESTIMATE ACCURACY VERSUS TIME

There are numerous methods with accompanying levels of accuracy for preparing project cost estimates. The information available determines which type of estimate is appropriate as a project evolves.

The information available is time dependent. Table 11.1 shows the approximate engineering progress for each category. These are not absolute percentages but only serve as a rough guide. Estimates are updated continually as better information becomes available.

For example, a semidetailed estimate is produced when engineering is roughly 30–50% complete. Part of the engineering may be complete whereas other engineering may be very preliminary. Semidetailed estimates then cover a wide range of engineering progress, and these estimates bridge a wide range of accuracy, all of which depends on the amount and quality of information available.

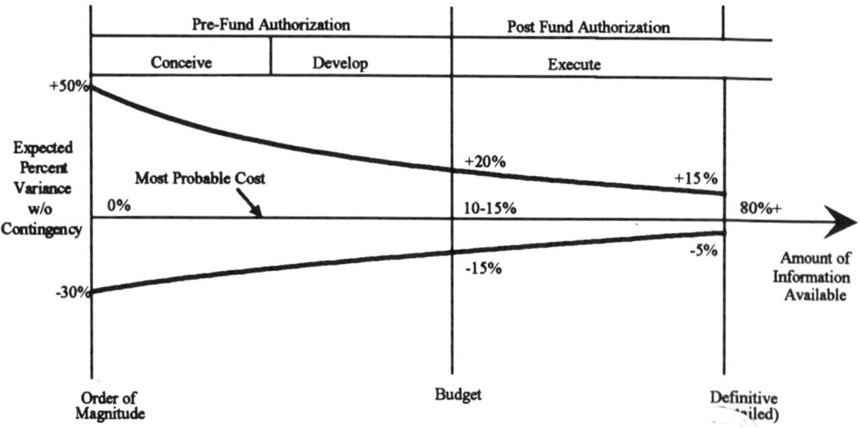

FIGURE 11.1 Estimate Accuracy Ranges (AACE Estimate Categories)

During the prefund appropriation phases (C and D) the amount of information grows to approximately 10–15% of the total project information. After funds have been authorized, engineering activity increases during the initial part of the implementation or execution (E) phase. Working drawings and specifications are the principle sources of information which define the project. This stage also produces most of the procurement documentation. Before construction begins, approximately 80–90% of the information has been generated. There are exceptions, such as in a fast-track project, but the point is that a considerable amount of information is generated during the execution phase.

The estimate accuracy ranges are shown in Figure 11.1. Note that the horizontal axis represents the amount of information available for estimating. The percentage figures are meant only to indicate very approximate percentages.

11.3 ESTIMATING METHODS

Estimates are required at all stages of a project with varying amounts of available information and, as a result, several methods have evolved. The following descriptions of these methods is intended to simply introduce this topic as an overview. However, the relationship between methods, information available, project stage, and estimate accuracy must be always kept in perspective by the user. An estimated cost is complete in meaning only with a stated range of accuracy.

Terminology used to label the many methods varies depending on the industries and authors. The following paragraphs elaborate on the previously presented estimate categories (i.e., order of magnitude, budget, and definitive), proceeding from the less detailed to the more accurate types.

11.3.1 Cost Indices

There are numerous cost indices available and most are intended to reflect the rate of inflation, that is,

$$C_2 = C_1 \cdot \frac{I_2}{I_1}$$

where

C_1 = known cost, time period 1
C_2 = estimated cost, time period 2
I_1 = index, time period 1
I_2 = index, time period 2

Numerous cost indices have been developed for specific niche market needs. Table 11.2 lists some indices with a brief statement of purpose.

Items 1–3 are output indices, that is, they are based on historical costs of actual construction. Items 4 and 5 are input indices which are compiled monthly by costing a constant package of resources that is intended to be representative of a building or heavy construction project. The ENR construction cost index (CCI) is the average cost for 20 U.S. (and 2 Canadian) cities for 7500 (1500) lb of structural steel, 1800 (1580 for Montreal and 1190 for Toronto) board feet of lumber, 6 (10) barrels of cement, and 200 hr of common labor.

The difference between the construction and building cost index (BCI) is the labor component. The CCI uses 200 hr; the BCI uses 68.38 hr. Since the indexes are computed using real prices, the proportion of a resource component in the index will therefore vary. In March 1993, the CCI index

TABLE 11.2 Some Inflation Indices

Item	Index	Purpose
1.	Chemical Engineering	Plant construction
2.	EPA–STP Treatment Plant	Complete primary and secondary waste treatment plants
3.	Nelson True-cost	Relative cost of constructing a barrel of refining or process capacity
4.	ENR Building Cost	Buildings
5.	ENR Construction Cost	Heavy construction

TABLE 11.3 ENR's Toronto Cost Index

Year	1Q	2Q	3Q	4Q
1985	4568.19	4810.03	5110.86	4769.78
1986	4807.65	4849.53	5101.61	4978.67
1987	5075.06	5256.94	5256.58	5253.34
1988	5279.99	5418.62	5584.94	5563.18
1989	5831.60	5710.50	6023.06	6023.06
1990	6018.10	6376.35	6401.54	6401.54
1991	6343.04	6554.50	6575.10	6537.05

consists of 77% common labor, 13% structural steel, 8% 2 × 4 lumber, and 2% Portland cement. Also, in March 1993 the BCI proportions are 61% skilled labor, 22% structural steel, 14% 2 × 4 lumber, and 3% Portland cement. Note that no adjustment is made for change in productivity, managerial efficiency, labor market conditions, contractor overhead and profit, and other less tangible cost factors. Both ENR indexes apply to general construction costs. The CCI should be applied when labor costs are a high proportion of total costs and labor is less skilled.

Of more value are the indexes produced for the various cities. Table 11.3 shows ENR's Toronto Construction Cost Index for each quarter. These indices are available for all major U.S. and Canadian cities, and many others around the world.

The effect of local conditions is reflected directly without the need to use location factors for adjustment from a base cost. The other indexes must be adjusted using location multipliers which account for cost differentials between geographic locations.

Example 11.1: Use of ENR Index. The cost of a warehouse completed in 1987 in Toronto was \$3,000,000. The estimated cost in 1991 is

$$\frac{6{,}537.05}{5{,}253.34} \times 3{,}000{,}000 = \$3{,}733{,}000$$

that is, an expected range between \$3,700,000 and \$3,800,000.

11.3.2 Limitation of Indices

An index must be relevant to the type of facility or construction, and this is the reason behind the many indexes available. Although the ENR indexes are relatively current, others will have some lag time. An index is a weighted average and is insensitive to short-term swings in the economy. Also, most will not reflect productivity changes, technological improvements or the competitiveness of the market.

In a screening estimate it is advisable to index only that part of the facility which could be considered constant for most sites. Special site conditions and foundations, environmental requirements, weather conditions, and site-specific special requirements such as long interconnecting pipeways must be considered as separate adjustments to the indexed cost of a facility.

11.3.3 Cost Capacity Factor Estimates (Proration)

Continuing with estimating for early project economics, cost capacity factors apply to changes in size of projects of similar types.

Costs for a proposed facility can be prorated using the following exponential expression

$$C_2 = C_1 \left(\frac{Q_2}{Q_1}\right)^m$$

when C_1 and C_2 are costs previously defined and

Q_2 = size of new facility
Q_1 = size of known facility
m = Cost capacity exponent which depends on type of industry. The values for m can be taken as 0.6 or selected from sources such as those shown in Figure 11.2

11.3.3 Component Proration for Equipment

Typically a new facility will have similar equipment components but of a different size than a previous installation. It is advisable to obtain recent

Process	Unit	Cost-Capacity Factor m	Capacity Range
Aluminum (from alumina)	Metric	0.76	20M-200M
Ammonia (by steam-methane reforming)	Tons/day	0.72	100-3M
Carbon Black	Tons/day	0.53	1-150
Ethylene	Tons/yr	0.72	20M-800M
Hydrogen (from refinery gases)	Cu ft/day	0.64	500M-10MM
Methanol	Gal/yr	0.83	5MM-100MM
Oxygen	Tons/day	0.72	1-1.5M
Power plants, coal, nuclear	Mw(elec)	0.88	100-1M
	Mw(elec)	0.68	100-4M
Styrene	Tons/yr	0.68	4M-200M
Sulfuric acid (100%)	Tons/day	0.67	100-1M
Water Treatment	Mgpd	0.67	1-100

M = million dollars

FIGURE 11.2 Example of Cost Capacity Factors (Adapted from F. C. Jelen, Cost & Optimization Engineering, McGraw-Hill Book Co., New York, 1970, p. 312

TABLE 11.4 Capacity Exponents

Equipment	Item	Capacity Exponent (m)
Horizontal	Pressure vessel	0.65
Centrifugal	Pumps and driver	0.52
Industrial	Boilers	0.5

price quotations for the required size. Otherwise an estimate of cost can be prorated using capacity exponents such as those shown in Table 11.4.

Example 11.2: Equipment Cost Proration. A field-erected 100,000 U.S. gal vertical API storage tank with a conical roof is required. What is the estimated price of this tank if a similar tank of 150,000 U.S. gal was purchased recently for \$2,000,000, foundations not included? Capacity exponent $m = 0.63$.

$$\text{Estimated cost } C_2 = C_1 \left(\frac{Q_2}{Q_1}\right)^m$$

$$C_2 = \$2{,}000 \times \left(\frac{100{,}000}{150{,}000}\right)^{0.63} = \$1{,}728{,}000$$

11.3.4 Equipment Factored Estimates

These estimates are categorized in the previously defined ''Budget'' category and, as the title suggests, are derived from the cost of the permanent equipment. The cost of each piece of equipment is determined and then multiplied by an appropriate module factor which has been developed from historical data. The total estimated cost is the summation of the factored cost of all pieces of equipment. Care must be exercised to include all items that are not equipment related but must be accounted for in addition to equipment-factored costs.

The factor is intended to include for the installation and supply of all components within an imaginary module.

The module contains items such as foundations, backfill, area grading, foundation and pipe support pads, steel platforms and ladders, increments of pipe rack and miscellaneous supports, fabricated vessel, carbon steel piping, underground sewer drain, electrical lighting and grounding, instrumentation, and paint and insulation.

Because of the lack of project definition at this stage, a high level of sophistication and background data is required to develop these factors, a sample of which is shown in Table 11.5.

In addition to the factored equipment costs, the complete estimate would include items that have not been considered in the equipment factors.

TABLE 11.5 Equipment Factors

Equipment	Factor
Electric motors	8.5
Heat exchangers	4.8
Centrifugal pump and motor	7.0
Tower columns	4.0
Furnaces (package units)	2.0
Centrifugal compressor and driver	2.0
Blowers and fans including motor	2.5

Example 11.3: Equipment Factored Estimate. The following is a factored estimate summary. Item 1 is the sum of all equipment factored costs (EFC):

$$\text{EFC} = \Sigma\, n_i\, C_i\, f_i$$

Where

i = a particular type of equipment
n = number of pieces
c = current cost of equipment type i
f = equipment factor (Table 11.4)

Other percentages and factors for this example are fictitious and could vary for different companies.

Example of Factored Estimate

Equipment	n_i	C_i	f_i	Cost	
1. Electric motors	20	2,000	8.5	$340,000	
Heat exchangers	10	20,000	4.8	960,000	
Tower columns	4	100,000	4.0	1,600,000	
Compressor	1	200,000	2.0	400,000	
Fans	4	5,000	2.5	50,000	
2. Major equipment cost					= $3,350,000
3. Miscellaneous equipment—10%					335,000
4. Equipment subtotal					$3,685,000
5. Total direct field cost (1.25 × 3,685,000)					= $4,606,250
6. Indirect field costs (0.26 × 3,685,000)					= 958,100
7. Total field cost					$5,564,350
8. Engineering and total office (0.20 × 3,685,000)					= 737,000
9. Total field + office cost					= $6,301,350
10. Escalation (3%)					189,050
11. Contingency (0.25 × 6,300,000)					= 1,575,000
12. Total plant cost					= 8,065,700
		Rounded off			= $8,100,000

11.3.5 Parameter (Elemental) Estimating

Several formats are popular for this type of estimate. Costs of a project are related to some parameters such as gross enclosed floor area, roof area, area of face brick, and others. Table 11.6 compares various formats.

Some sources for cost data are Mean's Square Foot Cost Data or Yardsticks for Costing by Hanscombe and Associates. Costs are based on square foot costs of a type of construction or on cost of assemblies (Means Assemblies Cost Data). An assembly for example would include all components of a masonry wall including brick insulation and interior finish. Parameter estimating requires schematic drawings in order to permit parameter quantity estimates. Occasionally ENR publishes examples of parameter costs for selected facilities and locations.

Figure 11.3 is an example of an elemental estimate using the format of the Canadian Institute of Quality Surveyors. Increasingly public authorities retain the services of professional cost consultants who are skilled at producing elemental estimates as a check on the A/E estimates prior to bid.

11.3.6 Definitive Estimates (Detailed)

Detailed estimates are made when the engineering is at least approximately 80% complete and the project is well defined. Lump sum bidding requires a

TABLE 11.6 Comparison of Elemental Estimate Formats

Means (Systems Estimate)	Engineering News Record (Parametric Costs)	Canadian Institute of Quantity Surveyors (CIQS)
1. Substructure	1. Site work	1. Substructure
2. Superstructure	2. Foundations	2. Structure
3. Exterior enclosure	3. Floor systems	3. Exterior cladding
4. Interior construction	4. Interior columns	4. Interior partitions and doors
5. Conveying systems	5. Roof systems	5. Vertical movement
6. Plumbing systems	6. Exterior wall	6. Interior finishes
7. HVAC systems	7. Exterior glazed openings	7. Fitting and equipment
8. Electrical systems	8. Interior wall systems	8. Services
9. Fixed equipment	9. Doors	9. Site development
10. Special foundations	10. Specialities	10. O/H and profit
11. Site construction	11. Equipment	11. Contingencies
12. General contingencies	12. Conveying systems	
13. Related costs	13. Plumbing	
	14. HVAC	
	15. Electrical systems	
	16. Special electrical	
	17. Markup	

Project: AL.C.B. – HERITAGE, EDMONTON Date: 24–Feb–91

Element	Elemental Cost		Elemental Amount		Rate per SF/M2		
	Quantity	Unit Rate	Sub–total	Total	Sub	Total	%
01. SUBSTRUCTURE				$46,800		31.33	8.3
(a) Normal Foundation	1,119 m2	31.33	35,058		31.33		
(b) Basement Excavation	m2		0		0.00		
(c) Special Conditions	31 piles	378.10	11,721		0.00		
02. STRUCTURE				$119,000		106.32	21.0
(a) Lowest Floor Constr.	1,119 m2	36.95	41,347		36.95		
(b) Upper Floor Constr.	m2		0		0.00		
(c) Roof Constr.	1,119 m2	69.37	77,625		69.37		
03. EXTERIOR CLADDING				$152,300		136.11	26.9
(a) Roof Finishes	1,119 m2	43.88	49,102		43.88		
(b) Walls Below Ground Fl.	m2		0		0.00		
(c) Walls Above Ground Fl.	683 m2	131.55	89,849		80.29		
(d) Windows	40 m2	220.00	8,800		7.86		
(e) Ext. Doors & Screens	5 No.	780.00	3,900		3.49		
(f) Balconies & Projections	33 m2	20.00	660		0.59		
(g) Skylite					0.00		
04. INTERIOR PARTITIONS				$22,500		20.14	4.0
(a) Permanent Partitions	346 m2	49.38	17,085		15.27		
(b) Movable Partitions	m2		0		0.00		
(c) Doors	9 No.	605.60	5,450		4.87		
05. VERTICAL MOVEMENT				$0		0.00	0.0
(a) Stairs	0 flgt	0.00	0		0.00		
(b) Elevators	0 No.	0.00	0		0.00		
(c) Escalators	0 set	0.00	0		0.00		
06. INTERIOR FINISHES				$54,100		48.36	9.5
(a) Floor Finishes	528 m2	81.67	43,122		38.54		
(b) Ceiling Finishes	1,119 m2	3.38	3,782		3.38		
(c) Wall Finishes	974 m2	7.40	7,208		6.44		
07. FITTINGS & EQUIPMENT				$1,200		1.07	0.2
(a) Fittings & Fixtures	1 sum	1,200.00	1,200		1.07		
(b) Equipment	1 sum		0		0.00		
08. (a) ELECTRICAL				$72,700		65.00	12.8
(i) Services & Distribution	1,119 m2	65.00	72,735		65.00		
(ii) Lighting & Power		incl.	0		0.00		
(iii) Systems		incl.	0		0.00		
08. (b) MECHANICAL				$56,000		50.00	9.9
(i) Plumbing & Drainage	1,119 m2	50.00	55,950		50.00		
(ii) Fire Protection		incl.	0		0.00		
(iii) HVAC		incl.	0		0.00		

FIGURE 11.3 Elemental Estimate—CIQS Method

09. OVERHEAD & PROFIT				$41,968	37.50	37.50	7.4
NET BUILDING COST				$566,600			100
10. SITE DEVELOPMENT				$68,200		60.97	
(a) General	1 sum	54,089	54,089		48.34		
(b) M & E site services	1 sum	14,135	14,135		12.63		
(c) Alterations							
(d) Demolition							
11. CONTINGENCIES				0			
(a) Design contingency			0				
(b) Escalation contingency			0				
(c) Construction contingency			0				
Total Building Cost				634,800			
Building Gross Floor Area (m2)				1,119			
Unit/ GFA				567.30			

FIGURE 11.3 (*Continued*)

complete scope definition, otherwise the propensity for claims for extras and disputes is high. Unit price bidding also requires detailed estimates although payment is based on measured quantities.

To prepare a definitive estimate, the usual steps are as follows:

1. Break the project into cost centers which can be physical elements such as direct costs for foundations, excavation and structure or non-physical elements such as indirect costs for insurance and bonding. The work breakdown structure and cost codes are used to organize the estimate into cost centers. Divisions 1-16 as shown in Figure 11.4 are a WBS based on the Masterformat index (see Appendix A also).
2. From the drawings, estimate the quantities of materials for each cost center. These are the direct materials such as concrete, lumber, structural steel, doors, windows and so on.
3. Price the quantities of materials using vendor quotations, suppliers catalogues or historical data. Include wastage.
4. The estimate "rolls off" the quantities, i.e. the labour cost is determined by multiplying the quantities by the appropriate productivity factor (production rate) for labour. For example, the labour for placing 100 cu. meters of concrete is multiplied by the productivity factor of 1.0 manhours per cubic meter. This would yield 100 manhours of direct labour for placing of concrete. All other material items are also multiplied by their applicable productivity. The result is the total manhours for the direct labour. Productivity data is obtained from the company's historical database if available, or from published data such as produced by the R. M. Means Co. or Hanscombe and Associates, Richardsons, and Gulf Publishing, to name a few.

CONDENSED
ESTIMATE SUMMARY

SHEET NO.

PROJECT | ESTIMATE NO.

LOCATION | TOTAL AREA/VOLUME | DATE

ARCHITECT | COST PER S.F./C.F. | NO. OF STORIES

PRICES BY: | EXTENSIONS BY: | CHECKED BY:

DIV.	DESCRIPTION		MATERIAL	LABOR	EQUIPMENT	SUBCONTRACT	TOTAL
1.0	**General Requirements**						
2.0	**Site Work**						
3.0	**Concrete**						
4.0	**Masonry**						
5.0	**Metals**						
6.0	**Carpentry**						
7.0	**Moisture & Thermal Protection**						
8.0	**Doors, Windows, Glass**						
9.0	**Finishes**						
10.0	**Specialties**						
11.0	**Equipment**						
12.0	**Furnishings**						
13.0	**Special Construction**						
14.0	**Conveying Systems**						
15.0	**Mechanical**						
16.0	**Electrical**						
	Subtotals						
	Sales Tax	%					
	Overhead	%					
	Subtotal						
	Profit	%					
	Contingency	%					
	Adjustments						
	TOTAL BID						

FIGURE 11.4 Estimate Summary Form

5. Price the labour by multiplying each class of labour by the prevailing average wage rate which includes a mix of craftsmen and apprentices. These calculations produce the estimated direct cost for labour.
6. Determine equipment requirements and price this account according to available rental rates or historical data.
7. Summarize the direct costs for material, labour and equipment.
8. Obtain specialty contractor's bids. A general contractor may perform only 20% of the work and subcontract the remainder which often include mechanical, plumbing, electrical, elevators, finishing trades, inspection services and any other specialities that the general contractor would not perform.
9. Calculate the indirect costs which would include overhead, payroll burdens, supervision, bonds, taxes, insurances, materials other than direct materials, small tools, rental of site facilities, site utility costs, and so on.
10. Add allowances that the owner wishes to include in order to include for those known items which are not adequately defined.
11. Add contingency, which is an unknown but expected cost.
12. Summarize the bid on a form such as shown in Figure 11.4.
13. Management will add profit to complete the bid.

Definitive estimates are produced for bidding purposes by a contractor or to produce an engineer's estimate for the owner. Both lump sum and unit price contracts require detailed estimates. In unit price contracts approximate quantities are calculated and payment is based on actual measured quantities. A unit price bid is essentially a lump sum bid and requires a detailed estimate in order to accurately calculate unit prices.

11.4 RANGE ESTIMATING

Range estimating is a probalistic method which can be applied to any category of estimates. An optimistic, pessimistic, and expected value for each item is assumed. Only those items with a high degree of uncertainty should receive a probablistic treatment. Those items that are estimated with a reasonable amount of certainty should be treated deterministically, as is done in the previously described methods.

A high, low, and expected value is established for each item. The formulation requires the generation of a random number and an assumed probability density distribution which for simplicity is often assumed to be skewed triangular. The estimated values generated by numerous computer runs are sorted into 100 equal increments in ascending order of values, that is, least value at the top of the list. The probability ($n\%$) that a cost will be equal to or less than a certain value is the nth value. For example 1000 runs are made and grouped into 100 equal increments. The sixtieth increment is a cost that has a 40% chance of being exceeded. Range estimating is useful in assessing risk and provides a method for evaluating the amount of contingency required. A mathematical treatment can be found in Chapter 16.

11.5 COMPUTERIZED ESTIMATING

Most construction estimating is still performed manually, augmented perhaps with spreadsheets and templates to facilitate calculation and organization. There is however a large array of estimating software packages such as Management Computer Controls ICE System, Timberline Software Corporation Medallion System or Estimating Lite, Bid Master, and many more. Most estimating software packages are PC based and have their own productivity and price databases. Others access the Means or Richardson Estimating Services Database which can be customized for specific company use.

The advantage of these programs is the consistent organization and format for estimating. The more sophisticated packages have the capability of estimating item by item or using assemblies. An assembly includes the cost and quantities for all the constituent components of a particular construction. For example, a particular wall assembly would include masonry, studs, drywall, insulation, and vapor barrier. A single cost per square meter of this wall would be used for the estimate, simplifying the estimating process.

The disadvantage of these computerized programs is that the format is rigid and very few shortcuts are permitted.

11.6 FACTORS THAT AFFECT ESTIMATE ACCURACY

Estimated costs are based on a combination of historical data and price quotations. Several influences contribute to the uncertainty, including the method of gathering the historical data. To some degree each project and the conditions under which the project will be executed are unique. The estimating process is in itself a forecast of the future, a process that is therefore risky.

Estimating errors are always a concern, but can be minimized with good procedures. However, there are other degrees of uncertainty present. The price of materials delivered to the site may be variable, but can be determined with a high degree of accuracy by soliciting quotations.

A construction estimate utilizes historical published data that reflect average or ideal conditions. The question always remains as to what factors will cause a variance. If one considers a totally new environment such as a new location, the factors that are most difficult to predict are those that affect labor productivity. Work study methods could be used to establish production rates, but these studies are seldom carried out during the estimating stage.

The factors that modify productivity are broadly grouped as those governed by the state of the economy and specific job circumstances.

General economic conditions include the local and regional business activity which determines the level of employment and available labor supply. A buoyant economy applies pressure to escalate prices and labor rates.

Labor skills, training, pay, quality of supervision, quality of communication, and management approach are a mixture of factors that affect productivity.

Job planning and schedule-related factors that affect productivity include work sequence, craft availability and mix, and overtime. A number of job characteristics that must be considered are job size and complexity, changes, rework, and site conditions.

Two other obvious factors to consider are climatic conditions and work methods.

As can be seen, numerous factors must be forecast in order to produce a reliable estimate. These same factors must later be considered for cost control.

11.7 CONTINGENCY

A specific provision must be included to account for unforeseen elements of cost. Contingency is therefore a legitimate and anticipated cost for unknowns. To put contingency in perspective one needs to decide how much of an overrun would be incurred if the job was bid with only what can be seen in the bidding documents. That amount is the contingency required, assuming the objective is not to lose money.

The most reliable assessment of contingency is based on the experience of the people involved. There are computerized methods to assist in evaluating the appropriate amount of contingency. A Monte Carlo simulation of risk can be performed.

Contingency is directly related to estimating accuracy, as was previously depicted by the "Accuracy Horn." This once again considers the type of estimate and the amount of information available.

National and local factors that must be considered include the degree of industrialization in that particular country, national political objectives, infrastructure, and the legal and regulatory systems. Local customs and work attitudes greatly influence the amount of contingency needed in the estimate.

A very important consideration is who assumes what portion of the risk. Whoever assumes the risk requires the contingency, whether owner or contractor. The type of contract used can transfer risk from one party to another.

11.8 ALLOWANCES

Occasionally the bidder is asked to include a specified sum of money for known items which have not been adequately defined so that an accurate estimate can be made. For example, the owner may require an allowance for rugs but the quality, which has a considerable bearing on cost, will be specified later. Cost adjustments are usually required to reflect actual costs or bid revisions. Allowances are included so that a total bid cost is available for financing, bonding, and insurance reasons.

Prime Cost Sums is another term which is equivalent to Allowances.

PROBLEMS

11.1 A 50-mgpd (U.S.) water treatment plant is proposed in 1993 for Ureka, Washington. A 60-mgpd plant using the same treatment process was built in 1986 in Vancouver, Washington. Make an order of magnitude estimate based on the following additional information and assumptions:

1. Poor site conditions in Vancouver required an additional $3,000,000 for foundations, which is not anticipated for Ureka.
2. Assume the size exponent $m = 0.67$.
3. Assume inflation will persist at an average of 4% per annum.
4. Assume the location index for Vancouver is 1.17 and 1.24 for Ureka.
5. A new high-pressure filter process will be added at Ureka at an additional cost of $4,000.000.
6. The construction period for the new plant will include winter construction. The Vancouver plant was not constructed through a winter season. Assume an increase in construction costs of 12%.

11.2 Make an equipment-factored estimate for the cost of a plant based on the following information:

Equipment	Cost	Cost Source	Equipment Factor[a]
E1	100,000	1986 in-house data	2.0
E2	200,000	Recent telephone quote	2.9
E3	300,000	1992 catalogue price	3.1
E4	400,000	1991 in-house data for 1/2 proposed size of E4	2.5

[a] Cost × factor = direct field cost (DFC).

Assume

1. Inflation = 8% per annum.
2. The original plant was completed in 1986. New plant completion is scheduled for 1993.
3. Construction support = 25% of DFC.
4. Engineering and office expense = 10% of direct field costs
5. Fee = 7%.

CHAPTER 12

COST BUDGETING

Project duration and cost are the two most important objectives of project management. It can be reasonably assumed that if these are met successfully by all participants, the quality objective will also be achieved. In the preceding chapters the discussion was focussed on planning methods and procedures followed for physical feasibility check and project scheduling. The following chapters will deal with schedule and cost plan implementation and control.

Unless the estimate is related to the schedule and converted into a project budget it is not very useful for project cost control. The basic question that arises is, how are the cost estimate and project schedule combined for use as a reference point, or baseline, for cost control?

Cost budgeting has the following objectives:

1. To provide analytical methods and procedures and to establish a reference point for monitoring and controlling project costs. The budget is the cost plan.
2. To serve as a basis for the owner for provision of funds as well as for disbursement of progress payments.
3. To provide a baseline from which forecasts and trends can be developed.

Cost budgeting tools that will be discussed in this chapter are the cost breakdown structure and cost coding. A study of these two topics will prepare the reader for implementation, monitoring, and control discussed in Chapter

14. The reader is referred to Chapter 2 for the interrelated topic of work breakdown structure, which is the basis for the project cost breakdown.

12.1 COST BREAKDOWN

On any project the contractor's organization and the owner's or project manager's organization have different reasons for breaking down costs.

The contractor views the cost of the project at the work item level. It is in the contractor's best interest to be involved from the very beginning with every aspect of the estimate in order to become intimate with all details of the job. This approach produces a more accurate bid and avoids "surprises" once the job is won and construction begins.

The contractor is interested in estimating the cost of such work items as formwork, reinforcing, and cast-in-place concrete. For the contractor's purposes, the cost accounting would be satisfied by lumping together all cast-in-place concrete on different parts of the project into one figure for the work item, although subdivisions of concrete items may be desirable in other instances.

There may be similar components which are not exactly alike in a functional element breakdown. These variations are not crucial because an owner is interested in the unit cost of a particular element without having to make a detailed item by item quantity take-off for each item of work. With reliable information of this type collected from several projects, the owners estimators or cost consultants will have good historical data for future estimates. This is the elemental estimating method discussed in Chapter 11. Figure 12.1 shows the contractor's viewpoint. For any given project the work items that comprise it are of most concern to the contractor. In arriving at the cost per unit quantity of work item, the contractor totals the contributing costs from all aspects of the job to come up with a single figure.

The owner, consultant, or project manager each view the project from quite different perspectives. To the owner the project consists of a series of functional elements. A functional element breakdown, as the name suggests, brings together components of the project that relate exclusively to a part of the project that serves a singular and unique function. On a building project, for example, some of the functional elements would be substructure,

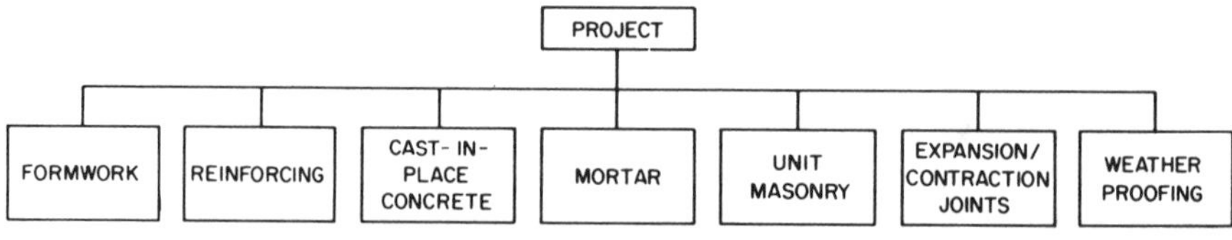

FIGURE 12.1 Work Item Levels—Contractor's Viewpoint

superstructure, exterior walls, interior finish, and so on. The components that contribute to these functional elements are not individually important cost-wise from the owner's viewpoint. Figure 12.2 illustrates the owner's view at the functional element level of a building project.

In order for any two differently oriented organizations to derive cost control information from a single database, there must be some degree of compatibility between their cost breakdowns. As already pointed out, the owner's and contractor's breakdowns approach a project from two very different standpoints which may, at first, appear irreconcilable. With a little cooperation and forethought, however, it is possible for the two independent systems to be meshed and for information to be obtained by both interests from the same basic data.

The technique for making the owner's and contractor's breakdowns agreeable is for the contractor to keep track of the components of the cost of work items, at the same time subdividing each work item so that the work done on each functional subelement can be tracked separately. By so doing, the contractor will be able to accommodate the owner, who wants to find the cost of any functional element, by summing up the dollars associated with the components of that element.

Another viewpoint that differs from either of these two is the perspective at the facility level, as shown in Figure 12.3. This view shows the project as seen by top management in a large construction company, where one project is only a fraction of their total concerns. A breakdown that satisfies both owner and contractor must, by its very nature, completely define the project with relation to the facility itself.

Figure 12.4 combines all of the viewpoints in a breakdown for one functional element, namely, substructure. Keep in mind that the partial list of functional elements shown belongs to only one of the various types of project breakdowns that are possible. In this case the functional breakdown is peculiar to building construction, largely a civil engineering undertaking. There

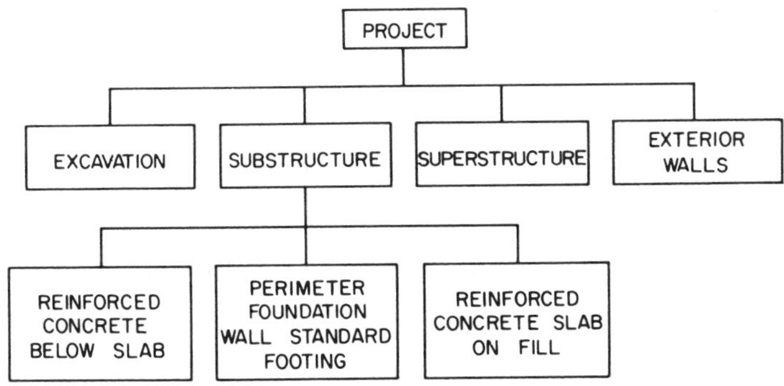

FIGURE 12.2 Functional Element Level of a Building Project

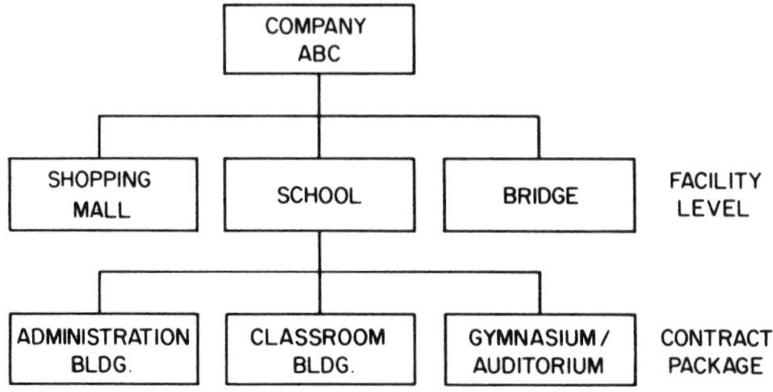

FIGURE 12.3 Facility Level Breakdown

are comparably detailed breakdowns for every kind of project—bridge construction, electrical substation construction, hydropower projects, nuclear power facilities, process and petrochemical plants, and so on. Each subelement of each project type has a breakdown of cost accounts which can be meaningful to both the owner and contractor and similar in detail to the one shown in Figure 12.3. In fact, every box of the cost breakdown structure represents a cost account.

A commonly used system for buildings is the Masterformat Index which was developed for specification categories and is shown in Appendix A. This system is useful for developing a work or cost breakdown structure and is widely used in building projects.

The reader is already familiar with the merging of the WBS and the ORC, the two different breakdowns from the project and organization viewpoints. Using the same concept, the cost account breakdown emerging from Figure 12.4 can be combined with the ORC so that a group of cost accounts can be monitored and controlled by an individual in the organization.

At this point the reader may well ask, why do we need a WBS and a cost breakdown structure (CBS)? In fact, for many companies, they are the same. However, it is best to think of the WBS as a planning tool, whereas the CBS is used for costing. Cost codes, which are discussed later, are the means by which WBS and CBS are brought into one costing system.

12.2 THE ENGINEERING PROJECT BUDGET

Determining a budget is the process of forecasting expected expenditures of money and materials during the course of a project's life. To arrive at a budget, it is necessary to know the predicted duration of the project and to account for the labor and materials to be directly consumed by the project. Since all organizations have different priorities and policies, the degree of detail required for budgets will differ from organization to organization. The

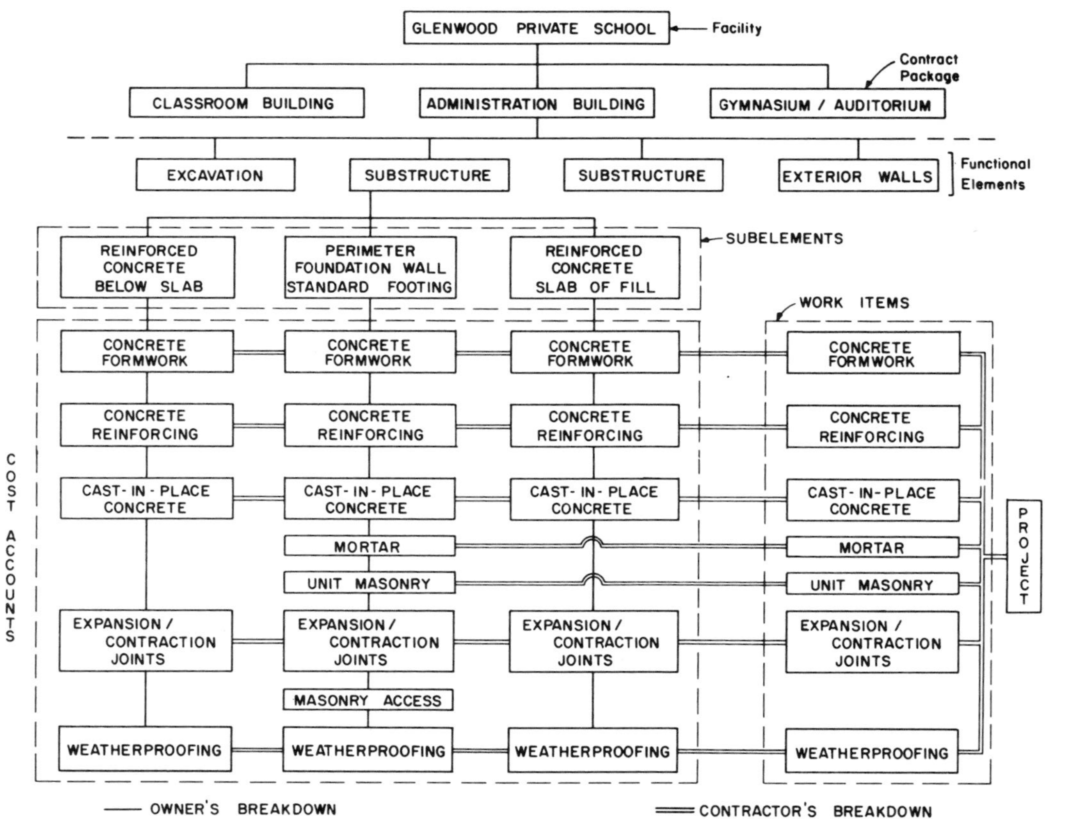

FIGURE 12.4 Partial Work/Cost Breakdown for a School Project

form of budget allocation is always in dollars but may also be in hours of labor, hours of equipment, dollars of supplies, and so on. Coinciding with these specific allocations, budgets are referred to by the facility to which they apply, such as power house, rockfill dam, and penstock for some of the facilities of a hydropower plant project.

Generally, each facility on a project will have its own budget, and project budgets are combined, if necessary (in a multiproject environment), to create department or summary project budgets. Each project or department-oriented division for which a budget is authorized and for which an individual has financial responsibility is a cost center, as shown for a multiproject environment in Figure 12.5. A detailed discussion on cost centers is presented in the following section. A cost account for each item of work is used for tracking of costs versus assigned budget, as illustrated in Figure 12.5.

At the lowest levels where budget figures originate, the estimator makes estimates of labor hours required by class (e.g., crane operator, carpenter, electrician) and converts their hours and rates into dollar figures. The estimator determines as well what materials are needed and what their costs amount to, and adds this sum to the figure for labor. This provides the basis for the budget.

The development and usage of budgets vary according to the requirements of the organization producing them. The owner needs a budget with at least two phases. Whether the owner manages the project himself or employs a project management firm, the first budget will be for the engineering man-hours spent on the design of the project. If the owner is controlling the entire project, once the detailed design, schedule, and estimate are completed, budgeting begins for the construction phase of the project. On an owner-managed project, the budget is usually broken down by monthly and by department to ensure complete control. If a project management firm over-

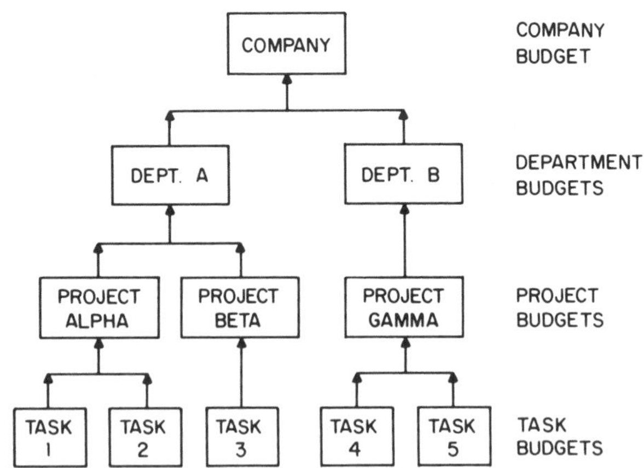

FIGURE 12.5 Budget Pyramid in Project Environment

sees the job, they will prepare a detailed design and engineering man-hour budget for the owner. Once bids are collected for the various component parts of the project and contractors are selected, the project manager is then able to firm up the budget for each contract which is then used for control purposes. These budgets must be passed along to the owner for approval and are used to forecast cash outflow.

The successful contractor must also establish a budget for its own work. Contractors usually use their bid schedule and estimate as a starting point for their project budget. A review of bid versus "issued for Construction" drawings must be made to capture the costs of revisions to the drawings that were made in the interim period. This provides an update which is needed to establish a final contract price prior to commencement of the work.

Whereas in the estimate the cost was arrived at for each work item, in the budget the contractor's interest in the cost will shift more toward a time base. That is, the contractor must determine what expenditures must be allowed for on a monthly basis and what portion of any and all activities occurs within each budget period, and then must total these figures to arrive at a monthly budget for each work item. Further breakdown of cost figures for manpower, materials, and equipment can be obtained as required for cost control and monitoring.

At this point the process by which budgets are obtained should be clear. Network plans permit scheduling; scheduling is followed by estimating; these cost estimates tied to a schedule are translated into budgets, or oftentimes the reverse; the budgets provide a basis of comparison for actual expenditures; and any discrepancy between budgets and expenditures provides input for the control process.

12.3 THE COST CENTER

Budgets are used for cost control and the smallest unit for which a budget is established and personal responsibility assigned is called a cost center. Expressed differently, a cost center is synonymous with the part of the company or project to which costs are to be assigned for budgeting and accounting purposes. Hence each engineering supervisor and manager in an engineering organization is responsible for a cost center. All are responsible for their expenses and attempt to perform their assigned tasks within the limit of their budgets. This helps motivate people by committing them to project objectives. The keys to commitment are the various WBSs that divide the project objectives into component elements of work, each having an associated budget and schedule. Commitment is made by specifically assigning these cost centers to engineering managers of small working groups and giving them an opportunity for input into the schedule and budget. The difference in the jobs of these engineering managers is reflected in the nature of the resources and costs of which they have control and the relative distribution of their effort between both areas of expenditure.

Besides the need to assign accountability for cost control, cost centers are created for a number of other special situations, such as large expenditure items and large outside services (i.e., subcontracts). Other reasons could be high uncertainty areas, which should be segregated for easier scrutiny, and areas that are potentially subject to many changes.

Cost centers are accumulated into summary level cost centers and this is readily achieved through the cost coding system. Successive summarizations result in the cost center for the entire company which monitors its performance by means of income statements that summarize revenues and expenditures.

12.4 COST CODING

Cost coding is the basic framework upon which the cost budgeting system is built. It provides a common language of identification and a means of communication to be used by all those concerned with project cost control.

The design and construction of a project is a multidisciplined, multiphased, complex activity requiring a high degree of interdependency among its various participants. Each organization generates data, most of which is required information for the other participating organizations. To facilitate this information exchange, it becomes necessary to develop a code that can be used by all project participants. Coding not only reduces confusion but also encourages all team members to speak the same language. It is a key to computer processing. If it is not simple, it will be subject to abuse. Data may be assigned to the wrong cost accounts, resulting in adulterated information.

This section discusses codes for subdivisions of the WBS (charge numbers) and work items, and for the various types of crafts, materials, and equipment.

A cost breakdown structure (CBS) is developed prior to estimating in a manner similar to the development of a WBS before preparing a schedule. It is necessary to ensure compatibility between the CBS and WBS.

The identification of each cost account in the CBS and the phase to which it belongs necessitates the use of numeric and/or alphameric coding to facilitate the use of computers for estimating and cost control. The codes assigned to the cost accounts, the CBS subdivisions, are called charge numbers. The length of a charge number depends on the number of levels in the CBS and the number of subdivisions at each level.

Any number of digits can be used for each level, but it is advisable to limit this to two digits if possible.

Figure 12.6 illustrates one possible charge number coding system. It can be used in a multiproject environment where the projects are of different natures and possibly located in different regions. Contract number, project phase, and extra or credit change orders are parts of the illustrated coding.

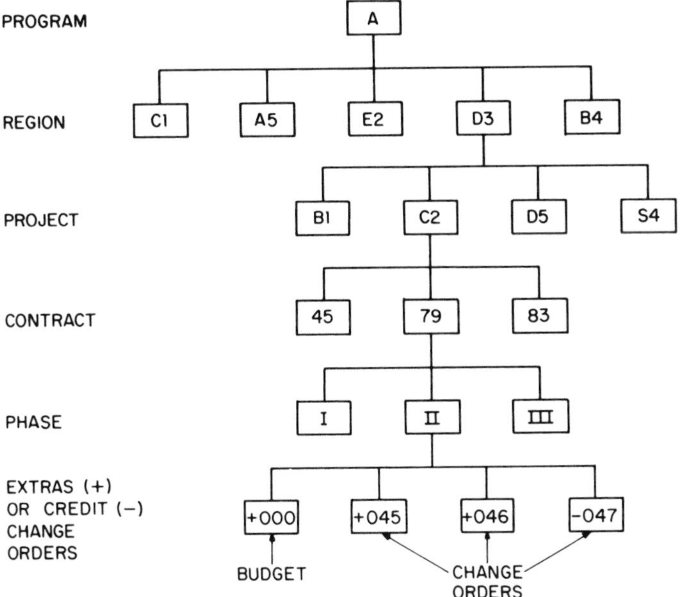

FIGURE 12.6 Example of Cost Coding

It should be clearly understood that the code conforms to the needs of the particular organization.

The lowest level element of the CBS is the cost account, which is comprised of the functional elements, subelements, and work items. Each cost account is assigned a costing code.

Also, purchase order and invoice records can be associated with the cost account in the manner shown in Figure 12.7.

The three subdivisions of the CBS, functional elements, subelements, and work items, are referred to by a costing code, as shown in Figure 12.8. Functional elements are assigned a single alphameric character followed by one alphameric character for the subelement and a five-digit numeric code for the work item. A single digit can differentiate between a labor, material,

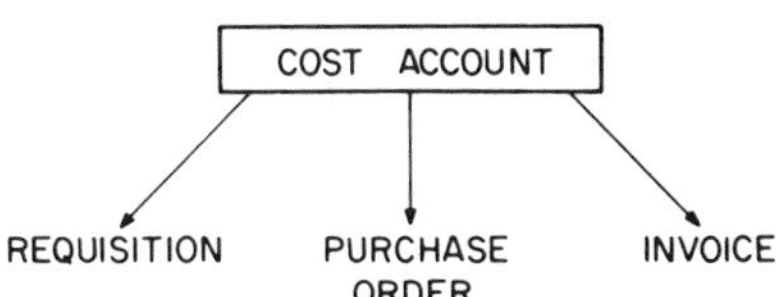

FIGURE 12.7 Procurement Cost Coding (The cost account suffix could include 1 for requisition, 2 for purchase orders, and 3 for invoice.)

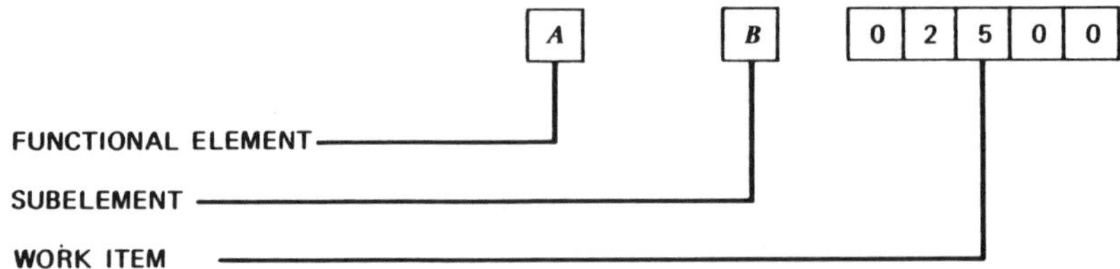

FIGURE 12.8 Example of Cost Coding Cost Code

expense, or rental item. The accounting period code can be simple. It consists of the accounting period's number and the year in which it falls, for example, 293 = second period of 1993.

The performing agencies, which are identified in the ORC, must also be assigned suitable codes. In the example shown in Figure 12.9 the company division for the project has a single alphameric character, and the performing agency has three characters in its field.

Coding is also necessary for different types of crafts, materials, and equipment. This is called the resource code and may be used on purchase orders and in assigning materials to cost accounts. For example, there may be up to 10,000 items in inventory for the construction of a nuclear power plant. Four numerals provide for almost 10,000 items (0001–9999), which theoretically is enough, but the divisions and sections of work and the need for flexibility may preclude the use of a large portion of the consecutive numbers, say from 1 to 9999. The number of items can be increased by using an alphameric code. It will be noted that the value could be in man-hours, cubic meters, machine-hours, square meters, tonnes, or dollars. This can be described by a single-character alphameric code. The type of cost, whether it is direct cost, indirect cost, budget, or commitment, can be described through another single-character alphameric code.

A coding system allows data to be collected and summarized according to a pyramidal system. Also, the different pyramids representing different coding systems can be merged in a manner similar to the merging of the WBS and the ORC, as discussed in Chapter 2.

At this stage it may appear that the coding system will generate enormous complexity. However, no cost code can be completely comprehensive, and

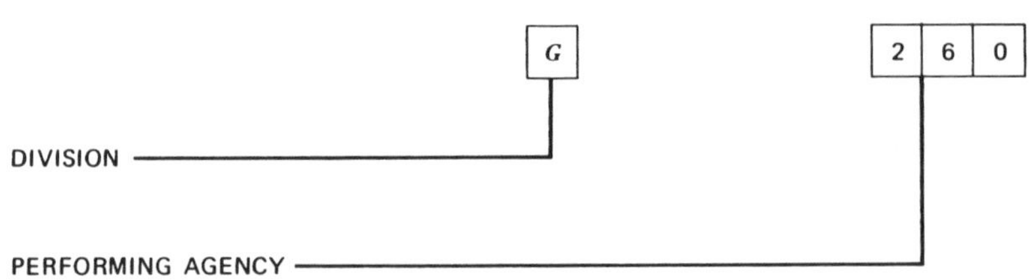

FIGURE 12.9 Coding Details

attempts to produce and use a standard, rigid, and very detailed code are usually doomed. Such a code will be too long and too inflexible, and it will eventually prove to be incomplete. On a daily basis only a few of the total number of digits are used. In fact, some labor will never use more than a very few digits. It is better to create a generic code with small beginnings and with a framework that has the potential for growth, so that with each job the code can grow to suit the work for which it is intended. This means that the estimator must not only use the cost code when estimating, but also work with it in creating code numbers for particular items. Computerized estimating software have a built-in generic system of cost codes such as the Masterformat Index System, which allows considerable flexibility and tailoring to suit the company's needs. Thus a code cannot be absolutely rigid; it must permit flexibility. Each cost code should be specially formulated for a construction company to reflect its work activities, none of which will be redundant. Appendix A shows the cost code for the Masterformat system popularly used for building projects.

Figure 12.10 shows the cost code used by a large EPC contractor. The facility consists of location, area, and facility identification. The Code of Accounts consists of the elements of work and cost. The elements of work can be the cost code numbers shown in the master format system, or these, which are specialty numbers for the process plant industry. Standard element work numbers are:

1000	Earthwork	5000	Machinery and Equipment
2000	Concrete	6000	Piping
3000	Metalwork	7000	Electrical
4000	Architectural	8000	Miscellaneous

Some examples of elements of cost are:

0	Labor	3	Subcontract
1	Burden	4	Expenses
2	Material		

We have discussed the main tools of cost budgeting, CBS, and coding. Bear in mind that cost budgeting is not simply taking the estimate figures

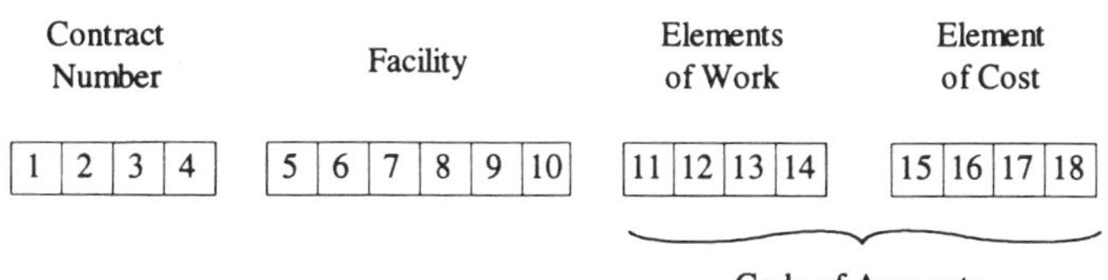

FIGURE 12.10 Cost Code for EPC Contractor

and plugging them into cost codes for the CBS, but instead relating these figures to the project schedule to meet the requirements of cost control during design as well as construction and commissioning.

PROBLEMS

12.1 A hydroelectric project has been planned for construction. Develop a breakdown that satisfies both the owner and contractor.

12.2 Using the Masterformat Index, create a standard cost code system for a midsize contractor in the general contracting business dealing with buildings and commercial developments.

CHAPTER 13

CASH FLOW FORECASTING

A project plan must be physically feasible; that is, it must be workable. In order to determine physical feasibility, the resources needed for the project and their availability must be checked. Procedures of resource allocation are therefore discussed in Chapter 9. Economic feasibility checks, as described in Chapter 10, determine if the solution obtained through resource allocation gives a project duration commensurate with the minimum total cost for the project. Cash flow forecasting is required to determine whether or not the funds to execute the plan are available, or in other words a financial feasibility analysis must be undertaken. This is the subject matter of this chapter.

13.1 OBJECTIVES OF FINANCIAL FEASIBILITY ANALYSIS

There are two objectives in financial feasibility analysis. First, contractors wish to find out if they will have the necessary funds required to carry out a project. They know the extent of their own investment and the progress payment expected at different stages in the life of the project. Their objective is to balance the expenditure on the project with the amounts available to them. Owners face a similar problem in determining their own ability to fulfill financial commitments to the contractors. Owners prepare their first cash flow forecast during the initial planning phase, followed by more sophisticated ones during investigative and control phases. When bids are invited, each contractor may be required to submit an estimate of progress payment requirements. This enables owners to update their cash flow forecast at the time of letting the contract.

Second is the case of financial feasibility. Contractors have other funds available to them besides their own investment and progress payments, which they expect to receive from an owner. To raise these funds, it is necessary for them to estimate their borrowings and produce a schedule for their bank or finance company. The advantage results from the accuracy attained in cash flow forecasting calculations so that maximum and average overdrafts, and the period for which they are required, are accepted with confidence. Of course, the user has to exercise judgment in using the procedure so that cash flow forecasts of the required accuracy are obtained.

A contractor's objective is to keep to a minimum the interest paid to financiers and to establish the financial feasibility of the project. In view of the high interest rates, interest during construction (IDC) can make the difference between a financially feasible and unfeasible project.

The two situations can be summarized as follows:

1. Balancing inflow against outflow.
2. Minimizing interest on borrowings.

13.2 CASH INFLOW AND OUTFLOW

If a contractor is to determine whether or not a project is financially feasible, he or she must study closely the estimated cash flow for the project. On any project there is an inflow and outflow of cash. Progress payments received by the contractor represent the inflow; payments made to subcontractors, suppliers, and others constitute the outflow.

An extimate of all income and expenditures and their anticipated dates based on the expected actual transfer of funds and billings is used to forecast cash flow. A positive cash flow indicates that the contractor has received more money to date than he or she has paid out; a negative cash flow indicates the opposite situation.

Many projects have a negative cash flow until the very end, when the final payment is received. This is a typical situation where the final payment consists of retention funds and where the retention percentage is greater than the profit percentage. However, there can be great variations in cash flow patterns. A contractor may achieve a positive net cash flow early in the project period. This is an attractive situation from the contractor's standpoint, since it not only eliminates borrowing or tying up organizational funds but makes new funds available for investment. Negative cash flow indicates the need to draw on the organization's working capital.

Outflow of funds is depicted in Figure 13.1 by an S curve. The top curve shows the outflow if the activities are started as soon as possible; the bottom curve represents the outflow associated with the late start of activities. In Figure 13.2 another, closer set of curves is drawn. The distance between the curves is indicative of flexibility in funding characteristic of a project.

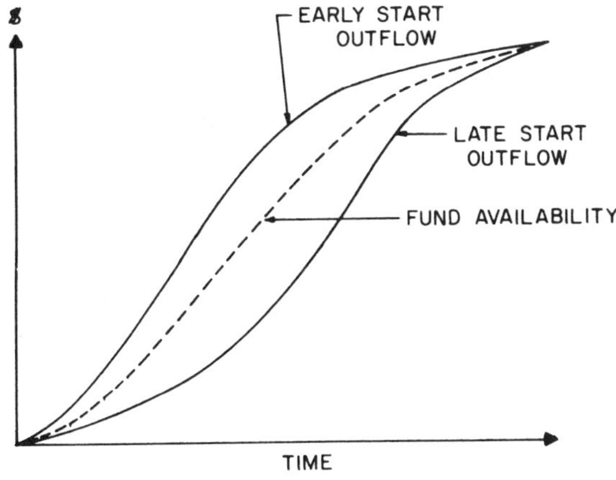

FIGURE 13.1 S Curves

The dotted curve shows the inflow of funds, including investment, loans, and credits. So long as the inflow curve falls within the two outflow curves, the project is financially feasible. The closer the inflow curve is to the late start curve, the higher the risk associated with the scheme of funding the project. This is reduced as the inflow curve moves closer to the early start outflow curve.

The inflow curve can be brought closer to the early start outflow by short-term loans available from banks and finance institutions. Suppliers and subcontractors also provide another source of financing, for there is a lag between the time when their invoices are received and the time when payment is made. This is called creditor financing. The inflow of funds is also affected by the profit earned on a project that is reworked into it.

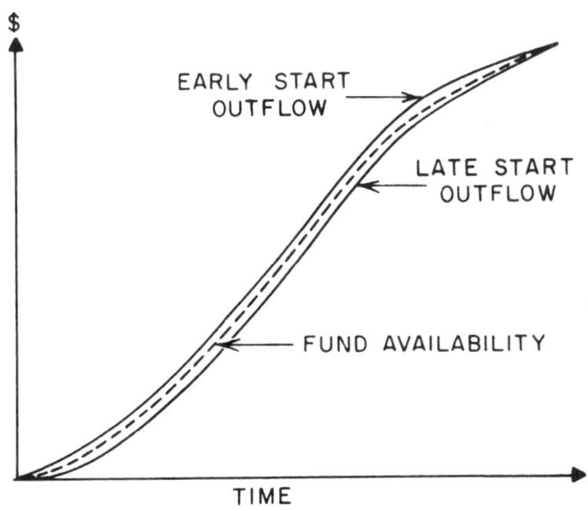

FIGURE 13.2 S Curves—Reduced Flexibility

13.3 EARNINGS, RECEIPTS, EXPENSES, AND DISBURSEMENTS

How cash actually flows on a project is illustrated on graphs for earnings, receipts, expenses, and disbursements in Figures 13.3 to 13.6. All graphs contain a plot of cumulative cost against project time in weeks. Figure 13.3 shows that progress payments lag behind earnings. Earnings are defined as the total value of work completed by a contractor. This value includes the profit, if any, earned by the contractor that he or she may not be allowed to collect at that time. The owner may like to hold back part of the value of the completed work as security so that the contractor will finish the project to his satisfaction. Thus the contractor may have earned more than the amount actually paid to him or her. If a contractor's progress payments are not equal to these earnings, the difference between earnings and receipts generally, but not essentially, indicates profit. Progress payments to the contractor represent cash outflow of the owner unless the contractor has other expenses chargeable to the project. The earnings of the contractor are the owner's commitments.

Figure 13.4 contains a plot of expenses versus time throughout the project. Expenses are the cost or cash outflow of the contractor. The contractor may have to make some disbursements. These may be for special services, certain materials, wages, loans, and so on; they are represented by amounts payable (solid line) and commitments (dotted line). Although the commitments made may be treated as expenses, the transaction takes place in the future. Thus the amount payable may lag behind the expenses incurred up to a certain time on a project. The commitment of today becomes the amount payable tomorrow and becomes a disbursement thereafter.

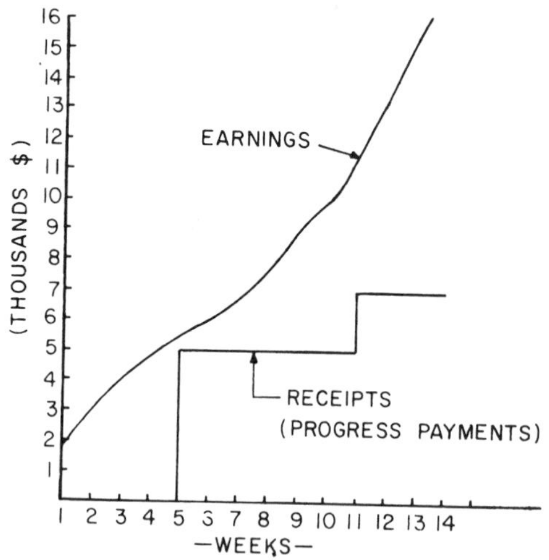

FIGURE 13.3 Earnings and Payments

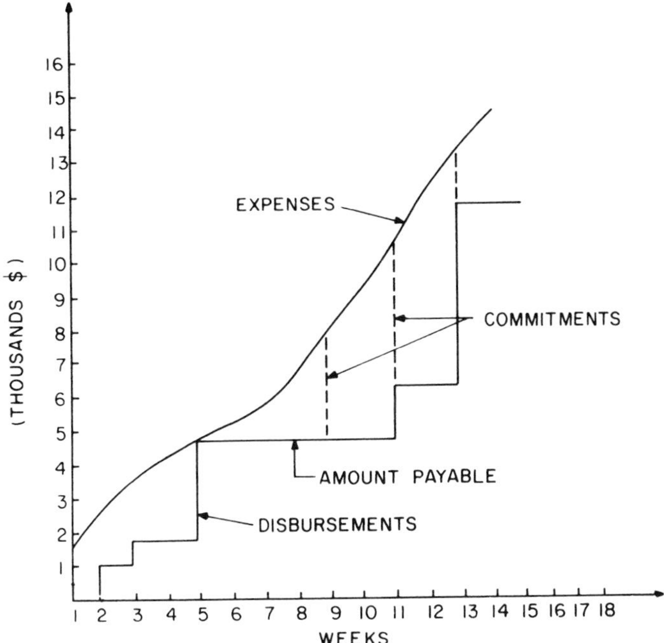

FIGURE 13.4 Expenses and Disbursements

Figure 13.5 is a plot of earnings (solid line) and expenditures (dotted line). The difference between the two normally indicates contractor's profit. The graph shows that the earnings of the contractor which form the value of work to the owner is in excess of the expenditure incurred by the contractor.

The shaded areas in Figure 13.6 show the difference between disbursements and receipts. These are the trouble spots where the contractor's receipts fall short of the required disbursements. Additional funds must be arranged to meet the deficit depicted by the shaded areas. The difference between disbursements by the contractor and progress payments received by him is covered by investments and loans. Financial feasibility analysis can show whether or not the investment is adequate; if loans are needed it can show how much and when.

Example 13.1. A sample project is shown in Figure 13.7. The number appearing above each activity arrow is the activity duration in weeks; the boxed number below the arrow is the estimated cost per week.

The contractor initially has an amount of $4500 to invest in a project undertaken by his forces. In addition, as work proceeds the contractor expects to receive the following progress payments: at event 5, $1000; at week

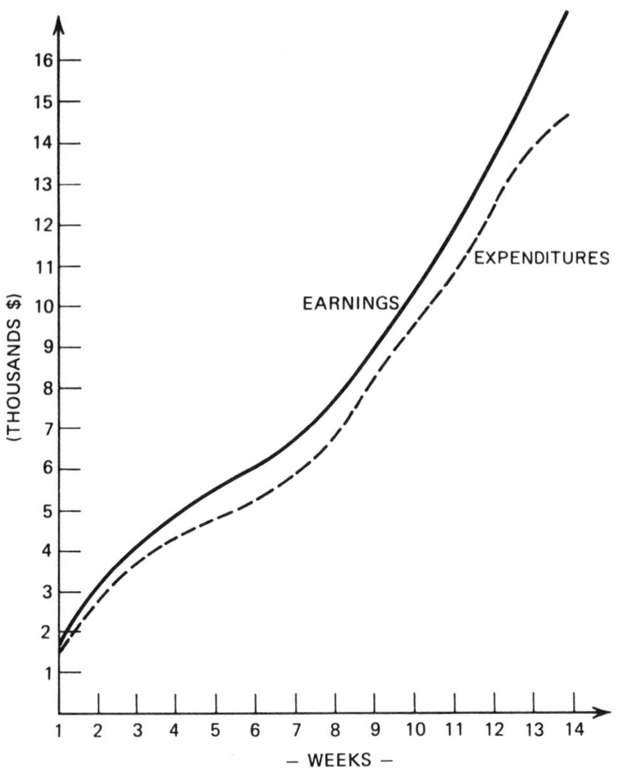

FIGURE 13.5 Disbursements and Receipts

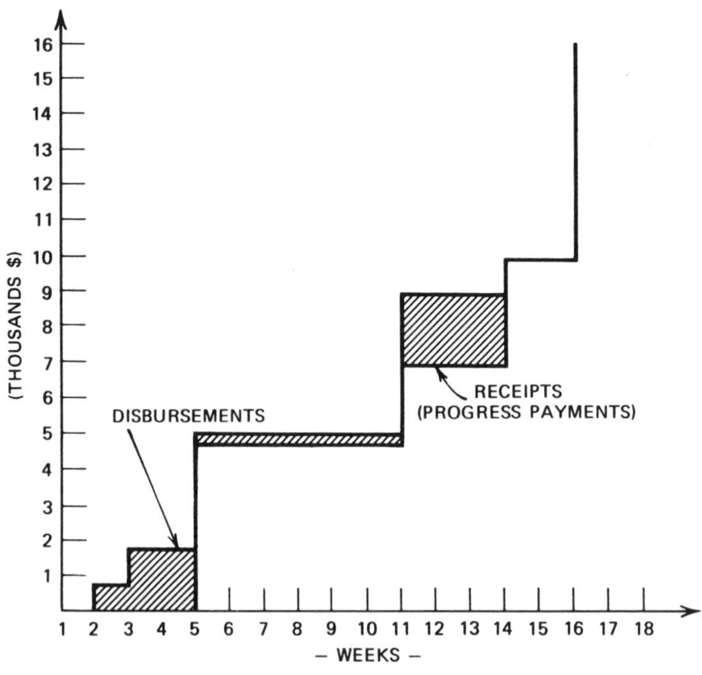

FIGURE 13.6 Earnings and Expenditures

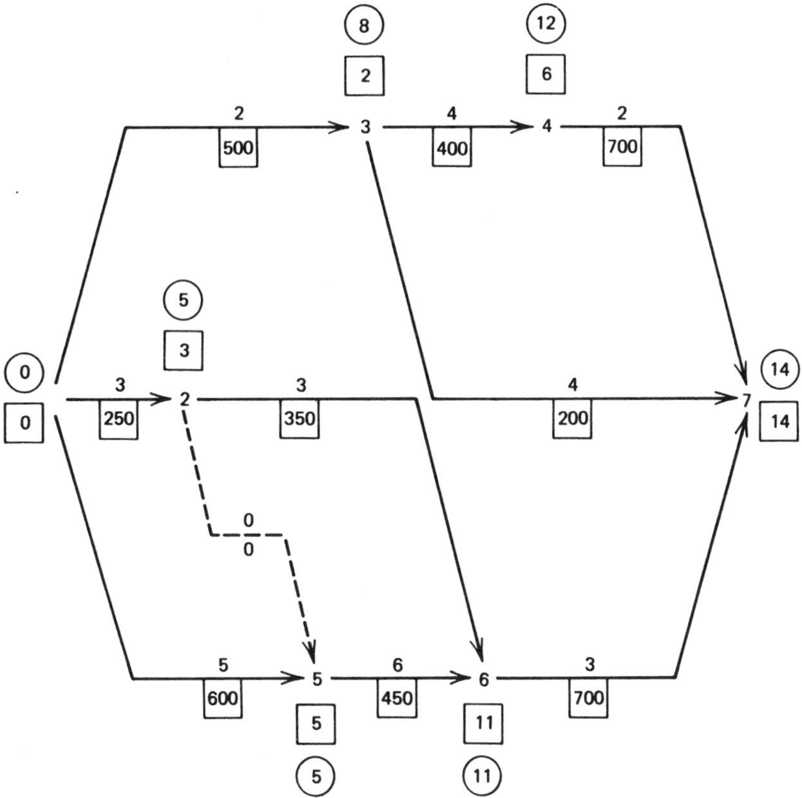

FIGURE 13.7 Example 13.1 Network

8, $5000; at event 6, $5000. Out of a total payment of $16,000 for the project, the balance, $5000, will be received one week after the project is finished. The contractor wants to know if the project can be funded without borrowing. The problem can be solved by using a simple resource allocation procedure, as described in Chapter 9. Instead of such resources as workers, machines, or materials, we now have money as a resource. We must keep track of the expenditures, the investments, the payments received, the amount available in the resource pool, and the project clock.

There are three possible ways of solving this problem: company financing (contractor's own investment), borrowing, and borrowing plus creditor financing. Each solution will be discussed separately.

Case 1: Company Financing

The solution is shown in Table 13.1. In this case the project must be completed using the contractor's own investment of $4500 and the progress

TABLE 13.1 Cash Flow

Activity	Duration	Weekly Cost	Activity Cost	Latest Start	Total Cost	Priority		1	2	3	4	5	6	7	8	9	10	11	12	13	14	15	16
						Loan																	
						Payment Received						1000			5000				5000				5000
						Commitment		600	1200	2050	2900	3750	4200	5150	5150	6450	7650	5050	10,100	11,400	13,000	14,400	
						Amount in Resources Pool	4500	3900	3300	2450	1600	1750	1300	350	5350	4050	2850	1450	5400	4100	2500	1094	6100
						Project Clock		↓	↓	↓	↓	↓	↓	↓	↓	↓	↓	↓	↓	↓	↓	↓	↓
1–2	~~3~~ ~~2~~ ~~1~~ 0	250	~~500~~ ~~250~~ ~~750~~ 0	~~2~~ ~~3~~ ~~4~~ 5	0	~~2~~ ~~2~~ ~~2~~			250	250	250												
1–3	~~2~~ ~~1~~ 0	500	~~500~~ ~~1000~~ 0	~~6~~ ~~7~~ ~~8~~ 9	~~0~~–1	~~2~~ ~~2~~ ~~2~~							500		500								
1–5	~~5~~ ~~4~~ ~~3~~ ~~2~~ ~~1~~ 0	600	~~2400~~ ~~1800~~ ~~600~~ ~~3000~~ ~~1200~~ 0	~~0~~ ~~1~~ ~~2~~ ~~3~~ ~~4~~ 5		~~1~~ ~~1~~ ~~1~~ ~~1~~ ~~1~~	600	600	600	600	600												

2–5	0	0	0 ~~700~~ ~~350~~	5	0	~~TOP~~						↕									
2–6	~~3~~ ~~2~~ ~~1~~ 0	350	 ~~1050~~ 0 ~~1200~~ ~~600~~	~~8~~ ~~9~~ ~~10~~ 11	0	~~3~~ ~~3~~ ~~3~~									350	350	350					
3–4	~~4~~ ~~3~~ ~~2~~ ~~1~~ 0	400	0 ~~1600~~ ~~400~~ ~~800~~ ~~400~~	~~8~~ ~~9~~ ~~10~~ ~~11~~ ~~12~~ 13	~~0~~–1	~~3~~ ~~3~~ ~~3~~ ~~3~~ ~~3~~										400	400	400	400			
3–7	~~4~~ ~~3~~ ~~2~~ ~~1~~ 0	200	0 ~~800~~ ~~200~~ ~~700~~	~~10~~ ~~11~~ ~~12~~ ~~13~~ 14	0	~~4~~ 2 ~~3~~ ~~3~~											200	200	200	200		
4–7	~~2~~ ~~1~~ 0	700	0 ~~1400~~ ~~2250~~ ~~1800~~ ~~999~~	~~12~~ ~~13~~ ~~14~~ 15	~~0~~–1	11															700	700
5–6	~~6~~ ~~5~~ ~~4~~ ~~3~~ ~~2~~ ~~1~~ 0	450	 ~~2700~~ ~~1350~~ ~~450~~ 0 ~~1400~~ ~~700~~	~~5~~ ~~6~~ ~~7~~ ~~8~~ ~~9~~ ~~10~~ ~~11~~ 12	~~0~~–1	~~1~~ ~~1~~ ~~1~~ ~~1~~ ~~1~~ ~~1~~ ~~1~~						450	450		450	450	450	450				
6–7	~~3~~ ~~2~~ ~~1~~ 0	700	 ~~2100~~ 0	~~11~~ ~~12~~ ~~13~~ ~~14~~ 15	~~0~~–1	~~1~~ 2 ~~3~~													700	700	700	
							600	600	850	850	850	450	950	0	1300	1200	1400	1050	1300	1600	1400	

payments received at event 5, week 8, and event 6. The allocation procedure is similar to the one described in Chapter 9. Although expenditure is incurred on activities as scheduled, disbursements are made only at the end of an activity unless otherwise specified. An activity is scheduled only if funds in the resource pool minus commitments are adequate for it.

The priority rules are as follows:

1. Latest start.
2. If a tie, minimum float.
3. If a tie, activity leading to a progress payment event.
4. If a tie, maximum total funds required.
5. If a tie, maximum daily funds required.
6. If a tie, *i–j* sequence.

Dummies are assigned top priority. The priority rules can be selected to suit project requirements.

It is found that the project is completed on week 15 and that no work can be carried out during week 8, as the necessary funds are not available either to start a new activity or to keep in progress those already started. The investment and the progress payment, which amount to $15,500, are spread over 16 weeks, and the amount of $1100 remains in the resource pool at week 15. The final payment of $5000 is received one week after completion of the project.

The cash flow for Case 1 is tabulated in Table 13.2. The expenditure and earnings are calculated for each week. The contract price is $16,000, and the total expenditure is $14,400, therefore the profit earned by the contractor is as follows:

$$\text{Profit} = \$16,000 - \$14,400 = \$1600$$

This profit will hold good only if there is no liquidated damages clause. Suppose a fine of $1000 damages is applied; the profit is then reduced to $600. Return on investment is 600/4500 = 13.33%.

If the network represents a resource adjusted schedule with fixed start and finish dates, restrictive company financing can (1) throw off the schedule, (2) cause equipment to sit idle, and (3) cause overhead to accumulate.

In the next two cases borrowed funds are available; the constraints imposed by a fixed project duration are met, but they lead to interest costs.

Case 2: Borrowing

Money can be borrowed to keep the already started activities in progress and to start all new possible activities. Despite the interest payable on

TABLE 13.2 Case 1

(1) Week	(2) Expenditure per Week	(3) Cumulative Weekly Expenditure	(4) Amount of Loan	(5) Interest	(6) Total Earnings	(7) Cumulative Total Earnings
1	$ 600	$ 600	$0	0	$ 667	$ 667
2	600	1200	0	0	667	1334
3	850	2050	0	0	994	2278
4	850	2900	0	0	944	3222
5	850	3750	0	0	944	4166
6	450	4200	0	0	500	4666
7	950	5150	0	0	1055	5721
8	0	5150	0	0	0	5721
9	1300	6450	0	0	1445	7166
10	1200	7650	0	0	1333	8499
11	1400	9050	0	0	1556	10,055
12	1050	10,100	0	0	1167	11,222
13	1300	11,400	0	0	1444	12,666
14	1600	13,000	0	0	1778	14,444
15	1400	14,400	0	0	1556	16,000

loans, a higher rate of profit can be earned than that in Case 1. The solution is shown in Table 13.3. Allocation for the first seven weeks is made out of the available investment. At the end of the seventh week only $350 is left in the resource pool, but activities 1–2 and 5–6 require $500 and $450, respectively, to be continued. Hence $600 is borrowed. There is nothing in the resource pool at the eighth week. At this time a progress payment of $5000 is received. At week 11, $5000 is received; therefore the loan plus interest is repaid. In this case the project is completed with approximately the same amount of expenditure as in Case 1; however, the project duration is reduced from 15 to 14 weeks. In order to achieve a duration of 14 weeks, interest of 12% must be paid. Thus the interest payable on the loan is the only additional cost.

The data for this case is tabulated in Table 13.4. Interest is calculated, for simplicity in this example, at the end of each week period.

At week 7 an amount of $600 is borrowed. At the end of week 11 interest is charged on this amount for a total of four weeks:

$$\text{Interest} = 600 \times \frac{4}{52} \times \frac{12}{100} = \$6$$

The return on investment is as follows:

$$\text{Profit} = \frac{100(16{,}000 - 14{,}406)}{4500} = \frac{1594}{4500} = 35.42\%$$

TABLE 13.3 Cash Flow

Loan (Repayment)														600				(606)					
Payment Received												1000			5000			5000			5000		
Commitment								600	1200	2050	2900	3750	4200	5150	6100	7300	8500	9900	11,200	12,800	14,400		
Amount in Resources Pool							4500	3900	3300	2450	1600	1750	1300	350	4400	3200	2000	5600	4294	2694	1094	6094	
Project Clock								↓	↓	↓	↓	↓	↓	↓	↓	↓	↓	↓	↓	↓	↓	↓	↓
Activity	Duration	Weekly Cost	Activity Cost	Latest Start	Total Float	Priority		1	2	3	4	5	6	7	8	9	10	11	12	13	14	15	16
1–2	~~3~~ ~~2~~ ~~1~~ 0	250	~~500~~ ~~250~~ ~~750~~ 0	~~2~~ ~~3~~ ~~4~~ 5	0	~~2~~ ~~2~~ 2			250	250	250												
1–3	~~2~~ ~~1~~ 0	500	~~500~~ 0 ~~1000~~	~~6~~ ~~7~~ 8	0	~~2~~ 2							500		500								
1–5	~~5~~ ~~4~~ ~~3~~ ~~2~~ ~~1~~ 0	600	~~2400~~ ~~1800~~ ~~600~~ 0 ~~3000~~ ~~1200~~ 0	~~0~~ ~~1~~ ~~2~~ ~~3~~ ~~4~~ ~~5~~ 6	0	~~1~~ ~~1~~ ~~1~~ ~~1~~ 1	600	600	600	600	600												

2–5	0	0	0 ~~700~~ ~~350~~	5	0	~~TOP~~						↕								
2–6	~~3~~ ~~2~~ ~~1~~ 0	350	~~1050~~ 0 ~~1200~~ ~~800~~	~~8~~ ~~9~~ ~~10~~ 11	0	~~2~~ ~~2~~ ~~2~~									350	350	350			
3–4	~~4~~ ~~3~~ ~~2~~ ~~1~~ 0	400	0 ~~1600~~ ~~400~~ 0 ~~600~~ ~~400~~	~~8~~ ~~9~~ ~~10~~ ~~11~~ 12	0	~~3~~ ~~3~~ ~~3~~ ~~3~~									400	400	400	400		
3–7	~~4~~ ~~3~~ ~~2~~ 1	200	~~800~~ ~~200~~ ~~700~~	~~10~~ ~~11~~ ~~12~~ ~~13~~ 14	0	~~4~~ ~~2~~ ~~3~~ ~~3~~											200	200	200	200
4–7	~~2~~ ~~1~~ 0	700	0 ~~1400~~ ~~2250~~ ~~1800~~ ~~999~~	~~12~~ ~~13~~ 14	0	~~1~~ ~~1~~													700	700
5–6	~~6~~ ~~5~~ ~~4~~ ~~3~~ ~~2~~ ~~1~~ 0	450	0 ~~2700~~ ~~1350~~ ~~450~~ 0 ~~1400~~ ~~700~~	~~5~~ ~~6~~ ~~7~~ ~~8~~ ~~9~~ ~~10~~ 11	0	~~1~~ ~~1~~ ~~1~~ ~~1~~ ~~1~~ ~~1~~						450	450	450	450	450	450			
6–7	~~3~~ ~~2~~ ~~1~~ 0	700	~~2100~~ 0	~~11~~ ~~12~~ ~~13~~ 14	0	~~1~~ ~~2~~ ~~2~~												700	700	700
							600	600	850	850	850	450	950	950	1200	1200	1400	1300	1600	1600

TABLE 13.4 Case 2

(1) Week	(2) Expenditure per Week	(3) Cumulative Weekly Expenditure	(4) Amount of Loan	(5) Interest	(6) Total Earnings	(7) Cumulative Total Earnings
1	$ 600	$ 600	$ 0	0	$ 667	$ 667
2	600	1200	0	0	667	1334
3	850	2050	0	0	944	2278
4	850	2900	0	0	944	3222
5	850	3750	0	0	944	4166
6	450	4200	0	0	500	4666
7	950	5150	600	0	1055	5721
8	950	6100	0	0	1055	6776
9	1200	7300	0	0	1333	8109
10	1200	8500	0	0	1333	9442
11	1400	9900	(600)	(6)	1555	10,997
12	1300	11,200	0	0	1443	12,440
13	1600	12,800	0	0	1777	14,217
14	1600	14,400	0	0	1777	15,994

Borrowed funds serve to increase the return on equity, as long as the contractor earns more on the money than he pays in interest. Loans help cash flows, often hurt borrowing capacity, and help supply cash for continuing project work.

Case 3: Borrowing Plus Creditor Financing

In Case 1 there was one week during which no work was performed. In Case 2, in order to avoid postponement of activities, the contractor borrows money where necessary. Here, besides borrowing money if and when needed, the contractor also has (1) four weeks creditor financing on all activities except activity 3–4 and (2) 2% cash discount on activities 1–2, 1–3, 5–6, and 6–7. This is the discount allowed by his creditors if he does not avail of the 4 weeks and pays his invoices on presentation.

A progress payment of $1000 is received at event 5, $5000 payable at week 8, and $5000 at event 6, and the balance out of $16,000 is received one week after the project is finished.

Besides the progress payments received, the contractor has commitments to meet and payments to make. Activity 1–2 begins on week 3 and terminates on week 5. It requires $250 per week; however, 2% discount is available on this activity. It is advantageous to use the discount option because of the comparatively higher interest cost of money. This is shown in Table 13.5. Also a 2% cash discount is available on activity 1–3. At week 5 the amount payable is $735; this is subtracted from the amount in the resource pool. The commitment of $3000 for activity 1–5 does not have to be paid until week 9. Similarly, other commitments are postponed four weeks where possible. The project is completed on week 14.

Table 13.6 contains the tabulated data for this case. At week 7 an amount of $538 is borrowed. At the end of week 11 interest is charged on this amount for a total of four weeks:

$$\text{Interest} = 538 \times \frac{4}{52} \times \frac{12}{100} = \$5$$

The amounts of cash discounts are computed as follows:

Activity			Discount
1–2	2%(750)	=	15
1–3	2%(1000)	=	20
5–6	2%(2700)	=	54
6–7	2%(2100)	=	42
		Total	$131

The total expenditure is $14,400 − $131 + $5 = $14,274. The rate of return is

$$\frac{(16{,}000 - 14{,}274)}{4500} = 38.35\%$$

There are other variations in timing of receipts and disbursements that can be added. Labor is usually paid twice a month. Such labor costs as the pension fund, unemployment insurance, and taxes are usually paid within two weeks of their due date. Subcontractors are paid, not when the amount becomes due to them but usually a month later. This is when the main contractor has collected a progress payment from the owner which generally includes 90% reimbursement for the materials on site. Owners on large projects sometimes allow additional money in early payments to help cover some of the contractor's mobilization costs. This amount is then recovered through deductions from the contractor's progress payments. Holdbacks are held until completion of the project. Field and office overheads are payable once a month as soon as they become due. The lag factor involved in payment to labor and subcontractors can be considered in the model on similar lines as the creditor financing from vendors.

TABLE 13.5 Cash Flow

Activity	Duration	Weekly Cost	Activity Cost	Latest Start	Total Float	Priority		1	2	3	4	5	6	7	8	9	10	11	12	13	14	15	16
Loan (Repayment)														538				543					
Payment Received												1000			5000			5000				5000	
Commitment								600	1200	2045	2890	3735	4176	5107	6038	7229	8420	9811	11,097	12,683	14,269		
Amount in Resources Pool							4500	3900	3300	2455	1610	1765	1324	393	4462	3271	2080	5684	4398	2812	1226	6226	
Project Clock								↓	↓	↓	↓	↓	↓	↓	↓	↓	↓	↓	↓	↓	↓	↓	↓
1–2	~~3~~ ~~2~~ ~~1~~ 0	250	~~500~~ ~~250~~ ~~700~~ 0	~~2~~ ~~3~~ ~~4~~ 5	0	~~2~~ ~~2~~ ~~2~~			245	245	245												
1–3	~~2~~ ~~1~~ 0	500	~~500~~ 0 ~~1000~~	~~6~~ ~~7~~ 8	0	~~2~~ ~~2~~							490	490									
1–5	~~5~~ ~~4~~ ~~3~~ ~~2~~ ~~1~~ 0	600	~~2400~~ ~~1800~~ ~~600~~ ~~3000~~ ~~1200~~ 0	~~0~~ ~~1~~ ~~2~~ ~~3~~ ~~4~~ 5	0	~~1~~ ~~1~~ ~~1~~ ~~1~~ ~~1~~	600	600	600	600	600												

2–5	0	0	0 ~~700~~ ~~350~~	5	0	~~TOP~~						↕									
2–6	~~3~~ ~~2~~ ~~1~~ 0	350	~~1050~~ 0 ~~1200~~ ~~800~~	~~8~~ ~~9~~ ~~10~~ 11	0	~~2~~ ~~2~~ 2									350	350	350				
3–4	~~4~~ ~~3~~ ~~2~~ ~~1~~ 0	400	0 ~~1600~~ ~~400~~ ~~600~~ ~~400~~	~~8~~ ~~9~~ ~~10~~ ~~11~~ 12	0	~~3~~ ~~3~~ ~~3~~ 3										400	400	400	400		
3–7	~~4~~ ~~3~~ ~~2~~ ~~1~~ 0	200	0 ~~800~~ ~~200~~ ~~700~~	~~10~~ ~~11~~ ~~12~~ ~~13~~ 14	0	~~2~~ ~~2~~ ~~3~~ 3												200	200	200	200
4–7	~~2~~ ~~1~~ 0	700	0 ~~1400~~ ~~2250~~ ~~1800~~ ~~900~~	~~12~~ ~~13~~ 14	0	~~1~~ 1													700	700	
5–6	~~6~~ ~~5~~ ~~4~~ ~~3~~ ~~2~~ ~~1~~ 0	450	~~2700~~ ~~1350~~ ~~450~~ 0 ~~1400~~ ~~700~~	~~5~~ ~~6~~ ~~7~~ ~~8~~ ~~9~~ ~~10~~ 11	0	~~1~~ ~~1~~ ~~1~~ ~~1~~ ~~1~~ 1						441	441	441	441	441	441				
6–7	~~3~~ ~~2~~ ~~1~~ 0	700	~~2100~~ 0	~~11~~ ~~12~~ ~~13~~ 14	0	~~1~~ ~~2~~ 2													686	686	686
							600	600	845	845	845	441	931	931	1191	1191	1391	1286	1586	1586	

TABLE 13.6 Data from Case 3

(1) Week	(2) Expenditure per Week	(3) Cumulative Weekly Expenditure	(4) Amount of Loan	(5) Interest	(6) Cash Discount	(7) Total Earnings	(8) Cumulative Total Earnings	(9) Receipts
1	$ 600	$ 600	$0	0	0	$ 673	$ 673	
2	600	1200	0	0	0	673	1346	
3	850	2050	0	0	5	947	2293	
4	850	2900	0	0	5	947	3240	
5	850	3750	0	0	5	947	4187	$1000
6	450	4200	0	0	9	495	4682	
7	950	5150	538	0	19	1064	5768	
8	950	6100	0	0	19	1064	6832	5000
9	1200	7300	0	0	9	1344	8176	
10	1200	8500	0	0	9	1344	9520	
11	1400	9900	(538)	(5)	9	1568	11,058	5000
12	1300	11,200	0	0	14	1456	12,544	
13	1600	12,800	0	0	14	1791	14,335	
14	1600	14,400	0	0	14	1791	16,126	

The examples discussed in this section illustrate the cash analysis of an extremely small project. In real life, restrictive borrowing with different interest costs, instead of borrowing as needed (as illustrated in these examples), can result in a longer project duration or different funding costs. A comparison of different alternatives is therefore profitable in deciding a financial plan.

It may be necessary in certain cases to juggle the activities so they conform to the progress payment schedule. Noncritical activities may be delayed until funds become available. When the project duration is exceeded, the network can be modified, and a financial feasibility check can be made on the new network. This approach succeeds only if the sum of progress payments, loans, creditor financing, and available investment up to the project finish exceeds the cost of the project.

This example has described cash flow forecasting. It should not be assumed that actual work will necessarily follow the planned cash flow. Indeed, by management decision, during the course of work the cash flow can be varied by activity shifting within available float, and by changing from one construction plan to another. This may become essential in order to eliminate or alleviate a critical work deficiency.

13.4 OWNER'S CASH FLOW

The previous section described three possible ways a contractor can finance a particular project. An investigation was made to determine which had the highest rate of return. In this section the financial feasibility of a project from an owner's standpoint will be studied.

The objective of an owner is to defer payment to a contractor, as long as it does not hamper the contractor's performance on the project. Before awarding a contract, the owner needs an idea of what the total cost for the project will be and when the progress payments must be made. Consider Case 2 of the example presented in the preceding section. The owner's progress payments to the contractor are shown in Table 13.3.

Since owners wish to get the most from the dollars they either have available or have to borrow, they must minimize the money on hand at any particular time. They will either invest their money in a short-term deposit or delay borrowing once they know the money is there when needed. A likely picture of the amount of money an owner will have on hand is shown in Figure 13.8.

Week 5 shows the owner obtaining $1000 to make the first payment to the contractor, followed by $5000 each for weeks 8, 11, and 15.

At weeks 1 to 4, 6 to 7, 9 to 10, and 12 to 14, the owner does not need funds to make the payment. This time could be one day, one week, or even one month, depending on the situation. In this example, one week was selected.

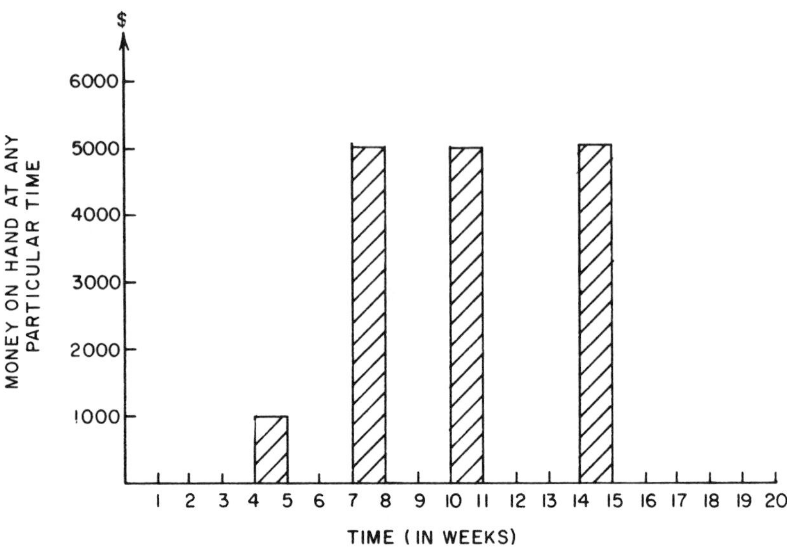

FIGURE 13.8 Money on Hand

13.5 MULTIPROJECT AND COMPANY CASH FLOW

Recall the three cash flow solutions to the problem in Figure 13.7. Any of these solutions could be used by a contracting company in performing the project. Now consider each of the three cases as three separate projects, each starting four weeks apart and controlled by one company. Figures 13.9*a*, *b*, and *c* contain plots of cumulative expenditure less cumulative payments received against project time in weeks for each of the three cases or projects. Figure 13.9*d* is a plot of income from other sources the organization has while carrying out the preceding projects. This money can be used to partly finance the projects. Figure 13.9*e* shows the overhead costs. Figure 13.9*f* contains a summary of all the other graphs of Figure 13.9. The cumulative expenditures less cumulative payments received for each project are summed in the plot marked $a + b + c$. The other plot, Figure 13.9*f* marked $a + b + c - d + e$, represents the overall organization cash flow from these projects and other sources. All of the data for these figures is organized in Table 13.7.

Projects are charged with rental cost for the company equipment. This amount is credited to the equipment reserve and is reflected as a part of the other income of the company shown in Figure 13.9*d*. The net cash inflow after meeting all repayments on equipment shows up in the company cash flow when the curve of Figure 13.9*d* is added to projects cash flow and company overhead cost curves to obtain the company cash flow as shown in Figure 13.9*f*.

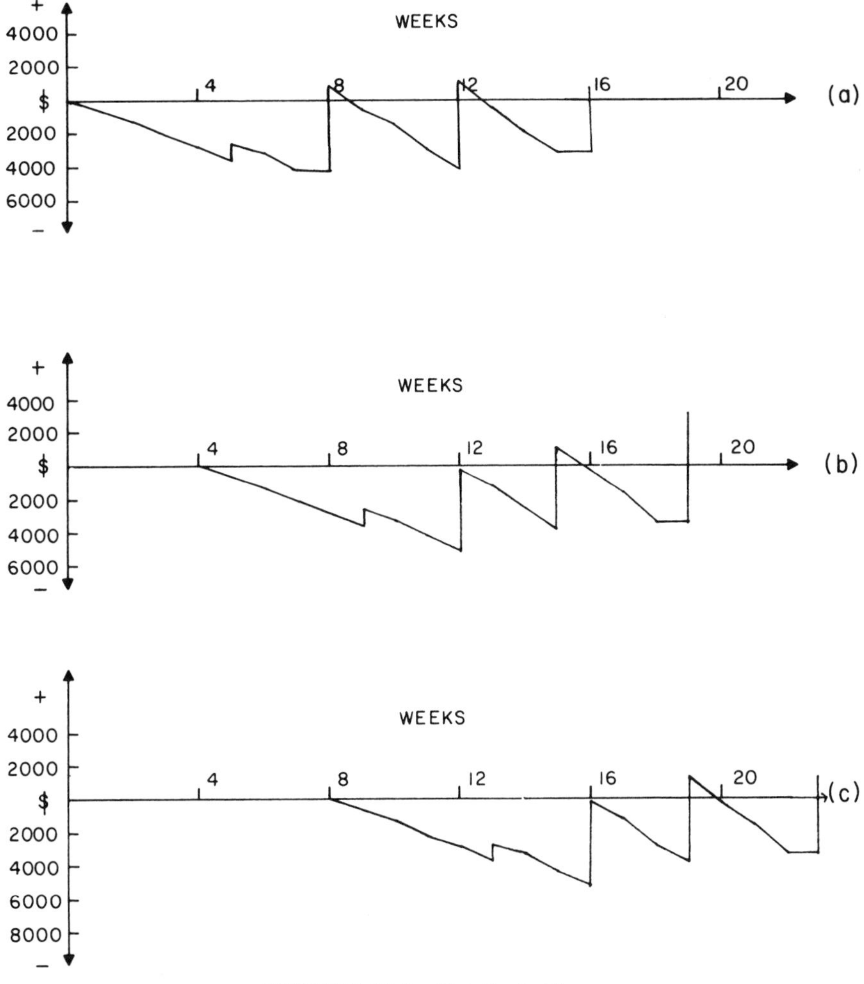

FIGURE 13.9 Net Cash Flows

The necessary input data for the company cash flow consists of project schedules, estimated cost of each activity, investments, progress payment schedules, financing terms, and time lag factors for the various income and expense components. Net project cash flow is obtained from this data for each project and then combined with the organization's income and expenditure from other sources and its overhead expenditure. A summary of this income and expenditure generates the cash flow forecast for the organization for a given period. Projects that have large negative cash flows should be staggered or rescheduled so that the cash outflows do not compound each

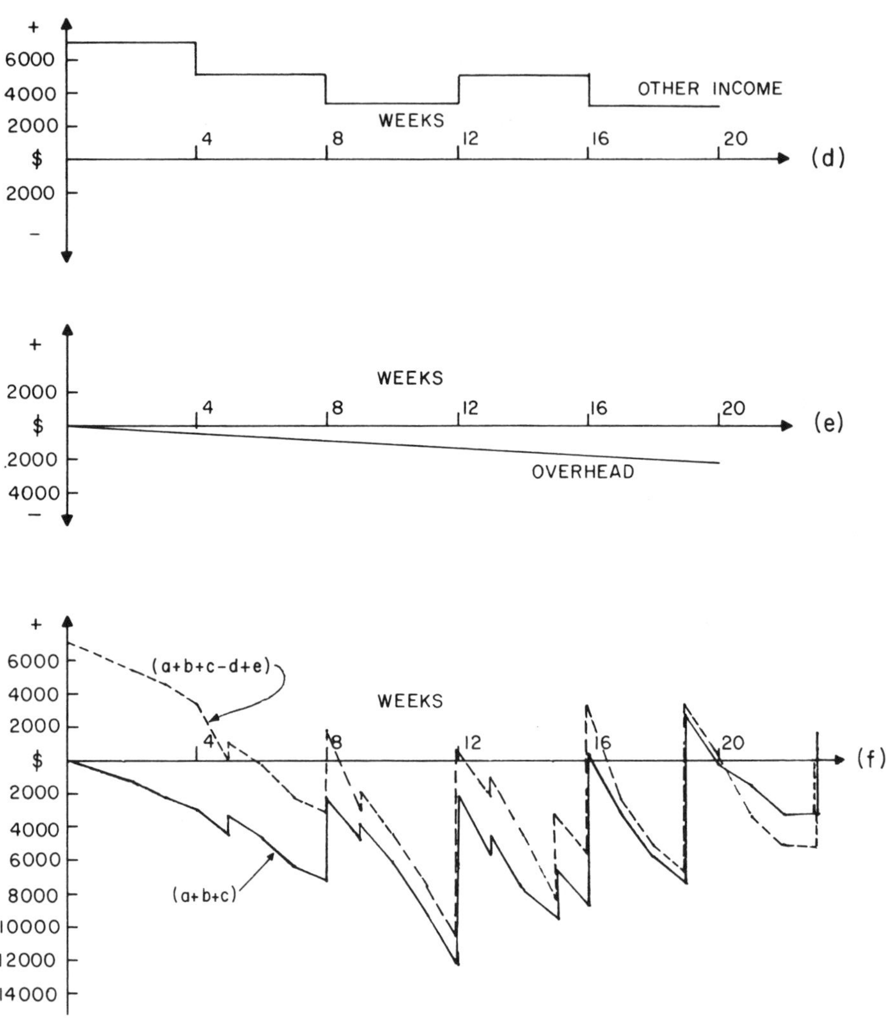

FIGURE 13.9 (*Continued*)

other. When surplus cash is available, efforts should be made to see that this cash is utilized, either by additional work or by short-term investments.

Using this method, all existing and planned projects can be related to overall liquidity situations for the entire organization. In this way a CPM-based cash flow forecast can help formulate realistic policies, especially concerning commencement dates for new projects, whether or not to bid on a new project, and construction rates to suit available working capital. Figure 13.9*f*, for example, may be used to indicate the estimated company cash commitment or working capital status for all the company projects at any time and for a specific project period. Committed expenditures for current

TABLE 13.7 Multiproject and Company Cash Flows

	(*a*) Project 1	(*b*) Project 2	(*c*) Project 3	($a+b+c$) All Three Projects	(*d*) Other Projects	(*e*) Overhead	($a+b+c-d+e$) Total
1	$ 600	$—	$—	$ −600	$7000	$ −125	$ 6275
2	−1200	—	—	−1200	7000	−250	5550
3	−2050	—	—	−2050	7000	−375	4575
4	−2900	—	—	−2900	7000	−500	3600
5	(−3750) −2750	−600	—	(−4350) −3350	5000	−625	(25) 1025
6	−3200	−1200	—	−4400	5000	−750	−150
7	−4150	−2050	—	−6200	5000	−875	−2075
8	(−4150), 850	−2900	—	(−7050) −2050	5000	−1000	(−3050) 1950
9	−450	(−3750) −2750	−600	(−4800) −3800	3000	−1125	(−2925) −1925
0	−1650	−3200	−1200	−6050	3000	−1250	−4300
11	−3050	−4150	−2045	−9245	3000	−1375	−7620
12	(−4100), 900	−(5100), −100	−2890	(−12,090) −2090	3000	−1500	(10,590) 590
13	−400	−1300	(−3735) −2735	(−5435) −4435	5000	−1625	(−2060) −1060
14	−2000	−2500	−3176	−7676	5000	−1750	−4426
15	−3400	(−3906), 1094	−4107	(−11407) −6413	5000	−1875	(−8282) −3282
16	(−3400) 600	−206	(−5038) −38	(−8644) 356	5000	−2000	(−5644) 3356
17	—	−1806	−1229	−3035	3000	−2125	−2160
18	—	−3406	−2420	−5826	3000	−2250	−5076
19	—	(−3406), 1594	(−3816) 1184	(−7222) 2778	3000	−2375	(−6597) 3403
20	—	—	−102	−102	3000	−2500	398
21	—	—	−1688	−1688	1000	−2625	−3313
22	—	—	−3274	−3274	1000	−2750	−5024
23	—	—	(−3274) 1726	(−3274) 1726	1000	−2875	(−5149) −149

and planned projects are drawn on this chart to indicate the total cash demand on working capital during a project period. In this way key decisions can be made regarding bidding for new projects, determining project duration and optimum start times, among other things, so financial crises in the company can be planned for, if not eliminated. Likewise, allowances can be made in the working capital budget for capital expenditures such as new equipment purchases.

13.6 MINIMIZING FUNDING

In the process of generating a feasible funding plan for a project, several alternatives are developed. In order to select the minimum cost alternative, the various factors that affect the cost of a project have to be considered.

On large projects the cost of financing a project is a major consideration. On a billion-dollar power project with an estimated duration of 10 years, the funds invested during the first year at 12% interest will approximately triple before the end of the project. It is therefore necessary to postpone the expenditure, as long as the postponement does not have a deleterious effect on the project duration.

On the other hand, there is an escalation cost that is very hard to predict. It is a part of any expanding economy. The earlier a component is obtained and installed, the cheaper it may be in the long run. Another method to guard against escalation may be to bring the components ahead of their requirement. This may involve inventory cost and possible damage in storage. Nevertheless, such decisions are important and have to be made. It is therefore necessary to estimate the cash flow for a project using the escalation factor for each year and the estimated inventory costs where advance buying is planned.

When part of the project is complete, income should begin to flow in. Income today is worth more than the same amount five years from now. Just as the income in five years is more valuable than the same income eight years hence. Thus the income from the project and the time of its receipt are important in financial feasibility analysis.

Although investment does not earn any interest for the contractor, it has the potential of earning interest. The progress payments received by the contractor on a project offer an opportunity income (i.e., the contractor may be able to invest the amount in a short-term bank deposit, in bonds, or in another construction project). Many organizations therefore charge interest on their own investment to the cost of a project. So long as this interest is not drawn by the company from the project, it does not affect the cash flow, but it definitely influences the total cost of the project.

Thus, when interest costs, escalation, inventory cost, income from the project, and interest on investment amounts to a considerable sum in a project, it is advantageous to determine not only the cash flow but also the true cost for each alternative funding scheme for the project and then select the minimum cost solution.

The procedure described in this chapter can be used to determine the basic cash flow on a project. To this can be added the interest, escalation, and inventory costs. Any income earned from short-term investments can be added and a plot obtained as shown in Figure 13.10. Compound interest tables can be used for computing present worth of the cash flow. The objective is to determine and minimize the total cost for each financially feasible plan. This method is useful for alternatives with similar project periods.

If a comparison is made between alternatives that do not have equal durations, the rate of return method may be used. An initial guess for i (the rate of return) is made, and the present worth of future profits is compared with the initial investment. If they are equal, then i is indeed the rate of return for the project. If the two are not equal (which can happen), then the results obtained from the comparison of the profits' present worth with the initial investment will be useful in selecting the next i to be tried. The process continues until the present worth is equal to the investment, which is when the rate of return is found. Present worth factors can be found in a standard textbook on mathematics of finance.

An alternative method for determining the rate of return, although based on trial and error as before, is to find the rate of interest that causes the present worth of all the receipts to be equal to the present worth of all the disbursements. A higher interest rate would cause the value of the disbursements to exceed the value of the receipts and indicate an unattractive investment.

Case histories of projects similar in magnitude, duration, and profit percentage may be compared. Although the profit percentage in relation to the contract amount may be the same, the rate of return on investment may be different. This could be because one project had a negative cash flow. A project with a predominantly positive cash flow needs very little investment.

The rate of return from alternative projects can also be used for comparing their merits and selecting the most beneficial project. A study of cash flow characteristics of such alternatives is helpful at the bid stage if the results are properly reflected in the pricing decision. This enables a contractor to

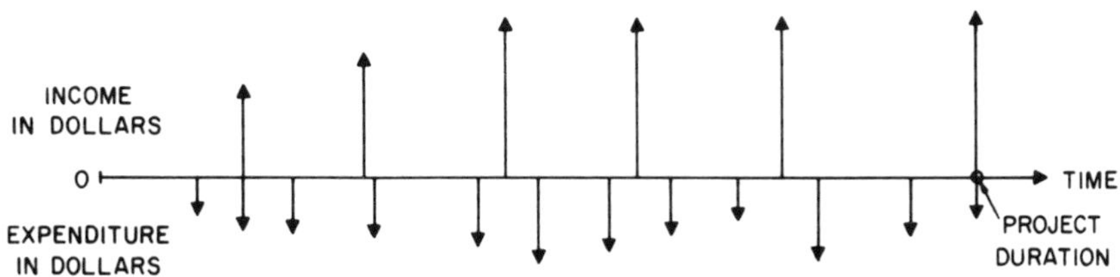

FIGURE 13.10 Cash Flow Diagram

bid more selectively, increasing his chances of obtaining the good jobs and leaving the poor ones to the undiscriminating competition or obtaining them at a proper price.

Resource allocation programs can generally be used for a cash flow forecast, provided such programs can accommodate negative resources. By treating cash inflow as a negative resource, the cash flow profile can be plotted. Drawing upon the discussion in this chapter, estimates of expenditures, disbursements, discounts, cash credits, and time lag in payments can be made and associated with activities of a summary network to make reliable cash flow forecasts that are then used in a resource allocation computer program. If interest calculations are desired to be made, a suitable program can be combined with the resource allocation program.

Financial forecasting procedures demonstrated in this chapter illustrate the effect that many variables have on cash flow. The method can be used for accurate detailed short-term forecasting. However, in the very early stages of the project financial forecasting does not require detailed CPM networks, which in fact are not even available at that time. Rather cash flow forecasts can be developed by using the bar chart method. Cost and duration probabilities can be associated with either method, and by using simulation procedures, a probability distribution may be generated for each time period's cash flow.

PROBLEMS

13.1 The researcher in the project depicted by the following network is given $5000 to start the work.

The estimated costs of the activities follow:

	Activities	Estimated Cost
1–2	Prepare research proposal	$4000
2–3	Await approval	0
2–4	Plan room	400
2–5	Design equipment	5000
3–4	Dummy	0
3–5	Dummy	0
4–6	Build room	4000
5–7	Order equipment	2500
5–8	Fabricate and install service pipework	1000
5–9	Write computer program	3000
6–7	Build founds	1000
6–8	Electrical wiring	1800
7–8	Install equipment	1500
8–9	Dummy	0

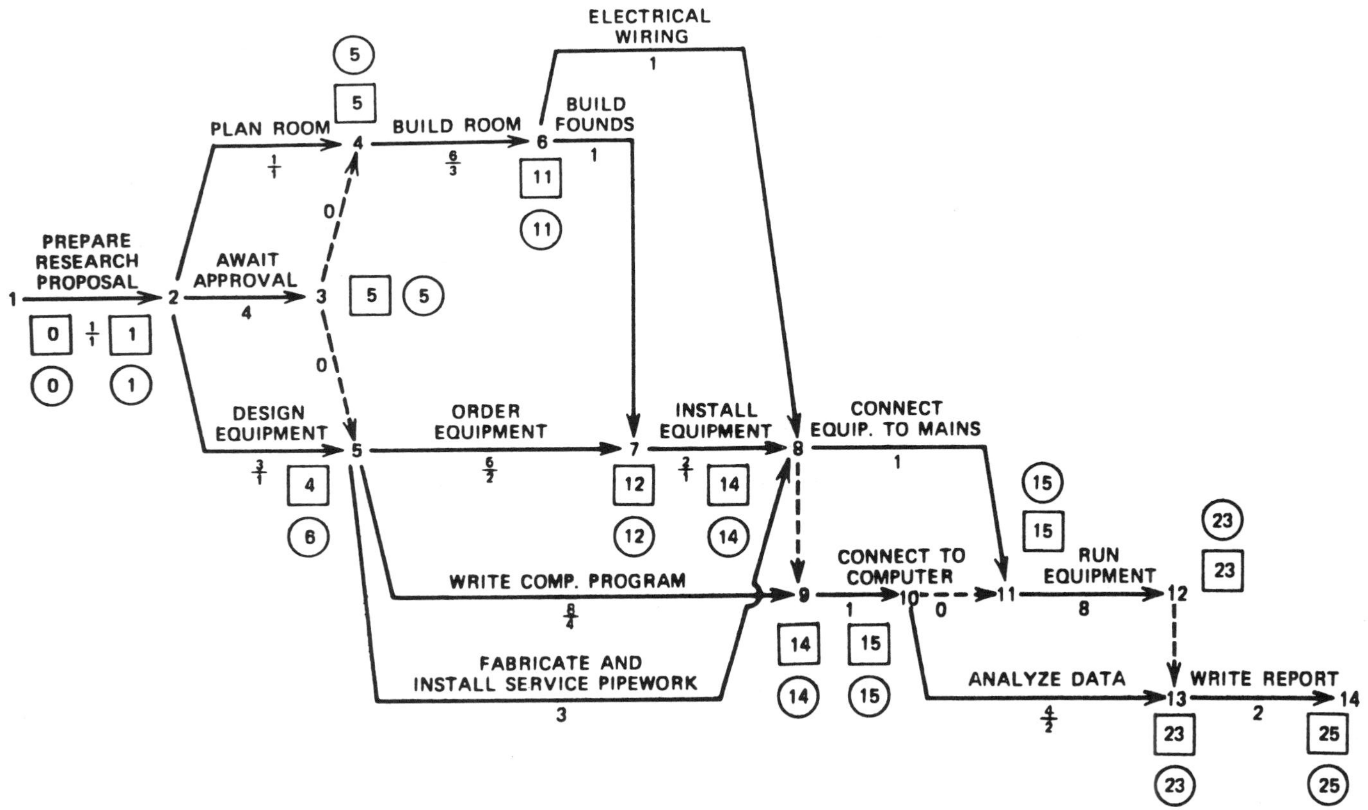

PREPARE RESEARCH PROPOSAL
AWAIT APPROVAL
PLAN ROOM
DESIGN EQUIPMENT
BUILD ROOM
ELECTRICAL WIRING
BUILD FOUNDS
ORDER EQUIPMENT
INSTALL EQUIPMENT
WRITE COMP. PROGRAM
FABRICATE AND INSTALL SERVICE PIPEWORK
CONNECT EQUIP. TO MAINS
CONNECT TO COMPUTER
RUN EQUIPMENT
ANALYZE DATA
WRITE REPORT

PROBLEM 13.1

	Activities	Estimated Cost
8–11	Connect equipment to mains	200
9–10	Connect to computer	1500
10–11	Dummy	0
10–13	Analyze data	1000
11–12	Run experiment	1600
12–13	Dummy	0
13–14	Write report	1000

The researcher is allowed by the company to borrow from the bank. Creditor financing is allowed on the following activities for a period of four weeks:

4–6 Build room
5–7 Order equipment
6–8 Electrical wiring

A 2% discount is allowed on activity 5–7 if cash payment is made as soon as it becomes due. Bank interest is computed at 1% every four weeks on the outstanding balance. The bank loan will be paid back out of the company's savings of $1000 per day, starting two weeks after the completion of the project.

Forecast weekly cash flow until such time that the loan is fully paid.

13.2 An entrepreneur has from a certain source a regular income of $1000 per month. He wants to build the gas station described in Problem 6.5. He has good credit in the market and can avail himself of four weeks credit or in lieu claim 2% discount. Interest on unpaid bills is charged at 12% per annum; he can borrow up to $10,000 from his bank at a prime interest rate of 12% per annum. He can also borrow from a finance company at a rate of 15% per annum. He assigns higher priority to a garage building under construction that is expected to be complete in five weeks and that needs an additional investment of $30,000. This is because of higher rate of return from the garage building. On its completion the entrepreneur expects an income of $500 per week, which he can invest in the gas station. There is a construction overhead cost of $150 per week on the garage and $200 per week on the gas station.

The following are the costs for each activity of the gas station. Each activity can be performed in the duration and cost noted against it. Determine from the cash flow for all the entrepreneur's projects the minimum cost duration for the gas station.

Activities	Normal Duration (in weeks)	Normal Cost
Excavate for sales island base	1	$ 200
Construct sales island base	1	400
Construct cash office	2	2000
Obtain pumps	16	2000
Install pumps	1	500
Connect pumps	2	150
Inspector approves pump installation	2	0
Obtain office furnishings	8	2000
Paint and furnish office and toilets	2	500
Connect office and toilet lighting	1	0
Excavate for office island	1	500
Construct office island base	1	250
Build offices and tiolets including all services	2	15000
Install burglar alarm	1	300
Connect up burglar alarm	2	200
Insurance company inspects burglar alarm	2	0
Electricity board installs meter	14	150
Connect main cable to meter	1	100
Install area lighting	4	2000
Mobilize	1	300
Set out and level site	1	500
Excavate trench and lay all underground services	1	600
Excavate for pipework and tanks	1	900
Construct concrete pit	1	1000
Install pipework and tanks	3	1000
Obtain pipework and tanks	2	3000
Obtain compressor	10	1200
Install compressor	1	300
Connect power to compressor	1	150
Inspection of compressor	2	0
Backfill and cover tanks	1	120
Construct concrete slab	2	0
Construct perimeter wall including air points	2	800
Connect air points	1	200
Demobilize and clean up site	1	400
Obtain approach road signs	4	500
Select site for approach road signs	12	200
Erect approach road signs	1	200
Inspection of pipework and tanks	2	0

CHAPTER 14

PLAN IMPLEMENTATION, MONITORING, AND CONTROL

We plan, organize, execute, monitor, and control; this is how we manage projects. As shown in Figure 1.1, this dynamic process can be remembered by the acronym POEM/C.

Costs need to be managed and controlled as well as schedule and quality. Each is as important as each leg of a three-legged stool. Although there may be some tradeoffs, the project objective is always to achieve maximum performance, which means producing the scope with the required quality, on time, and within budget.

Once a project has been planned, it is management's responsibility to implement it in such a manner that the project objectives are attained. This is achieved by monitoring progress and expenditure, comparing it with planned objectives, and if necessary, taking corrective action. The constantly changing environment of a project makes unceasing efforts by management absolutely essential, or else, instead of management controlling the project, the project controls management, and there is movement from crisis to crisis without any plan. The plan implementation, monitoring, and control of a project is the subejct matter of this chapter.

14.1 IMPORTANCE OF FEEDBACK IN PLAN IMPLEMENTATION

Implementation of a project plan must be planned. The plan should describe the schedule and cost targets, gradual build-up and arrangement of the organization (as described in Chapter 3), training programs, and procedures. It often happens that owing to the unpredictable environment of a project (e.g., inclement weather, uncertain labor productivity, labor shortages), execution takes longer than planned. Likewise, the actual cost of the project often differs from the estimated cost. Also, as time progresses, the needs of the project many change, and on any project change orders are not an unusual occurrence. Key personnel may move to other projects, resulting in a change of procedures reflected in project performance. The original plan can therefore be rendered inoperative by many influencing factors. Among these are:

1. Changes in time objectives for completion.
2. Changes in cost objectives for the project.
3. Changes in operating policies.
4. Changes in the technical specifications of the projects.
5. Changes in construction methods.
6. Changes in needs.
7. Revised activity time estimates.
8. Inaccurate planning of activity relationships.
9. Failure of suppliers or contractors to deliver on time.
10. Reassessment of resource requirements for individual activities.
11. Inability to utilize resources as originally planned.
12. Unexpected technical difficulties.
13. Unexpected environmental conditions (strikes, weather, etc.).
14. Unexpected market fluctuation.

The foregoing list is not all inclusive but illustrates an important aspect of project management, that is, the management of change—in scope and in conditions. Change control is discussed in the next section.

Obviously, labor strikes or inclement weather cannot be eliminated, but most situations can be compensated for by efficient management control. The control of any system necessitates adequate response to the changing conditions in its environment. Obtaining feedback from the output and comparing it with the designed performance level is an essential feature of the control process. With adequate feedback on progress and expenditure, the project team can work together to exercise control and design compensatory corrective action for any probable occurrence.

Information generally flows efficiently through organizations. However, bad news is consistently impeded in flowing upward. Beware of this filtering of information as it passes from lower to upper levels of management. Hence it is often the case that people at the top make decisions as though times were good, while people at the bottom know that the project is in a chaotic state. Such is the communications problem.

Communications managment is critical if a project team is to act in unison for project control. The objectives are to maintain a balance between reality and one's perception of reality. Programs, policies, and procedures cannot work effectively unless all members of the team know about them and why they exist, and regularly receive performance evaluation feedback. If the team members are not kept fully informed, there may be misinterpretations of the project's objectives, affecting performance and productivity of the team. Information must be passed on to the person who needs the information for a specific task.

To overcome human communication problems requires a tactical plan designed to assure free flow of information throughout the organization without loss of valuable production time. The design of this information flow should provide for all information to flow toward the project control services group where it should be filtered and forwarded on to those who need to know all or parts of it. These details of a project management information system are discussed in Chapter 19. Chapter 18 includes examples of reports used for reporting and monitoring of a project. Next we shall examine aspects of change control and information necessary for control of schedules, followed by information relating to cost control.

14.2 CHANGE CONTROL

Changes arise for many reasons and during all phases of a project. To control changes a project manager must be familiar with the causes and plan to avoid situations which give rise to changes. During the conceptual and development stages, many changes are made in the name of design development. At these early stages the ability to influence cost is greatest and unless control is exercised, the financial ability of the owner or the economic viability of the proejct may be exceeded. Owners are often the most difficult to control. For example, in a processing plant, the operators of each area will attempt to get a facility which is the most maintenance free and least costly to operate. A tradeoff often exists between the best that can be built and that quality which is necessary for an economic operation. A life cycle cost approach is necessary to assess planned expenditures.

The cost of changes is readily evident on lump sum contracts because the contractor will submit a claim for each change. In cost reimbursable contracts, changes may be more readily accepted because of the nature of the contract itself.

Some changes are necessary and others may be discretionary. Changes may be required to correct errors or omissions. Changes in scope may be necessary because of a reevaluation of the project for either economic or functional reasons. A project manager must establish a formal procedure for the control of changes. Figure 14.1 shows a flowchart which outlines the procedure required before implementing a change.

Changes should not be acted upon unless the change has been authorized by the recognized authority. An authorized change becomes a change order, which will affect the cost of the project. Figure 14.2 shows a change order

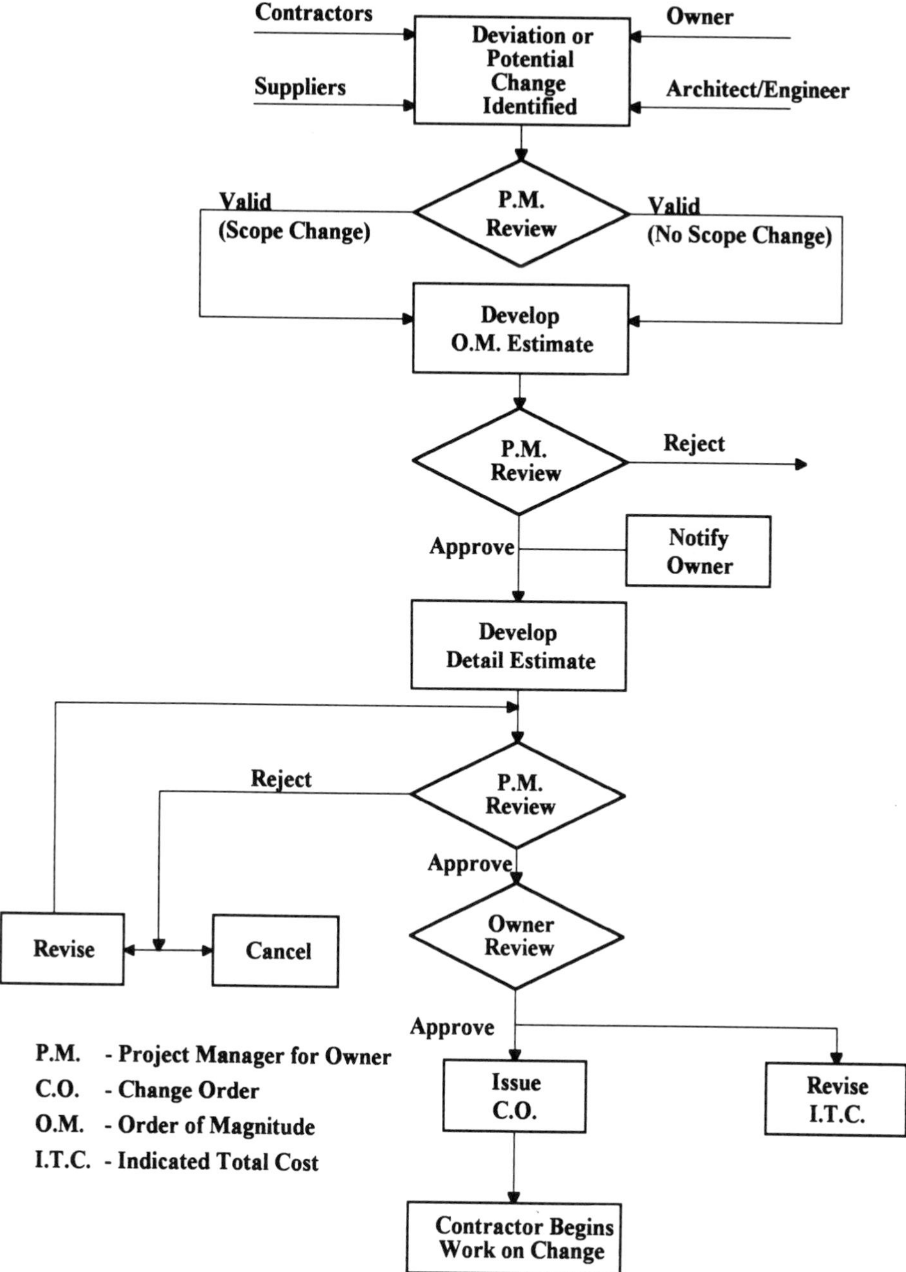

FIGURE 14.1 Change Control Procedure

JONES AND SMITH
PROJECT MANAGERS
GANDER, NFLD.

CONTRACT CHANGE ORDER

Project: *Airport Terminal Building* Change Order No. *1*
For: *Dept. of Transportation* Date July 27, 19 __
To: *The Labrador Construction Co., Ltd.*
Churchill, NFLD.

Revised Contract Amount

Previous contract amount	*5,762,634.00*
Amount of this order	
~~(decrease)~~ (increase)	*5,478.00*
Revised Contract Amount	*5,768,112.00*

An ~~(increase)~~ ~~(decrease)~~ (no change) of ______ days in the contract time is hereby authorized.

This order covers the contract modification hereunder described:

Providing and installing 50mm dia. copper pipe as shown and described by Supplemental Drawing GB 25 attached hereto

The work covered by this order shall be performed under the same terms and conditions as included in the original construction contract.

Changes Approved

____
(Owner)

by ______________________

____
(Contractor)

by ______________________

Jones and Smith, Project Managers

by ______________________

FIGURE 14.2 Contract Change Order

completed with a description and a revised contract amount. This revised amount is either a firm quotation or based on an estimate. Also, note the requirements for authorized signatures.

Contractors undertake considerable risk if a change is implemented without written authorization. If written authorization has not been received, and this is often the case, the contractor should confirm the owner's intent by taking the initiative to document the proposed change and to inform the owner accordingly. A FAX is a convenient and speedy method for this purpose. Also, disputes can be avoided if the intent and cost of the change are clearly outlined prior to starting the work.

14.3 SCHEDULE CONTROL

Time is one resource that we manage and control; it is nonrenewable. Schedules are a graphical representation of time management on a project.

Control of project progress is an ongoing activity. Progress must be marked on the plan for everyone to see; it should be reported from site and must be supplemented by the report on expediting procurement activities so that reliable updated reports can be prepared at regular intervals.

14.3.1 Marking Project Progress Information

Progress information can be displayed on a project network or CPM-based bar chart, as shown in Figure 14.3. A thick line can be used above an activity line to show the extent to which the activity has been completed. Since the bars are drawn to scale, the length can show the status of activities on a

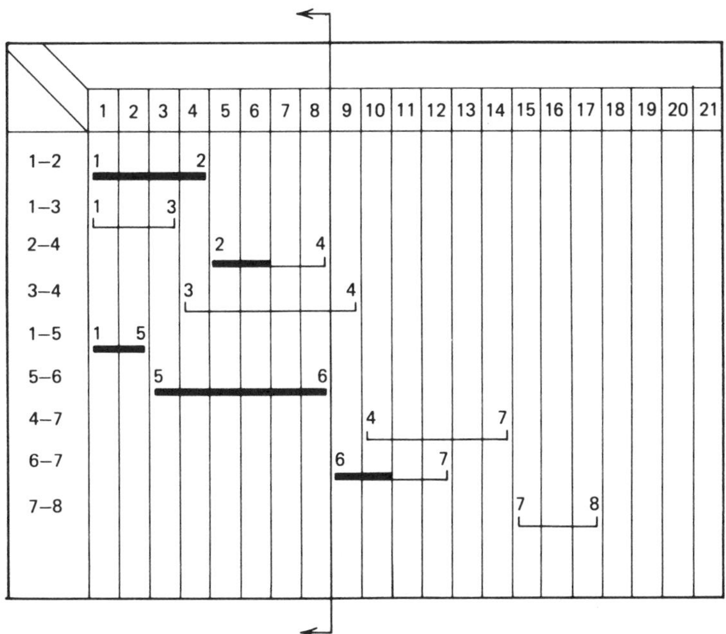

FIGURE 14.3 Bar Chart Displaying Progress

certain day. Notice in Figure 14.3 that activities 1–3, 3–4, and 2–4 are behind schedule. Activity 6–7, on the other hand, is 50% complete and two days ahead of schedule. On the day of update, a vertical line is drawn through the network to indicate how far the project should have progressed. Other considerations necessary for updating a CPM bar chart are similar to those for updating networks. The use of such control information in a format that enables easy visualization of the activity status enables all members of the project team to predict future events more accurately, whether favorably or unfavorably.

The following is intended to include an illustration of a computerized status. Software packages commonly have the facility for showing the status of a project, as illustrated in Figure 14.4. Progress has been shown only for two reporting periods. However a careful study of Figure 14.4 will reveal a considerable amount of performance data.

Most computer software packages can output multicolored plots of schedules, and this feature provides an effective means of presentation. For exception reporting a color such as red can be used to vividly indicate missed or late dates. A tabular computer output such as shown in Figure 14.5 can also be highlighted with color. The reader is encouraged to explore the multitude of plotters and software options that are available.

14.3.2 Estimating Progress Status

Numerous techniques can be used to determine the progress or status of activities on a project:

1. Quantities of work units in place are physically surveyed and compared to those shown on the drawings.
2. Elapsed time is compared to the estimated activity on project durations.
3. Resource usage is plotted versus expected requirements for labor, materials, and equipment.
4. Judgment by an experienced construction supervisor is applied to estimate the percentage complete on individual activities.

Each of these methods has its advantages or disadvantages. For example, field measurements may be more accurate than a "guesstimate" of the percentage complete, but it is expensive to use a survey crew to obtain these data. Guesswork, in turn, can reflect qualitative factors not evident in statistical data alone.

Percentage complete can be estimated from previously estalished thresholds, such as preliminary design or specification release, or by comparing drawings of monthly estimates with drawings of work completed, monthly estimates of pipe in meters with pipe actually installed, and so on. Despite any effort to be objective, the percentage complete estimate will always

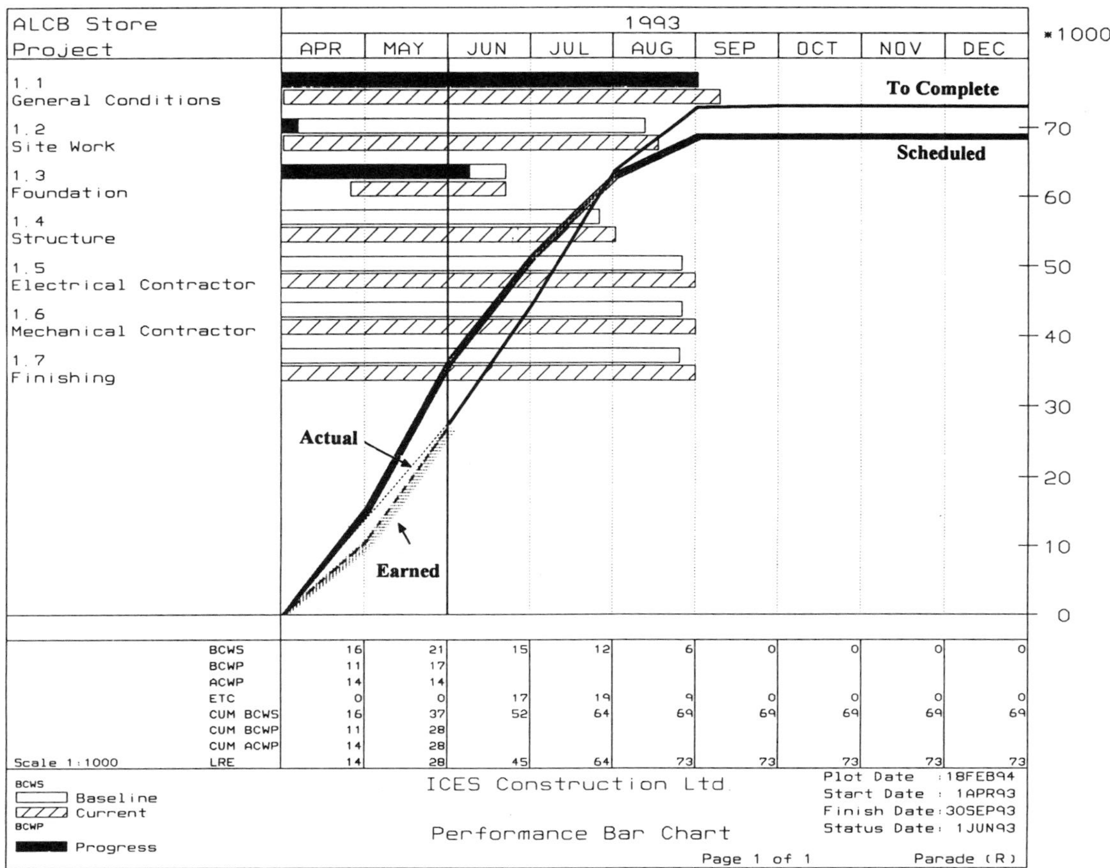

FIGURE 14.4 Sample Schedule Showing Progress to Date

ACTIV-ITY	DURA-TION	ACTIVITY	EARLIEST START	LATEST FINISH	TOTAL FLOAT	PER-CENTAGE COM-PLETE	RE-MAINING DURA-TION
1–2	6	Mobilize	01 MAR 93	09 MAR 93	0		
1–5	12	Mobilize mech.	01 MAR 93	12 APR 93	18		
1–8	12	Mobilize elec.	01 MAR 93	24 MAY 93	48		
2–3	6	Excavate	09 MAR 93	17 MAR 93	0		
3–4	6	Form work	17 MAR 93	25 MAR 93	0		
4–5	12	Pour foundations	25 MAR 93	12 APR 93	0		
5–6	15	Steel structure	12 APR 93	03 MAY 93	0		
5–7	6	Rough-in mech.	12 APR 93	11 MAY 93	15		
6–7	6	Forming slab	03 MAY 93	11 MAY 93	0		
7–8	9	Pour slab	11 MAY 93	24 MAY 93	0		
8–9	12	Masonry I	24 MAY 93	09 JUN 93	0		
8–14	6	Rough-in elec.	24 MAY 93	29 JUL 93	42		
8–15	6	Roofing	24 MAY 93	29 JUL 93	42		
9–10	12	Windows	09 JUN 93	25 JUN 93	0		
9–11	3	Doors	09 JUN 93	25 JUN 93	9		
10–11	0	Dummy	25 JUN 93	25 JUN 93	0		
11–12	18	Masonry II	25 JUN 93	21 JUL 93	0		
12–13	6	Insulation	21 JUL 93	29 JUL 93	6		
12–15	0	Dummy	21 JUL 93	29 JUL 93	0		
13–16	12	Heating	29 JUL 93	16 AUG 93	0		
14–15	0	Dummy	01 JUN 93	29 JUL 93	42		
15–16	12	Electrical	21 JUL 93	16 AUG 93	6		
16–17	6	Sanitary fittings	16 AUG 93	01 SEP 93	6		
16–18	12	Paint	16 AUG 93	01 SEP 93	0		
17–18	0	Completion	24 AUG 93	01 SEP 93	6		

FIGURE 14.5 Schedule for a Building Project

be subjective. Thus difficulties will always be encountered in reaching an agreement on it, particularly if the individuals have conflicting interests. A case in point is an owner and contractor having to agree on the percentage complete status for activities on a project before a progress payment is to be made to the contractor. Since the information is to be used for management control, one solution may be to allow the first-line supervisor to give a definition of percent complete within the guidelines of the prespecified "thresholds;" the manager will soon enough find out what each supervisor means by 50% complete.

A common understanding of what percent complete means can be facilitated by the use of a previously established and agreed upon earned credit for the activities. The following example of a common item such as formwork will illustrate the method. Formwork can be credited as follows:

Manufacture forms	50%
Erect forms	10%
Reinforcing steel	20%
Dismantle formwork	15%
Move to next location or store	5%

14.3.3 Activity Status Sheet

Progress status reports must be put together by on-site field supervisors at regular intervals and relayed to management. Consider the project schedule shown in Figure 14.5, which gives a schedule for the construction of a building starting on March 1. It is hoped the project will be completed by late August. The project is to be carried out on a six-day workweek basis.

The activity status sheet shown in Figure 14.6 indicates that all the activities preceding node 8 are complete, whereas the activities following that node are in progress or have not yet begun. The activities in progress on May 24 provide several start points for the project in the updated network. Each start point has the same start date when the plan is updated, which is May 24 in the present case. The durations of the activities that were in progress are now changed to the durations required to complete these activities, as indicated in the activity status sheet Figure 14.6.

In reporting the status of activities, it is important to think in terms of effective completion rather than total completion. For example, an activity for roofing is not considered incomplete because one small roof hatch is

ACTIVITY	DURATION	EXPECTED/ ACTUAL	FLOAT TOTAL	PERCENT COMPLETE	REMAINING DURATION
8–9	12	24 MAY 93	0	0.0	15
8–14	6	24 MAY 93	42	67.0	2
8–15	6	24 MAY 93	42	0.0	6
9–10	12	09 JUN 93	0	0.0	12
9–11	3	09 JUN 93	9	0.0	3
10–11	0	25 JUN 93	0	0.0	0
11–12	18	25 JUN 93	0	0.0	18
12–13	6	21 JUL 93	0	0.0	6
12–15	0	21 JUL 93	6	0.0	0
13–16	12	29 JUL 93	0	0.0	12
14–15	0	01 JUN 93	42	0.0	0
15–16	12	21 JUL 93	6	0.0	12
16–17	6	16 AUG 93	6	0.0	6
16–18	12	16 AUG 93	0	0.0	12
17–18	0	24 AUG 93	6		0

FIGURE 14.6 Activity Status Sheet

shipped late and requires half a day's work to install while made temporarily tight with plastic sheeting and mastic.

It is not sufficient to record only the progress achieved in an activity up to the cutoff period. A useful part of the information is the start time of an activity. The time between this start date and the cutoff date is the elapsed part of the duration, which may sometimes exceed the original duration. The remaining duration is an estimate of the time that this activity is expected to take to complete, and this estimate is furnished for all "in progress" activities at every update. If it is known that a future activity, such as the delivery of materials, approval of drawings, or an inspection, will not take place on its scheduled date, this information must be included. If such information is not available on the site, it should be added in the office. Likewise, the completion dates of activities that are the responsibility of management can be entered.

14.3.4 Expediting

The expediting function succeeds procurement and supplements project monitoring and control. It ensures that the materials requested are delivered at the specified time and place. As already discussed, early and late deliveries can have profound effects on the schedule, the condition of the materials, and the costs to the owner. For those reasons, once a delivery date is determined that minimizes these hazards, everything possible must be done to assure that the materials will arrive when and where needed.

In order for such stringent deadlines to be met, the delivery dates must be realistic. Otherwise, the procurement schedule will be nothing more than an invitation for disaster. There is a marked tendency for purchasers to anticipate the tardiness of a supplier and order materials too early.

Often overlooked is the vendor data that must be supplied for timely completion of working drawings. Expediting this data is as important as expediting the actual purchased item. For example, it is to no avail to deliver on time a pump when the foundation pad and anchor bolts have not been completed because of the lack of approved vendor drawings which outline the size of pad and location of anchor bolts for the pump. Schedule problems often occur at the interface between suppliers and user, and this interface must be carefully managed, especially in a just-in-time mode for delivery of equipment and materials.

The fabrication and delivery phase is the most critical stage of a material's life cycle because the largest amounts of time can be lost or recovered at this stage. The time involved depends on the type of material and the degree of transformation required. The contractor must therefore take great pains to ensure materials are monitored throughout this process. The delivery of materials is discussed in more detail in Chapter 18.

The expediting function must ensure that the decided delivery dates for any and all materials and equipment are met. To do this, the individual acting as the expediter must know what goods are ordered, when they are expected, and the ramifications of their untimely delivery. If the delivery date is changed for any reason, the supplier must be informed immediately. The supplier who appears to be falling behind schedule must be pressured to meet his or her commitment. Should the supplier fail to meet the requirements altogether, another source must be found to supply the needed materials.

14.3.5 Updating

When the progress report has been received from the site, it is necessary to compare it with the original schedule. Although the duration of each activity can be compared with its planned duration, this does not give an accurate picture of actual performance. For a clear understanding of what a delay on an activity means to the complete project plan, it is necessary to perform an update. In effect this involves entering the progress information into the network plan and analyzing the network with this added information.

Updating is carried out to accommodate configuration changes, to assign a new target date instead of a previously planned target date, and to reflect remedial action designed to correct deviations in order to predict their effect. For example, in the construction of a wharf for loading and unloading crude oil and its products, suppose information regarding the postponement of a delivery of loading arms is received. Moving forward this delivery date would affect the project completion date. Remedial action is necessary to speed up certain activities if it is crucial to meet the delivery date. When the information about the late delivery is received, updating will show the results of this remedial action ahead of time. Sometimes it becomes necessary to change the logic to accommodate more economical construction methods. Such changes should be incorporated as they become needed.

The project schedule of Figure 14.5 is updated according to the progress report appearing in Figure 14.6. The reader may observe that the project shown in Figure 14.7 is delayed by three days. Instead of being finished on September 1, 1993, it will now be finished on September 6, 1993. If this project completion date is now acceptable to management, the scheduled completion date causes negative slack for the activities that need to be expedited so that the project completion time can be met. After such activities are selected and negative slack is eliminated, event times are computed once again and a project schedule is drawn up.

It may be necessary to reallocate or crash the schedule of some activities. Possibly a more economical solution can be obtained by further rearrangement of activities.

14.3.6 Frequency of Updating

Most frequent updating is required on a complex project where many contractors are coordinated than on a similar project with few performing agencies.

ACTIVITY	DURATION	ACTIVITY	EARLIEST START	LATEST FINISH	TOTAL FLOAT	PERCENTAGE COMPLETE	REMAINING DURATION
8–9	15	Masonry	24 MAY 93	24 JUN 93	0		
8–14	2	Rough-in elect.	24 MAY 93	03 AUG 93	40		
8–15	6	Roofing	24 MAY 93	24 AUG 93	45		
9–10	12	Windows	14 JUN 93	30 JUN 93	0		
9–11	3	Doors	14 JUN 93	30 JUN 93	9		
10–11	0	Dummy	30 JUN 93	30 JUN 93	0		
11–12	18	Masonry II	30 JUN 93	26 JUL 93	0		
12–13	6	Insulation	26 JUL 93	03 AUG 93	0		
12–15	0	Dummy	26 JUL 93	03 AUG 93	6		
13–16	12	Heating	03 AUG 93	19 AUG 93	0		
14–15	0	Dummy	01 JUN 93	03 AUG 93	45		
15–16	12	Electrical	26 JUL 93	19 AUG 93	6		
16–17	6	Sanitary fittings	19 AUG 93	06 SEP 93	6		
16–18	12	Paint	19 JUL 93	06 SEP 93	0		
17–18	0	Completion	27 AUG 93	06 SEP 93	6		

FIGURE 14.7 Updated Schedule

On a complex project, one small change can affect the entire project, influencing not only project duration, but also many target dates for different contractors. In special cases, such as defense projects, or startup and commissioning on large projects, very short updating intervals may be necessary, a day-to-day progress check may be required by management; whereas on similar, less intensive projects, monthly updates would be sufficient.

The frequency of reporting should be considered from another point of view. If on a project it takes a week to collect and process information, a weekly report may be meaningless and a monthly report more logical. Ideally, the reporting period should be such that the reports are received by management in time for job meetings, which may be held weekly, fortnightly, or monthly. Lower management is interested in the immediate future whereas higher management is concerned with future trends. Reports to lower management should include the activities on which work is to be done in the immediate future, and higher management should receive summary reports showing progress trends and the probability of achieving the target.

The updating procedure itself involves certain costs for both manual work and computer analysis. This cost aspect must also be considered in determining the frequency of updating.

14.4 COST CONTROL

All costs must be managed and most costs can be controlled. An important question is which costs are controllable and by whom? Some costs are not directly controllable and yet a project manager must manage within a given environment, such as the state of the general and local economy.

There are three cost categories—direct, indirect, and overhead costs. Direct costs are those that can be related to the production, such as the cost of labor and material inputs that remain as part of the permanent facility (e.g., concrete supply and labor for placing concrete).

Indirect costs include labor, material, and expenses that are incurred but cannot be readily apportioned to a particular part of the project. They are usually applied as a percentage of direct costs, and include items such as general supervision, daily subsistence allowances, temporary roads and facilities, snow removal, licenses, and permits, insurance, bonding, and first aid facilities.

The third general category is general overhead costs. These are home office costs that are charged to a project on a predetermined basis such as man-hours of direct labor and hours of equipment usage. General overhead costs include costs for executive management, home office facilities and insurance, and other costs required to carry out the normal course of company business.

The costs incurred on a project can be measured best by field personnel.

The starting point for data collection is at the grass roots level through the use of time cards, invoices for materials, material requisition forms, and equipment utilization sheets. Reliability of the output information from the cost system depends on the accuracy and completeness of the input. Presently on most projects, data is generated manually, but new electronic notepads and similar devices are making it possible to gather and transmit the data electronically, thereby increasing the speed and accuracy of the transaction record.

14.4.1 Labor Cost

On all projects it is necessary to obtain the number of hours spent on each part of the job by each employee. Although the man-hour distribution reports used by different organizations may vary somewhat, they all contain basically similar information. The man-hour distribution sheet is usually prepared by the foreman, except for costing, which is done by the office. The information on this sheet is summarized from the daily time check report in which the foreman records each worker's time against the work items on which the individual works. Traditionally it is the foreman's responsibility to distribute the worker's time to items of work. The activity code column on the daily time sheet, as shown in Figure 14.8, and the man-hour distribution sheet

DATE: June 24/94

JOB NO. 132

DAILY TIME SHEET & WORKER-HOUR DISTRIBUTION

REPORT NO. 32

PAGE NO. 1

OCCUPATION	EMPLOYEE NAME	EMPLOYEE NO.	ACTIVITY CODE	PERF. AGENCY	WORK ITEM HOURS BA03100		BA03100		BA03300		BA03200		DA04200		DA04400		REGULAR		OVERTIME		AMOUNT	
					R	O	R	O	R	O	R	O	R	O	R	O	HOURS	RATE	HOURS	RATE		
LABOR	JAMES OLSEN	32	15-32		4		4	2									8	12.00	2	18.00	132	00
CARP.	J. M. STILL	135	16-18		4		4	2									8	15.00	2	22.50	165	00
IRON WORKER	H. KELLEY	240	20-43								8						8	16.50			132	00
MASON	SAM ZEBAC	315	25-50										5		3		8	16.00			128	00
"	FRANK PARKER	320	35-55										4		4		8	16.00			128	00

FIGURE 14.8 Example of Daily Time Sheet

have this function. The daily time sheet should show if special difficulties have been experienced on an activity for a certain work item resulting in the use of more than estimated man-hours. Man-hours are related to activities by using work item codes and activity codes. Two separate code numbers are shown on this time sheet, whereas often a single code number is used. Although the task of relating man-hours to work items may seem tedious, it is necessary for project control. In so doing, the foreman must refer to an estimate to note the costing code for a particular work item. This helps to overcome the general inclination to ignore the estimate and to bury these costs in other work items.

Daily time sheets and man-hour distribution records are totaled weekly, as well as the costs for each activity and each work item. As discussed in the chapter on information management, the use of computerized relational databases removes the drudgery of summing and transferring costs for further reporting.

Full use of man-hour information cannot be made without knowing the interim quantity of work. It is of little value to know the man-hours expended on an incomplete work item if the quantities of work done are not known. The quantity report, as shown in Figure 14.9, is used to measure the work done on different work items in the preceding period. It provides space for noting the budgeted and actual quantities of work done. The forecast quantity is normally estimated, taking into account any changes.

A labor report as shown in Figure 14.10, enables management to keep track of man-hours worked. Such reports generally give a comparison of work item in terms of estimated hours for each craft with the actual hours to date and indicate any overrun or underrun. The actual hours are taken off the daily time sheet and man-hour distribution (Fig. 14.8). The estimated man-hours for the current period are for the quantity of work completed during the current period and are obtained from progress reports on various activities. From the current period data for each craft it is possible to study learning curve effects and to project anticipated man-hours to complete the activity and plan for increased productivity. Total estimated and actual man-hours for each craft within each work item are listed. The difference between

CHARGE NO.	AS OF DATE			SHEET		
	WORK ITEM BA 03300 WORK UNIT m^3			WORK ITEM BA 3100 WORK UNIT m^2		
ACTIVITY	BUDGET	ACTUAL	LATEST REVISED	BUDGET	ACTUAL	LATEST REVISED
15–32	750	400	800	200	125	220

FIGURE 14.9 Foreman's Quantity Report

CONTRACT DESCRIPTION:		RESPONSIBLE ORGANIZATION	CONTRACT NO.			REPORT DATES			
MUNICIPAL GARAGE		AREA ENGINEER IV	MG 25			FROM ______		TO ______	
STRUCTURAL METAL FRAME						CUTOFF DATE RUN DATE		25 JAN. 83 01 FEB. 83	
IDENTIFICATION			CURRENT PERIOD WORKER-HOURS			TOTAL WORKER-HOURS			
RESOURCE	PERFORMING AGENCY	WORK ITEM	ACTUAL	ESTIMATE	(OVERRUN) UNDERRUN	ACTUAL TO DATE	ESTIMATE	FORECAST TO COMPLETE	PROJECTED (OVERRUN) UNDERRUN
Crew supervisor	G115	Structural metal frame	205	200	(5)	875	2000	2500	(500)
Structural steelworker			850	800	(50)	3387	8000	8864	(864)
Welders			475	400	(75)	1745	4000	3800	200
Crane operator			198	200	2	830	2000	2280	(280)
Light equipment operator			495	400	(95)	1884	4000	3800	200

FIGURE 14.10 Example of Worker-Hours Summary Report

the forecast to complete and the original estimate is computed and listed as the projected overrun or underrun, depending on whether the balance is positive or negative.

14.4.2 Materials Costs

Material cost feedback is generated mainly through a purchase requisition control procedure. All materials used on a project are requisitioned by the home office (project manager) or the field office (construction superintendent). A good definitive estimate and a bill of materials provide excellent control documents by which the material cost can be kept in check on a project.

For control to be exercised at the appropriate time, it is essential that a record of purchases be maintained by the organization. If the quantity and cost of materials for a particular work item do not match its estimate, the cost engineer must determine the reasons for the discrepancy and report to the project manager. Variance reporting will be discussed in this chapter. Each requisition must be sent to the purchasing department to procure the materials. The completion of a purchase order constitutes a commitment of funds.

The materials received on the site must be recorded on a material receiving report or in a similar manner. The data may be gathered in a number of ways. For instance, on a concreting job, this individual may collect tickets from the concrete trucks, stating the time of arrival and the quantity delivered. The materials receiving report is useful for inventory control and for keeping track of the status of purchase orders received at the site.

Materials are charged to work items by means of the costing code, and every order, invoice, and delivery ticket should bear both a job name and number and a costing code for the work item. With many distinctive materials and components, this is easily done. A costing plan spells out in detail which cost codes are to be used for every material item, as well as labor and expenses. The problems arise with the ordinary materials, such as ready-mixed concrete and construction lumber, which are used in many items of work and are sometimes also used in indirect cost items such as temporary fences and barriers. If no effort is made to record the use of the basic materials and to allocate them to specific items, the looseness and lack of knowledge concerning them will be perpetuated, and the misplaced costs of several hundred cubic meters of concrete or several tonnes of rebar will make attempts at accuracy elsewhere in the estimating and cost control futile. It therefore needs to be emphasized that materials should be charged to work items as soon as they are received on site.

For the materials that cannot be charged in this way, an inventory of materials on site is taken at the end of the week. After deducting these quantities from the materials received, the materials used can be determined. These are distributed over the volume of work performed in each work item, as shown in Figure 14.11. Such materials are costed at the average purchase price for the period. The unit cost of material for each work item can be determined from the material distribution sheet for comparison with estimated unit cost.

To check on the use of a certain material on a project, it is useful to generate a material consumption report separated by work items using a format similar to the man-hours report described in this section.

14.4.3 Equipment Cost

Equipment cost must also be charged to work items just like man-hours and materials cost. To do this, a record of number of hours per work item and the hourly rate for each piece of equipment is required. The number of days that equipment is assigned to a project can be derived from checking-in and checking-out procedures. Hours of operation can be accumulated from equipment time cards, as shown in Figure 14.12.

These cards can be turned in weekly, indicating daily hours of usage. The activity column, as in man-hour reports, helps to identify the reason for using equipment longer than the estimated hours.

Idle time should be distributed to items of work and prorated to the distributed working time or captured in a separate idle time account. Such equipment as hoists and cranes, which have many uses and therefore hardly ever have idle time, should nevertheless have their total idle and working times reported daily. Time should be distributed to specific activities. Thus, when a crane is used for such items as placing forms, steel rebar, and concrete, the time can and should be segregated and identified. This may require more effort than a company deems as warranted and often an item such as this crane cost is aggregated into one account for all activities requiring crane usage.

When use of equipment is not heavy on a project, the equipment costs can be applied to all work items as well as to all activities as an indirect cost. In such a case operators and operation costs are charged directly to work items, along with the manpower costs.

From the foreman's quantity reports, unit equipment costs for each work item can be calculated and compared with the estimated unit costs.

Equipment reports, which are generally concerned with comparing budgeted with actual equipment-hours, are similar to man-hours reports. With

WORK ITEM	PERFORMING AGENCY	MATERIAL	UNITS	QUANTITY USED	RATE	AMOUNT
BA 03300	G115	Concrete	m^3	600	100.00	60000.00

FIGURE 14.11 Material Register

PERFORMING AGENCY	ACTIVITY	WORK ITEM	MON. (hr)	TUES. (hr)	WED. (hr)	THURS. (hr)	FRI. (hr)	SAT. (hr)	TOTAL (hr)	RATE	AMT.
AECCO	4–5	AA02200	4	2	5	—	—	—	11	20	220
	Total										

FIGURE 14.12 Weekly Equipment Record

some modifications to title and column headings, the format of the man-hours report given in Figure 14.10 can be used to generate the equipment-hours report.

The major change in the report is in the resource code column. Instead of indicating a craft, the code indicates each type of equipment. The report includes estimated and actual information on equipment-hours used.

Not all data in the worker-hour, equipment and material sheets exhibited in this and the two preceding sections have to be produced manually. In a computerized environment, conversion to money units is not required from the reporting personnel.

14.4.4 Lump-Sum Contract Costs

Bids solicited for lump-sum contracts need not detail man-hours, equipment time, or materials used. Lump-sum contracts may refer to the main contract or a subcontract. However, the cost engineer's control starts with an analysis of all bids in terms of original estimates.

Each item in the project estimate represents a work item that has a definite meaning and scope. The bids received must be compared with this detailed estimate and the associated specifications; then the cost engineer must determine whether any additional work will be needed to complete the work described in the contract documents. A good example of this may be seen in the installation of a compressor, including all the small piping required for its operation. If the bid price of the contractor does not include the cost of piping, the cost engineer must recognize that an expenditure beyond the bid price will be required to complete the installation. Thus the bid price received can have a very different meaning from what might appear at first glance. By detecting incongruities of this type, the cost engineer will have provided the means for cost control of contract items before a contract is signed, and not after the fact when it can give rise to claims. The need for such observations is not limited to lump-sum construction contracts alone. It is equally important in unit price contracts and purchase of materials or equipment for the project.

Suppose that on a lump-sum contract a bid price breakdown is requested of the contractor. The only way it can be obtained with minimum difficulty to all concerned is by including the cost breakdown in the specifications, making it obligatory for the contractor to follow the project breakdown format. The project planning group should also use this format in preparing their estmates. In this way the prices obtained are easily analyzed, side by side, by the project manager. If the breakdown price is not acceptable, necessary adjustments are negotiated. When an acceptable breakdown price has been negotiated, it becomes the basis for progress payments and cash flow forecasts.

Whether the cost associated with activities is linearly distributed over the duration of the project or associated with certain events, the only information needed as feedback from the site is an updated progress report on the activities and the completion of events. Progress reports are used to determine the amount payable to the contractor. The exactness of this amount depends on the level of detail of the price breakdown and the care exercised in preparing it.

In the case of a lump sum contract, control over manpower, equipment, and material costs is the contractor's concern. The progress payment to the contractor is made according to the price breakdown submitted by the contractor and accepted by the owner.

Progress payment is directly related to the completion of activities in the project. As the activities are completed, they are recorded in the progress reports by the project staff on the job. From the activities completed, the percentage completed of unfinished activities in a progress period, and the accepted price breakdown, the amount payable is determined. Then a progress payment is released to the contractor.

An example of a progress payment summary report is shown in Figure 14.13. In this report all the various subcontractors are listed, and the total value of the contract is recorded. The completed value to date is the value of work completed up to the payment period, and this value less the holdback is the amount due the subcontractor for work completed to date. In the example shown, the previous payment made is listed under "paid to date." It is deducted from the amount "payable to date" and the balance is shown under the heading "amount payable." This is the amount payable to the subcontractor for this progress payment period.

14.4.5 Other Contract Costs

Because there are many types of contracts other than the lump-sum contract, a variety of methods can be used for cost control. Unit price and cost-plus contracts stand out as the two basic types; other contracts are variations on these two. The present discussion therefore will be confined to costs from the two other contract types.

Progress payment under the unit price contract is based on the amount of work completed in a payment period, that is, on measured quantities. In the initial bid for the job contractors put forward a unit price for each item of work by trade. An example is the $40/m unit price for installing cast iron pipe. During the pay period the amount payable is determined from the number of meters installed. A progress payment report is prepared that compares (1) quantities and costs estimated to actual quantities up to the previous period and total quantity to date, and (2) amounts estimated to actual payment up to the previous period and total payable to date. This report is useful for maintaining a check over quantities as well as costs for each work item on the project.

There is no standard form for a report on a cost-plus (cost reimbursable) contract. This is because the design of a reporting system is dependent on

DESCRIPTION	CONTRACT AMOUNT	COMPLETED VALUE TO DATE	HOLDBACK	PAYABLE TO DATE	PAID TO DATE	AMOUNT PAYABLE
C100 Holden Construction Co. Ltd.	2,070,000.00	1,002,640.00	50,132.00	952,508.00	752,400.00	200,108.00
C101 Crosbie Construction Ltd.	856,000.00	249,800.00	12,490.00	237,310.00	126,200.00	111,110.00
C102 Patiala Construction Ltd.	595,000.00	443,100.00	22,155.00	420,945.00	335,477.00	85,468.00
C103 Green Masonry Ltd.	291,000.00	21,000.00	1,050.00	19,950.00	9,940.00	10,010.00
C104 Del Drywall & Decorating Co. Ltd.	125,000.00	4,000.00	200.00	3,800.00	3,800.00	0.00
C105 Whitten & Clarke	550,000.00	200,000.00	10,000.00	190,000.00	100,000.00	90,000.00
C106 Steel Structures	90,000.00	50,000.00	2,500.00	47,500.00	2,500.00	45,000.00
C107 Avalon Roofing Ltd.	96,000.00	96,000.00	4,800.00	91,200.00	46,200.00	45,000.00
C108 Seaboard Electric Ltd.	100,000.00	50,000.00	2,500.00	47,500.00	45,000.00	2,500.00
C109 Wight & Associates	1,439,800.00	1,054,800.00	52,740.00	1,002,060.00	894,456.00	107,604.00
Project totals	6,212,800.00	3,171,340.00	158,856.67	3,012,773.00	2,315,973.00	696,800.00

FIGURE 14.13 Progress Payment Summary Report

mutual trust between the owner and the contractor. An owner may accept all the expenses incurred by the contractor subject to periodic auditing. He or she may institute preauditing control over all expenditures incurred on manpower, equipment, materials, and overhead. In such cases the owner's control is exercised through the audit reports. The owner may require the contractor to periodically submit manpower, equipment, materials, and overhead expenditure reports and exercise control by comparing the expenditure figures with the estimates. Reports on cost-plus contracts could also be combination of the different reports described here.

To avoid disputes it is prudent to spell out clearly and carefully the items that are allowed as chargeable costs. These can be restricted to direct costs; most of the indirect costs can be included as part of the hourly direct labor costs. For example, the contractor can estimate the required costs per direct labor hour for site office expenses.

14.4.6 Indirect Costs

Indirect costs for items such as access roads, camps, project supervisory staff, telephones, transportation, and so on are controlled by periodically comparing a report of actual expenditure with estimated costs. (This report is similar in format to a man-hours report.) A standard charge rate as a percentage of the actual direct cost of work is added to the estimated costs for items such as small tools and supplies. Unit costs are obtained by dividing total amount of work items by the quantity of work. Final adjustments to the unit costs are made when the actual indirect costs are determined at the completion of the project.

Indirect costs include costs of mobilization and demobilization, field staff, equipment for a project with low equipment usage, tools and plant, temporary construction, guards and watchmen, cleanup and housekeeping, permits and licenses, bonding and insurance, safety and transportation, and so on. Feedback on such costs is in the form of a monthly statement comparing expenditure on these work items with provisions in the budget.

14.4.7 Overhead Costs

Salaries of administrative, design, and engineering staff and office, stationery, telephone, traveling, and similar expenditures incurred by headquarters are collected under overhead costs as different work item subheadings. These are distributable over several projects.

Standard overhead percentage is derived from the ratio between the overhead costs and normal business volume. The overhead cost for a project is determined from this percentage and direct project costs. Periodically the overhead cost is distributed over the total value of work done on all projects; adjustments for the difference between the standard percentage and the actual overhead are applied to the projects in proportion to the work performed, and an overhead percentage chargeable to projects is determined for the next accounting period.

Like indirect costs, the feedback for control of overhead consists of a monthly statement comparing actual expenditure incurred on overhead work items with their budget provisions.

14.4.8 Work Items Cost

The experience gained in charging resource costs to the various work items, thus arriving at the full cost of work items, can be used in planning projects in the future. However, man-hours, equipment hours, and materials must be converted to dollars.

A work item cost status report, shown in Figure 14.14, must be generated. The report also lists separately the estimated and actual man-hours to date and man-hours for the total project as well as the overrun or underrun. It also shows to date and totals at completion estimated cost, forecast to complete, and the projected overrun or underrun, depending on whether the forecast to complete exceeds or falls short of the estimated cost. The dollar cost includes cost of man-hours, equipment-hours, and materials. In case of a serious overrun, corresponding man-hours are given under "to date" and "totals at completion." If the reported man-hours do not adequately explain the cause of the overrun, man-hours by craft, equipment-hours, or the materials report, a variance analysis is made. Normally, the work item cost report is sufficient for reporting status and cause of overrun. Man-hours, equipment-hours, and materials reports serve as backup reports when required. When functional subelement costs must be reported, the work items are grouped together in a vertical column as a subelement of the cost breakdown. A report similar to the work item cost report is obtained.

14.5 PROJECT PERFORMANCE CONTROL

Monitoring the actual progress and cost individually against schedule and budget does not alone ensure performance will proceed according to plan. Even when progress is ahead of schedule, the project may be running over budget, resulting in below par performance. It is therefore necessary to make the following comparisons for performance control:

1. For project and budget performance. Percent expenditure on the project versus the percent completed; remaining budget versus forecast requirements.
2. For production rate. Planned versus current production rates.
3. For productivity rate. Planned versus current productivity versus forecast at completion.

CONTRACT DESCRIPTION	RESPONSIBLE ORGANIZATION	CONTRACT NO.	REPORT DATES
MARINE LABORATORY	AREA V	ML45	TERM (SPAN) CUTOFF DATE: 30 MAY 93 RELEASE DATE: 03 JUNE 93

SEVERINO CONSTRUCTION

WORK ITEM IDENTIFICATION	WORKER-HOURS TO DATE		TOTAL WORKER-HOURS AT COMPLETION			TOTAL COST* TO DATE		TOTAL COSTS AT COMPLETION		
	ESTIMATE	ACTUAL	ESTIMATE	FORECAST TO COMPLETE	PROJECTED (OVERRUN) UNDERRUN	ESTIMATE	ACTUAL	ESTIMATE	FORECAST TO COMPLETE	PROJECTED (OVERRUN) UNDERRUN
BA03200	1234	1210	1824	1868	(44)	514	504	760	778	(18)
BB03200	931	931	1320	1347	(27)	321	315	455	464	(9)
BC03200	1123	1126	1474	1480	(6)	416	417	546	548	(2)
CB03200	856	838	1221	1180	41	276	270	394	388	6
FA03200	997	993	1271	1300	(29)	302	301	385	385	0
Total	5141	5080	7110	7175	(65)	997	1807	2540	2563	(23)

* In thousands of dollars

FIGURE 14.14 Work Item Cost Status Report

14.5.1 Earned Value Concept

There are many ways of measuring progress, each requiring a certain amount of effort. The earned value approach requires a reasonably accurate measure of progress. The better the measure, the better the results.

The earned value is based on the measured amount of work completed. This quantity is multiplied by the productivity rate used for estimating. For example, if 100 m^2 of masonry wall has been completed and the productivity value used for estimating was 2 man-hours per m^2, the earned value is $100\ m^2 \times 2$ man-hours/m^2 = 200 man-hours, irrespective of the actual man-hours expenditure. A variance occurs if the earned value is different from the actual expenditure.

Rules of credit can be predetermined for partially completed items, and credit is taken only after completion of an interim milestone. For example, credit can be taken for electrical work when wire has been pulled (50%), terminated (30%) and tested (20%).

14.5.2 Project and Budget Performance

To evaluate project performance effectively, management needs the budgeted cost of scheduled work, committed cost for work performed, and earned cost for work performed, which are broken down in terms of the major parts of the project. Commitments and obligations are also added to the costs. When an order is placed for material or equipment, a commitment is made to the supplier and is recorded in a Commitment Register. When work has been done, or material and equipment has been supplied, the payment becomes due although it is not actually made. This is an obligation or incurred cost. A sum of these costs is entered under the actual cost. A Project Status Report, as illustrated in Figure 14.15, is prepared at regular intervals.

Analyzing the data presented in Figure 14.15, each cost center is considered separately, with actual cost of work performed compared against budget allocations up to the time of the report. The report also shows overruns and underruns and predicted costs at completion.

In Figure 14.16, thirteen possible combinations are shown, using x, $x + y$, and $x + y + z$ as all positive real numbers such that $x + y + z > x + y > x$. A comparison is made among the three qualities in terms of the effect on schedule and cost. This interpretation is noted on each line.

Similarly, the estimated forecast cost at completion can be compared with the budgeted cost at completion leading to the interpretation noted in Figure 14.17. The schedule and cost information from a project is plotted in Figure 14.18.

The cost variance is the difference between the budgeted cost of work performed minus the actual cost of work performed. The schedule variance represents the difference between the budgeted cost of work performed and

Page 1A of 1B　　Page 1B of 1B

PARADE (R) Primavera Systems, Inc.

Title:

COST PERFORMANCE REPORT - WORK BREAKDOWN STRUCTURE

Start Date: 1APR93　Period End Date: 31MAY93

Report Date: 15DEC93

Report Period: 2

ITEM	CURRENT PERIOD					CUMULATIVE TO DATE					AT COMPLETION		
	BUDGETED COST		ACTUAL COST WORK PERFORMED	VARIANCE		BUDGETED COST		ACTUAL COST WORK PERFORMED	VARIANCE		BUDGETED	LATEST REVISED ESTIMATE	VARIANCE
	Work Scheduled	Work Performed		Schedule	Cost	Work Scheduled	Work Performed		Schedule	Cost			
General Conditions 1.1	.0	.0	.0	.0	.0	10.0	10.0	12.0	.0	-2.0	10.0	12.0	-2.0
Site Work 1.2	6.5	.0	.0	-6.5	.0	9.4	1.2	2.2	-8.2	-1.0	26.2	26.2	.1
Foundation 1.3	14.5	16.5	13.7	2.0	2.8	17.7	16.5	13.7	-1.1	2.8	19.5	17.2	2.2
Structure 1.4	.0	.0	.0	.0	.0	.0	.0	.0	.0	.0	7.7	7.7	.0
Electrical Contractor 1.5	.0	.0	.0	.0	.0	.0	.0	.0	.0	.0	1.6	1.6	.0
Mechanical Contractor 1.6	.0	.0	.0	.0	.0	.0	.0	.0	.0	.0	1.9	6.3	-4.4
Finishing 1.7	.0	.0	.0	.0	.0	.0	.0	.0	.0	.0	2.5	2.5	.0
COST OF MONEY	.0	.0	.0	.0	.0	.0	.0	.0	.0	.0	.0	.0	.0
GEN AND ADMIN	.0	.0	.0	.0	.0	.0	.0	.0	.0	.0	.0	.0	.0
UNDISTRIBUTED BUDGET	XXXXXXXX	XXXXXXXX	XXXXXXXX	XXXXXXXX	XXXXXXXX	XXXXXXXX	XXXXXXXX	XXXXXXXX	XXXXXXXX	XXXXXXXX	N/A	N/A	XXXXXXXXXX
SUBTOTAL	21.1	16.5	13.7	-4.6	2.8	37.1	27.7	27.9	-9.4	-.2	69.3	73.5	-4.2
MANAGEMENT RESERVE	XXXXXXXX	XXXXXXXX	XXXXXXXX	XXXXXXXX	XXXXXXXX	XXXXXXXX	XXXXXXXX	XXXXXXXX	XXXXXXXX	XXXXXXXX	.0	.0	.0
TOTAL	21.1	16.5	13.7	-4.6	2.8	37.1	27.7	27.9	-9.4	-.2	69.3	73.5	-4.2

(ALL ITEMS IN 1000)

FIGURE 14.15 Example of a Project Performance Report

Combin-ation	Budgeted Cost of Work	Actual Cost of Work Performed $	Earned Cost of Work Performance $	Interpretation	
				Schedule	Cost
1	x	x	x	At Par	At Par
2	x	x + y	x	At Par	Overrun
3	x	x	x + y	Ahead	Underrun
4	x	x + y + z	x + y	Ahead	Overrun
5	x	x + y	x + y + z	Ahead	Underrun
6	x	x + y	x + y	Ahead	At Par
7	x + y	x	x	Behind	At Par
8	x + y	x + y	x	Behind	Overrun
9	x + y	x	x + y	At Par	Underrun
10	x + y	x + y + z	x	Behind	Overrun
11	x + y	x	x + y + z	Ahead	Underrun
12	x + y + z	x + y	x	Behind	Overrun
13	x + y + z	x	x + y	Behind	Underrun

FIGURE 14.16 Actual vs. Earned Cost Comparisons

the budgeted cost of work scheduled. The project overrun is denoted by the difference between the estimated cost (forecast) at completion and the budgeted cost at completion. Besides cost, the graph also shows the new completion date and therefore the projected slippage.

This graph will be useful as long as both the budgeted rate of expenditure and production rate are in proportion to the man-hours used, which is not always the case. Hence it is necessary to consider production rate.

14.5.3 Production Rate and Productivity

Most projects have one or more repetitive activities that are very critical to the timely completion of the project. For instance, rock fill in a dam construc-

Budget Cost at Completion $	Estimate of Cost (forecast) at Completion $	Interpretation
x	x	Forecast on Budget
x + y	x	Forecast Underrun
x	x + y	Forecast Overrun

FIGURE 14.17 Budget vs. Forecast Cost Comparisons

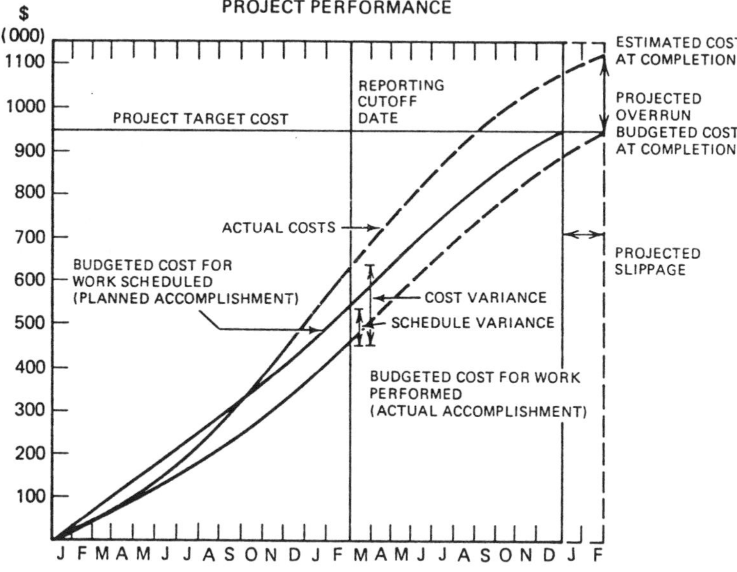

FIGURE 14.18 Performance Curves

tion project, pipe installation in a processing plant, cast-in-place concrete in a concrete structure building project, and so on. A monthly plot of actual cubic meters of rock fill, meters of pipe installed, or cubic meters of cast-in-place concrete against the estimated quantities gives a very quick, but fairly reliable, indication of whether or not the project will finish on time. If the man-hours are being monitored, the plot also indicates whether the project will finish on schedule.

The project performance curve in Figure 14.18 is for dollars versus time on a cumulative basis. Another useful technique for selected items is to review the incremental units versus time. For example total labor, or labor for a key activity such as labor for formwork, concrete pouring, and so on, can be plotted for each week or month.

Figure 14.19 shows such an incremental plot for the quantity of cast in place concrete in cubic meters budgeted and actually poured every month on a construction project. When studied in conjunction with the project performance curve, such plots give a picture not only of actual status with respect to cost but also progress. These plots provide an indication of short- and long-term trends, which are important control tools.

A healthy production rate in fact may have an overrun in the man-hours estimate. This can only be detected if production is divided by man-hours expended for each item of work to determine current productivity. A forecast of productivity is then made, which considers the learning effect that occurs during repetitive operations. This can be done regularly at specified intervals.

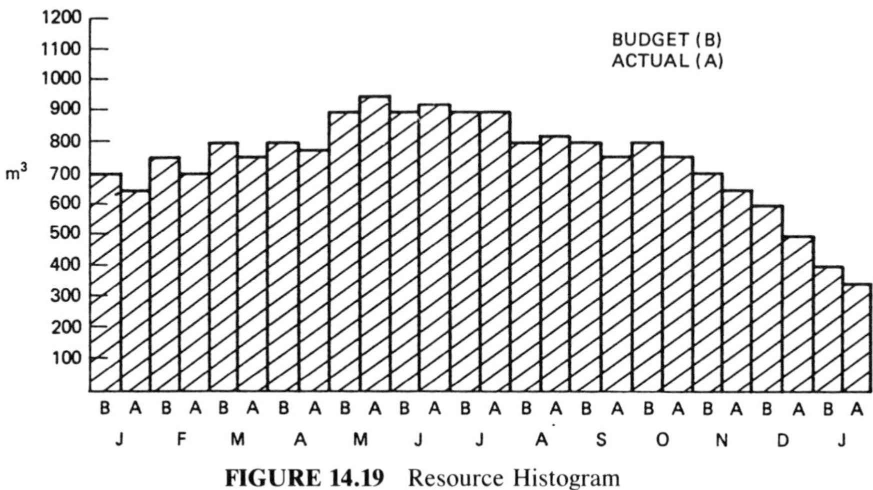

FIGURE 14.19 Resource Histogram

14.6 PROJECT STATUS REPORT

The cause of overruns or underruns is not always readily apparent. To exercise control requires an analysis of the contributing factors.

Material costs are equal to the product of quantity, that is usage, times price. A material cost variance is attributable to a variation in price and/or quantity used. The following example illustrates material variance analysis. (Similar analyses can be made for variances in labor and overhead costs.) Total variance is \$2900 overrun; Budget Price = \$100 m^3; Actual Price = \$98/m^3; Budget Volume = 1000 m^3; and Actual Volume = 1050 m^3.

Budget Vol. × Budget Price	Actual Vol. × Budget Price	Act. Vol. × Act. Price
1000 × 1000 = 100,000	1050 × 100 = 105,000	1050 × 98 = 102,900

Volume Variance = +5000 Price Variance = −2100

Net Variance = +\$2900

Two factors that affect labor costs are labor rates and efficiency. Labor rates include wages plus the many components of payroll burden which can give rise to variances from budget.

Estimators utilize an average labor rate which can be affected by the wage itself and the mix of tradesmen, apprentices, and unskilled workers.

Labor efficiency, which is a measure of output to input, is affected by many factors categorized as follows:

1. Morale related and motivational—which include quality of management, supervision, communication and information, level of training, absenteeism, and the frustration effect of changes.

2. Regional—such as experience level of the workers, quantity and quality of labor supply, and local customs such as coffee breaks.
3. Work scheduling and planning considerations—include the realism of schedules and deadlines, use of overtime and shiftwork, sequence of work and material flow. Craft mix affects efficiency if the mix of apprentices to craftsmen is not at an optimum.
4. Environmental and climatic conditions.
5. Job characteristics—such as size of project, complexity, and repetitiveness.
6. Changes—as they affect job rhythm, disruptions, and acceleration.

These categories of factors are not intended to be exhaustive, but they do show the many factors that could impact labor efficiency and contribute to variances.

14.7 FINANCIAL CONTROL BY OWNER

Project performance can be guided by controlling progress, labor, equipment, material, overhead, and other expenditures, as discussed in the preceding sections. An owner might show relatively little hesitation in awarding a lump-sum contract for a short project in which his or her liability would be fixed with no additional costs. However, a different situation prevails if the owner's project extends over a very long period, as in a hydropower project. In projects with a long time span, there are several contract packages passing through different phases, each getting its share of escalation and additional cost resulting from a change in scope. Because the owner can obtain progress reports at selected milestones for any contract package, as well as for the entire project, progress control is not difficult. The same, however, may not be said for financial control. Here the owner needs control over engineering design, engineering supervision, construction, and contract package costs, and the owner must keep the total project within budget without getting involved with details that fall under the project manager's jurisdiction. Control must also be exercised over escalation and contingencies to minimize project cost. Among the questions that arise are the following: How is the project budget controlled? What is the best procedure for financial control? These questions are discussed in this section.

14.7.1 Owner's Control Over Project Budget

A project owner is vitally concerned with budgeted costs, not only in monitoring but also in controlling them. The owner's prime concern is control of the overall project budget. This necessitates periodical revisions of various estimates to ensure that contract package costs at actual market prices do not exceed the original estimate plus authorized changes. Thus the owner's sights must be firmly fixed on the project budget, and his or her management actions must focus on keeping the project cost within this budget.

The owner's project budget comprises the base capital cost and the escalation, contingency, design, and engineering management costs for the different parts into which a project is divided. The base capital cost is based on current estimates and quotations without provision for contingency and escalation. Consider, for example, a regional development program where facilities such as water supply and sewage disposal constitute the different parts of the project. Each facility has a capital cost estimate based on the current dollar value at the time of estimating. The escalation allowance is calculated for the time span over which work is expected to be performed on this facility. Contingency, is calculated as a percentage of the base construction estimate. Engineering design and management are also calculated on a percentage basis, but costs for these items are better determined by detailed estimates whenever possible.

As the work for each facility is parceled into contract packages, the facility base cost estimate is distributed over these packages. Escalation is assigned relative to the time span of each contract package. Any saving from escalation or excessive demand on its results in a corresponding addition to or subtraction from the contract package contingency. These differences in contingency provision for contract packages can add up to a positive or negative total for the project contingency account, indicating an overrun or underrun in the current estimated project cost as compared with the original budget. Engineering design and engineering management costs are prorated or assigned as required to the associated contract packages of each facility.

Since the contract packages are worked on progressively, the estimated cost of each package is ultimately transformed into actual cost when work is complete. Before this takes place, there is an intermediate stage when work on a contract package is in progress; its cost is partly committed and partly an estimated forecast of cost to complete. The total project budget at this stage is a combination of estimates, commitments, and actuals for the contract packages in design, in progress, and already completed. The breakdown into base construction cost, escalation, contingency, engineering design, and engineering management contract packages is maintained throughout the life of the project.

Figure 14.20 illustrates a project budget status report for a regional development program. The original estimated cost consists of base construction, escalation, contingency, engineering design, and engineering management. Since the project estimate is periodically reviewed, all figures except budget cost and base construction cost are susceptible to change. Initially, all contract packages are in the design stage, so a project budget contains their estimated cost only. Later on, the contract packages are divided into three groups: work not started, work in progress, and work completed. Estimated cost figures are used for work not started, commitments and forecasts for work in progress, and actual costs for completed work.

CONTRACT PACKAGES	ORIGINAL BUDGETED COST	BASE CONSTRUCTION COST	ESCALATION	CONTINGENCY ORIGINAL PROVISION	CONTINGENCY (+)	CONTINGENCY (−)	ENGINEERING DESIGN	ENGINEERING MANAGEMENT	REVISED ESTIMATED COST
(1)	(2)	(3)	(4)	(5*a*)	(5*b*)	(5*c*)	(6)	(7)	(8)
Work not started									
Water intake	6.65	5.0	0.65	0.25	0.30		0.20	0.25	6.35
Sewage treatment plant	19.60	15.0	3.0	0.75		0.50	0.60	0.75	20.10
Work completed									
Sewage transmission	26.20	20.0	3.0	1.00	0.40		0.80	1.00	25.80
Water transmission	32.55	25.0	3.75	1.25	0.30		1.00	1.25	32.25
Work completed									
Reservoir development	1.73	1.20	0.06	0.06	0.30		0.05	0.06	1.43
Outfall	1.12	1.00	0.03	0.03		0.03	0.04	0.05	1.15
Total cost	87.85	67.20	10.49	3.34	1.30	0.53	2.69	3.36	87.08

FIGURE 14.20 Project Budget Status Report
(all costs in millions of dollars)

Every month the status of the project estimate is reviewed, and escalation is assigned on the basis of the next expected time frame. Contingency is estimated from the perceived scope changes. Then the revised estimate is worked out.

When a job is in progress, no more changes to its estimates of escalation, engineering design, and engineering management are necessary except when their scope changes. Such changes in turn necessitate changes in the revised estimated cost and consequently changes in the contingency provisions (columns 5*b* and 5*c*). A revised estimate for work in progress is the forecast of final cost.

No more changes are necessary in the cost of completed jobs, and the final revised estimate is equal to the actual contract package cost. Similarly, actuals are entered under engineering design, engineering management, escalation, and contingency.

The project budget status report presents for comparison the total revised project estimate and the original project budget as well as revised and original estimates on an individual contract package level. The difference between the revised estimate (column 8) and the original budget (column 2) is the savings or overrun in the original budget cost. This difference is subtracted or added to the contingency fund, as shown under columns 5*b* and 5*c*. In the report presented here, the total original estimated project cost is $87.85 million. The present revised total project cost estimate stands at $87.08 million. The difference of $0.77 million between the two estimates is the savings in the contingency fund [$1.30 million (column 5*b*) − $0.53 million (column 5*c*)]. The costs for contract packages are shown in the same way.

Contingency is an expected cost and the fund provision contingency must be carefully managed. Owners with assistance from contractors must periodically assess the risk and the amount of contingency required. As the project components are completed, the amount of contingency required is reduced, except for major potential developments. During the course of the project, contingency should be "drawn down," that is reduced to zero except for provision for latent defects.

14.7.2 Procedure for Financial Control

Since, as is evident from the preceding discussion, the revised estimate is subject to constant change, comparing actual cost with a changing baseline may be misleading. Hence the question arises whether the actual cost should be controlled against the original budget, which does not change. Consider a contract package with an original estimate of $2.0 million. A pretender revision shows that this contract package can be finished for $2.1 million. The lowest bid received on this contract is for $1.8 million. Obviously, project management will be authorized a budget of $1.8 million and told by the owner to administer this contract within the budget. This indicates the necessity of setting up a system of appropriations by contract packages.

Before design of a contract package is started, appropriations are made by the owner, usually based on preliminary or less than complete information. Expenditure has to be kept within the appropriation for the contract package. When a lump-sum contract is awarded or a purchase order is signed, an appropriation is being made for a definite commitment. Because most parts of project work are of a contractual nature, the owner can exercise effective control over cost by judicious control over the multitude of monetary commitments involved. Once any commitment is made, monitoring of work and the recommendation of progress payments become strictly a project management function. Where work is being done by in-house forces or on a cost-plus basis, besides appropriation control, a closer monitoring of progress and expenditure by the owner is required to avoid runaway costs.

Requests for appropriations are received from project management when contract package work is about to become a firm commitment. The estimate for engineering design of a contract package is established when the contract package scope is firmed up. On a request from the project manager, the owner appropriates the amount to be spent on specific contract packages. If this design is contracted out by the project manager, the appropriation is equal to the contract commitment. If it is carried out by in-house engineering departments of the project management organization, only the expenditure incurred, as determined from progress and manpower reports, becomes the commitment, with the balance being the forecast to complete.

When a tender package is ready, bids are invited and compared with the pretender estimate for the contract package. On recommendation by the project manager, appropriation is made by the owner from the base construction cost and, if the situation warrants it, from escalation and contingency. Unlike a lump-sum contract whose total contract amount becomes the commitment, in a unit price contract the estimated quantities and accepted unit prices are treated as individual cost commitments. In a cost-plus contract the escalated base construction cost is the committed cost. By the time of tender award the request for engineering management is received and an appropriation is made. Thus, separate appropriations are made for each contract package. These are for engineering design, construction, and engineering management costs. Besides these additional appropriations are made for change orders and major equipment purchase orders. These appropriations are also assigned to contract packages.

The project manager becomes responsible for reporting expenditure separately against these appropriations. Here the dichotomy between the project control and financial control becomes distinct. A project manager controls expenditure against appropriations. The owner controls appropriations against budgeted cost. By exercising control over appropriations, the owner retains control over any possible savings from the project estimate.

Viewed from another perspective, an appropriation can be considered as a budget for a contract package allocated out of the project budget by the owner. It is a budget for the project management team responsible for controlling expenditures within this budget. Since several appropriations are in force at any time, a forecast can be made for the funds required during the year, which is termed the budget for the year.

As the project progresses, invoices, progress payments, and forecasts to complete for each contract package are reported against its appropriations. This reporting is carried out periodically for the benefit of the owner and is presented in the format such as shown in Figure 14.21. The information contained in this owner project status report is partly derived from the status reports prepared for the project management, an example of which was given previously in Figure 14.15. Figure 14.21 compares estimated cost and appropriations for each contract package. When all appropriations are made, the column for unappropriated funds is blank. It also compares the forecast of final cost with the corresponding appropriation. Column 5 shows commitments. As invoices and progress bills are processed, commitments become incurred costs.

For a contract package the sum of columns 3 and 4 in Figure 14.21 is equal to column 7. That contract package overrun/underrun is the difference between the estimated cost and the sum of the appropriated and inappropriate funds: (2) − (3 + 4), as shown in Column 8. The difference between the forecast to complete and the appropriation (3 + 4) − (7), is shown for each appropriation in column 9. The report also shows the status of the project by contract package totals at the bottom.

Change orders affect the scope of work and therefore change project cost. The management of changes is a major part of every project. Control of changes is of paramount importance and is discussed earlier in this Chapter.

14.8 COLLECTING INFORMATION FOR PROJECT CONTROL

Project implementation strategy must include procedures for collecting information on project performance which is vital for monitoring and control. Information requirement therefore must be considered thoroughly vis-a-vis the project organization. On a lump-sum contract the owner may require only a record of progress payments and change orders. For projects executed by in-house forces, information on manpower, equipment, materials, overhead costs, and change orders is necessary. On unit price contracts information on the quantities of work performed is essential. For a project organization having a mix of different types of contracts and in-house work, a thorough study of information requirements before starting the works prevents future chaos. Information collection for project control is therefore described in general so that it can be designed specifically, combining elements required for a certain project organization.

On any project the first task is to develop a sound design. Time and cost planning functions are carried out concurrently followed by control. The elements involved in control are schedule and estimate; progress reports,

CONTRACT PACKAGE	ESTIMATED COST	UNAPPROPRIATED	APPROPRIATION	COMMITMENT	INCURRED COST	FORECAST OF FINAL COST	CONTRACT PACKAGE (OVERRUN) UNDERRUN (2) − (3 + 4)	APPROPRIATION (OVERRUN) UNDERRUN (3 + 4) − (7)
(1)	(2)	(3)	(4)	(5)	(6)	(7)	(8)	(9)
Work not started								
Water supply	**6.65**	**6.15**	**0.20**		**0.15**	**6.35**	**0.30**	
Sewage treatment plant	**19.60**	**17.00**	**3.10**	**2.50**	**0.40**	**20.10**	**(0.50)**	
Work in progress								
Sewage transmission	**26.20**	**3.95**	**21.85**	**20.05**	**11.68**	**25.80**	**0.40**	
Construction	24.40	3.95	20.00	20.00	10.42	23.91		0.04
Engineering design	0.80		0.80		0.75	0.75		0.05
Engineering management	1.00		1.00		0.51	1.09		(0.09)
Change order			0.05	0.05		0.05		
Water transmission	**32.55**		**32.25**	**25.00**	**17.09**	**32.25**	**0.30**	
Construction	30.30		30.00	25.00	15.40	29.82		0.18
Engineering design	1.00		1.00		1.05	1.05		(0.05)
Engineering management	1.25		1.25		0.64	1.38		(0.13)
Work completed								
Reservoir development	**1.73**		**1.43**		**1.43**	**1.43**	**0.30**	
Construction	1.62		1.26		1.26	1.26		
Engineering design	0.05		0.08		0.08	0.08		
Engineering management	0.06		0.09		0.09	0.09		
Outfall	**1.12**		**1.15**		**1.15**	**1.15**	**(0.03)**	
Construction	1.03		1.08		1.08	1.08		
Engineering design	0.04		0.03		0.03	0.03		
Engineering management	0.05		0.04		0.04	0.04		
Total	**87.85**	**27.10**	**59.98**	**47.55**	**31.90**	**87.08**	**0.77**	

FIGURE 14.21 Regional Development Program—Owner Project Status Report

manpower, equipment, and material costs; lump-sum contract costs; other contract costs; indirect costs; overhead costs; and change orders. They are interrelated through the project control function.

Each element of the project must be monitored by the project manager, who is responsible for project control. The project manager must have adequate feedback on the project so that corrective action can be applied when and where it is needed. For instance, if manpower costs escalate, material deliveries are delayed, equipment or overhead costs increase, subcontractors fail to meet deadlines, or the safety factor is affected in some way, it may be necessary to revise project plans. This is accomplished by the issuance of change orders. These change orders go through the project control function and subsequent action is taken on all other elements. In effect, then, the various elements of a project are all interrelated through the project control function, and each element must be linked with this function through the cost coding structure and adequate information flow channels. Besides these functions, project control is also responsible for quality, safety, human relations, and many other functions, all of which comprise part of the project plan.

The benefits of free flow information are many: (1) removal of misconceptions about the objectives and policies of project management, (2) improved interpersonal relationships, (3) improved interaction and cooperation between the various disciplines involved, and (4) improved morale and attitudes leading to increased performance and productivity.

The importance of reporting promptly and accurately needs to be emphasized to all contributors. This reporting function appears to divert effort from the main task of constructing the project and management may be required to remind the personnel of the value of good reporting. Errors in costing occur because of incorrect application of cost code numbers, and occasionally due to manipulation of data in order to mask the actual costs. This misinformation is misleading and distorts both the present control and future estimating purposes.

It should be remembered that the feedback loop in construction projects is not electronic but human.

The scheduling and cost engineers cannot passively depend on progress meetings or written progress and cost reports to obtain crucial information on project performance. They are expected to have prior knowledge of the problems; reports and meetings should only confirm their assessment of the problems. Actively appraising the developments and continuously updating the project plan in light of the latest information are signs of a successful information-gathering effort.

There are numerous sources from which scheduling and cost engineers can obtain their information. These are the project engineers, design engineers, estimators, quantity surveyors, site resident engineers, consultants, contractors, subcontractors, and vendors. They must also have up to date knowledge about the project and its environment. They must be involved in the analysis of progress reports and have first-hand knowledge of bid data tabulations, material takeoffs, and claims settlement. They must remain in contact with the accounting department and be informed about expenditures incurred and in-house charges reversed; contact with procurement personnel is necessary to find out about commitments made through purchase orders, work orders, contracts awarded, and contracts bound by letters of interest. All of this is necessary for monitoring and controlling the project.

14.9 HISTORICAL DATA

It is also necessary to collect data for projects to be carried out in the future. Suppose a project involves the driving of 500 piles. If it takes an estimated 8-hour day to drive one pile, the estimated project duration is 500 days. On the other hand, if it takes two days for one pile, total estimated duration is 1000 days. If the construction engineer does not have a log of a previous project and has to estimate from memory, the tendency is to remember the worst experience and to use the most pessimistic estimate. The planner may seek guidance from the past experience of others, as published in handbooks, but at best the planner obtains the mean estimate, which may not be relevant to the project. The most suitable data can be obtained from a record of the work actually performed by the organization if procedures exist for automatically storing experience from one project for use on future projects.

Although the estimator is not responsible for the execution of a job, the daily record of each work item should be accessible to this person. When the work is finished, the estimator has a record of man-hours, equipment usage, material consumption, overhead expenditure, comments about special difficulties, the slow progress in the beginning, the learning curve, and normal progress. The following data for all work performed by in-house forces will aid the estimator greatly in estimating the cost and duration of similar work items on future projects:

1. Crew composition and man-hours per unit measure.
2. Equipment usage per unit measure.
3. Materials consumed per unit measure.
4. Overhead expenditure on various items as a percentage of the total job cost.

Historical data are collected on the assumption that the information will be used on a similar project. But every project is different, and no project can be an example of perfect control. Valuable lessons can also be learned from a project's distinct nature. Therefore an analysis should be carried out not later than one month after project completion, which should include the following:

1. An "as-built" CPM network of the entire design through construction stages showing the actual work sequence and activity durations used.
2. An outline of problems encountered.
3. An evaluation of consultants', contractors', and subcontractors' performances in relation to the schedule.
4. A statement comparing subelement cost in the design estimates with the actual amounts of completion.

This information is useful for reliable scheduling and accurate estimating of projects in the future. Monitoring and control have no meaning without this information.

PROBLEMS

14.1 Describe the variables that affect productivity.

14.2 How will you enhance the reliability of your schedule after an update in view of the following situations?

(a) A possible labor strike. (You are working for an owner and the work is preformed by a contractor).

(b) Abnormally bad weather. (You are working for a contractor.)

(c) A hazard such as overtopping of a dam. (You are working for a utility company and the work is performed by a contractor.)

CHAPTER 15

PROGRAM EVALUATION AND REVIEW TECHNIQUE (PERT)

The two network methods described in the previous chapters (CPM and PDM) do not provide a measure of uncertainty associated with the estimate of a particular milestone in a project. They are deterministic tools. The duration of each activity can assume only one value, which is assumed to be realized in the field. Consequently the two methods provide the project manager with one estimate of the project completion time (or any milestone event).

The time required to complete a construction activity depends on many factors. Consider for example the activity "formwork placement for continuous footings in a building" on a given project. The time required to complete this activity given a certain resource configuration is 4 days. If the weather is bad or the labor productivity encountered is low, the activity may take up to 8 days. When the conditions are favorable as few as 3 days may be required, but not any less. In fact the 4-day time requirement assumed for the CPM analysis was only the most likely duration to take place. The program evaluation and review technique (PERT) is about incorporating this uncertainty of the duration estimate in the overall network analysis. If each activity in the network is no longer deterministic (i.e., one value) but rather probabilistic, we can better estimate the uncertainty associated with achieving a given schedule.

The PERT method was first used in planning the development of the Polaris Weapon System (USA Navy, 1958), where schedules were continuously missed or considerably differed from the estimated ones. The concept was based on breaking the project down into individual components (activities), probabilistically estimating the times required to complete the work component and defining the precedence relations between them, performing a simple network analysis, and estimating the project completion time with an associated probability distribution.

The PERT method is a simplification of the risk analysis procedure covered in the next chapter under simulation methods. In essence, the duration of each activity will be estimated using a three-time estimate reflecting the pessimistic, optimistic, and most likely values of the duration. Once this is accomplished the mean and variance for each activity time is estimated and used to find the mean and variance of the project completion time. Knowing these parameters will enable one to approximate the probability of completing the project in a particular time frame and to estimate other "risk assessment" measures.

15.1 AN OVERVIEW OF PERT

Consider the network given in Figure 15.1. The CPM analysis described in Chapter 5 and shown on the network of Figure 15.1 illustrates that the project will be completed in 42 days. This is the case assuming that the activity durations indicated on the network were exactly realized during the construction of the project.

To illustrate the application of PERT, suppose that bad weather was encountered during the first week, extending the duration of activity 1–2 (Clear site) from 10 to 16 days. The updated network analysis shows that the project will now require 46 days to complete. Likewise one can envision many scenarios yielding many different estimates of the project duration. In risk analysis we attempt to measure the uncertainty content of a given estima-

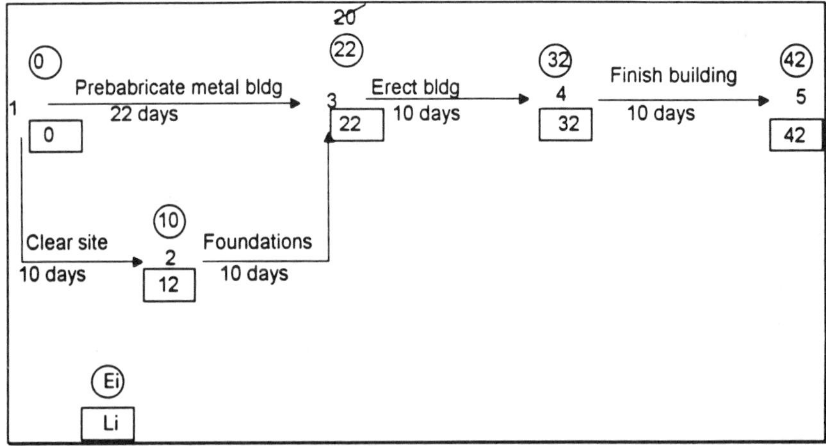

FIGURE 15.1 Sample CPM Network of a Convenience Store

tor, in this case the duration of the activity. During planning one may envision that productivity on activity 1–2 could be hindered by bad weather or other effects and, therefore, its duration estimate is associated with some degree of uncertainty. In order to quantify this, PERT allows the scheduler to estimate the duration of an activity using three values rather than one deterministic estimate. These will include an estimate of the duration if all conditions are favorable (i.e., an optimistic estimate), a duration with all conditions unfavorable (i.e., pessimistic estimate), and one that reflects the most likely conditions (i.e., most likely estimate). For activity 1–2 this would be in the form

$$\text{Duration} = 8,\ 10,\ 20$$

This implies that the site clearing activity (1–2) may take from 8 to 20 days to complete with a most likely value of 10 days depending on the conditions encountered. Now assume that we repeated the same procedure with all activities in the network of Figure 15.1, resulting in the duration estimates given in Table 15.1.

It should be obvious that once the computations for project completion time are carried we will end up with more than one estimate for the project time. If the most likely value of each activity time is used, the project will consume 42 days. Should the pessimistic estimate of each activity be used, the project will take 56 days to complete, and 37 days if the most optimistic values are used. In fact, the durations may be combined in many ways, thus making it virtually impossible to analyze the project duration without using statistical methods.

TABLE 15.1 Three-Time Estimates for Sample Project

Activity	Source of Uncertainty	Optimistic Estimate	Most Likely Estimate	Pessimistic Estimate
1–3 (Order/prefabricate mtl bldg)	Past experience with supplier	20	22	25
1–2 (Clear site)	Weather	5	10	15
2–3 (Underground and foundation)	Unforeseen conditions underground may be unfavorable	5	10	15
3–4 (Erect prefab bldg)	Possibility of cold weather, precipitation, etc.	8	10	20
4–5 (Finish interior)	Labor productivity	9	10	11

15.2 PERT METHOD

Modeling a project schedule using PERT is very similar to the CPM method discussed in Chapter 5. The method may be summarized in three steps, as discussed below.

15.2.1 Development of the PERT Network Model

A project is first divided into its component activities. Activity definition, precedence relations, and network building is similar to the CPM method. An activity is represented by an arc connecting two nodes. The nodes represent events in time denoting the completion of certain activities and the start of others. The project network may be arbitrarily connected by one start node and one terminal node.

Many variations for network representations exist. It is possible to use activity on node to represent the network without any loss of significance. We will adhere to the activity on arrow representation in the spirit of the original work on PERT. Networks may be also represented using various simulation modeling tools, as will be demonstrated in the following chapters.

15.2.2 Estimating Activity Time Distributions

In order to represent the stochastic (random) nature of activity times, PERT assigns a statistical distribution model for the time consumed in accomplishing a given activity. To facilitate hand computation procedures certain assumptions were adopted reducing the distribution representation for an activity i–j to its mean μ_{ij} and its variance σ_{ij}^2. These are approximated as shown in Equations 15.1 and 15.2.

$$\mu_{ij} = (D_{\text{opt}} + 4D_{\text{ml}} + D_{\text{pes}})/6 \tag{15.1}$$

$$\sigma_{ij}^2 = [D_{\text{pes}} - D_{\text{opt}})/6]^2 \tag{15.2}$$

where D_{opt} represents the optimistic, D_{ml} the most likely, and D_{pes} the pessimistic estimate of the activity time.

15.2.3 Network Computations

Forward Pass Calculations. Consider the network shown in Figure 15.2. The algorithm for manual computations of event times on the forward path can be summarized as follows:

Set the first event at an arbitrary time (conveniently set at zero). Any event time E_j may then be computed as being the maximum time of all

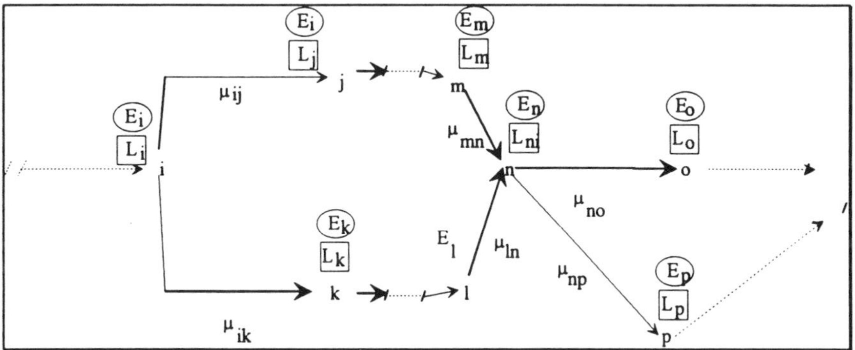

FIGURE 15.2 Generic Portion of a PERT Network

preceding events added to the appropriate mean activity time. For a simple connection *i–j* as shown in Figure 15.2, the calculation are then given by

$$E_j = E_i + \mu_{ij}$$

For a merge node like node *n* in Figure 15.2 the maximum of the event times computed from all incident nodes will be used as follows:

$$E_n = \text{MAX } \{(E_m + \mu_{mn}), (\text{E}_1 + \mu_{1n})\}$$

The mean project completion time corresponds to the latest terminal event (maximum event time if more than one terminal node). It will be the longest path connecting the origin node and the terminal node. In other words, if there are *n* paths T_i connecting the origin and terminal nodes, each path will have the following computed values:

$$T_1 = \mu_{ij} + \ldots \mu_{mn} + \mu_{no} + \ldots$$

$$T_2 = \mu_{ik} + \ldots \mu_{1n} + \mu_{no} + \ldots$$

$$T_i = \ldots$$

And the mean project time is given by

$$\overline{T} = \text{Max } T_i$$

The path that determined this maximum value $\overline{T}$ is termed the critical path. The sum of the variances of activities comprising the critical path is the variance of the project time.

Backward Pass Calculations. First the latest event time of the terminal node is set equal to the project mean time or any other desired project completion time. The late event times are computed on the backward path (refer to Fig. 15.2) as follows. Given the burst node n, the late event time L_n is determined by

$$L_n = \text{MIN}\ \{(L_o - \mu_{no}), (L_p - \mu_{np})\}$$

This corresponds to the least event time computed from all activities connected to node n.

For a simple connection like m–n there is only one event time to consider while computing L_m, namely that of L_n since node m is connected by one activity only (m–n), therefore $L_m = L_n - \mu_{mn}$.

Once all event times are computed and the mean project completion time is determined, its corresponding variance will be the sum of all variances on the critical path.

15.3 PERT CALCULATIONS AND THE CENTRAL LIMIT THEOREM

So far we have simply accepted the fact that the mean project time is the sum of the means of all activities on the critical path and its variance is the sum of variances on that path. The Central Limit Theorem is used to validate this assumption. The theorem is given in Hahn and Shapiro (1967) as follows:

> The distribution of the mean of n independent observations from any distribution, or even from up to n different distributions, with finite mean and variance approaches a normal distribution as the number of observations in the sample becomes large—that is, as n approaches infinity. The result holds irrespective of the distribution of each of the n elements making up the average.

Now assume that the n activity durations that make up the critical path of the PERT network are independent. Further assume that each has a given distribution D with some mean and variance known to be finite (e.g., in a beta distribution). If the number of activities on the path is large, the distribution of the mean of the activity times approaches a normal distribution according to the Central Limit Theorem.

In other words, if n is large enough the distribution of project completion time T can be approximated with a normal distribution with mean $\bar{T}$ and variance V^2 as follows:

$$\bar{T} = \sum_{xy=1}^{m} \mu_{xy} = \mu_{ij} + \mu_{jx} + \ldots + \mu_{mn}$$

such that ij belongs to the longest path leading to the terminal node.

$$V^2 = \sum_{xy=1}^{m} \sigma_{xy}^2 = \sigma_{ij}^{\ 2} + \sigma_{jx}^{\ 2} + \ . \ . \ . + \sigma_{mn}^{\ 2}$$

such that i belongs to the longest path leading to terminal node.

Since $\overline{T}$ and variance V^2 are assumed to follow a normal distribution, statistical analysis may be performed to calculate probabilities and percentiles based on this assumption. This is discussed in another section of this chapter.

15.4 EXAMPLE OF PERT NETWORK COMPUTATIONS

We will illustrate the network calculations by referring to the example of Figure 15.1. First estimate the mean and variance of each activity from the three-time estimates given in Table 15.1. For example, the mean and variance of activity 1–2 are given according to Equations 15.1 and 15.2 as follows:

$$\mu_{1-2} = (5 + 4 \times 10 + 15)/6 = 10 \text{ days}$$

$$\sigma^2_{1-2} = [(15 - 5)/6]^2 = 2.8 \text{ days}$$

The remainder of the activity mean and variance values are similarly calculated and are summarized in Table 15.2.

Now we proceed with the PERT event calculations as described earlier. The mean and variance of each activity are placed on the network for convenience, as shown in Figure 15.3. Note that while calculating the mean event time μ the same analysis as discussed in CPM regarding the burst and merge nodes is followed. Therefore, $\mu_3 = 22.2$ is found from comparing 22.2 from path 1–3 and 20 from path 1–2, 2–3 and the maximum value is used. Once the mean is found the variance is simply the sum of variances of activity durations on that path (note that it is not found by comparison of the sum of variances on all possible paths as the mean). This is simply reflecting the uncertainty associated with the estimate of the mean event time and must correspond to the path that determined that mean.

The mean project completion time corresponds to the last event (i.e., 5)

TABLE 15.2 Mean and Variance Estimates for Activities

Activity	D_{opt}	D_{ml}	D_{pes}	μ_i	σ_i^2	σ_i
1–3 (Order prefabricate mtl bldg)	20	22	25	22.2	0.7	0.8
1–2 (Clear site)	5	10	15	10.0	2.8	1.7
2–3 (Underground and foundation)	5	10	15	10.0	2.8	1.7
3–4 (Erect prefab bldg)	8	10	20	11.3	4.0	2.0
4–5 (Finish interior)	9	10	11	10.0	0.1	0.3

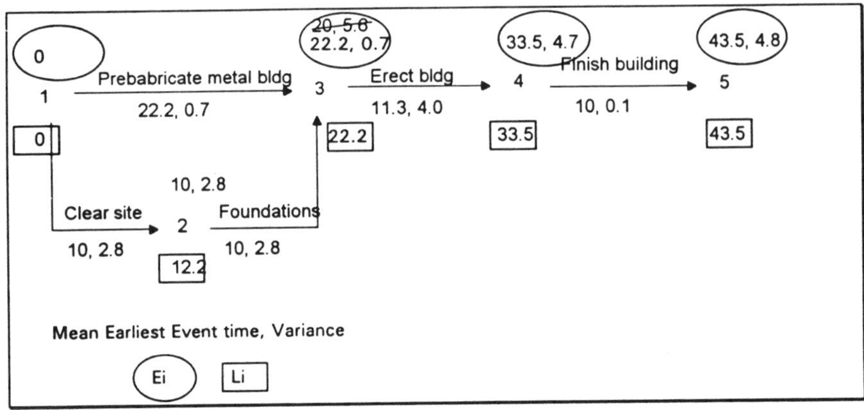

FIGURE 15.3 PERT Analysis on Convenience Store Example

and was found to be 43.5 in Figure 15.3. The variance of this estimate was 4.8, corresponding to the path 1–3–4–5.

The backward pass calculations are also shown on the network given in Figure 15.3 and should be obvious to the reader at this point.

15.5 ANALYZING UNCERTAINTY WITH PROJECT SCHEDULES

Uncertainty is analyzed by assuming that the project completion time (or any arbitrary milestone event) follows a normal distribution with the mean $\overline{T}$ and variance V^2. We will first review the properties of the normal distribution and then discuss how to estimate probabilities associated with the attainment of given events.

15.5.1 Normal Distribution

The normal distribution probability density function (pdf) with the mean μ and variance σ^2 is given by

$$f(x) = \frac{1}{\sigma \sqrt{2\Pi}} \exp \left[\frac{-(x - \mu)^2}{2\sigma^2} \right] \tag{15.3}$$

It has the general shape depicted in Figure 15.4.

The normal distribution is unbound, it has a symmetric shape where its mean coincides with the mode and the median (fiftieth percentile).

The cumulative distribution function (cdf) of a normal distribution with the mean μ and variance σ^2 is given by

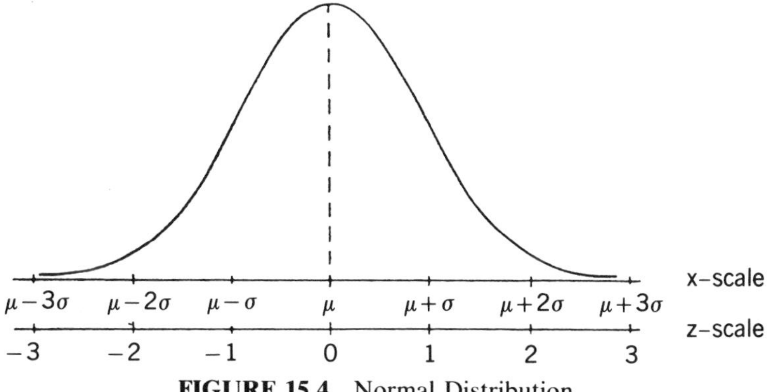

FIGURE 15.4 Normal Distribution

$$\Phi(x) = \int_{-\infty}^{x} \frac{1}{\sigma\sqrt{2\pi}} \exp\left[\frac{-(z-\mu)^2}{2\sigma^2}\right] dz \tag{15.4}$$

This is graphically depicted in Figure 15.5.

The cdf is important because it provides the means for determining the probability of a certain observation or range. For any given value x the probability of having a number less than or equal to x is obtained from Equation 15.4 (or by mapping a line from the x axis to the cumulative function and reading the probability on the y axis as illustrated in Figure 15.5). This requires that the mean μ and variance σ^2 are known and x is given. Since there is no closed form solution for $\Phi(x)$, it is approximated and usually given in tables.

Given that the mean $\overline{T}$ and variance V^2 of the project completion time have been estimated, the normal distribution function given in Equation 15.4 is fully defined and the probability of attaining any given scheduled time x can be estimated by solving Equation 15.4 or reading the probability value from a normal distribution table. Normal probability distribution is presented in tables with a mean of zero (0) and a variance of one (1.0). In order to use such tables, the following transformation is required:

If x is normally distributed with mean μ and variance σ^2 then the random variable

$$z = \frac{x - \mu}{\sigma}$$

is normally distributed with mean $\mu = 0$ and variance $\sigma^2 = 1$. Then z is referred to as a standard normal variate. Therefore, to read the probability of completing the project in x time units or less, one would simply read the probability corresponding to the value z from Table 15.2.

Consider again the convenience store example with project mean time

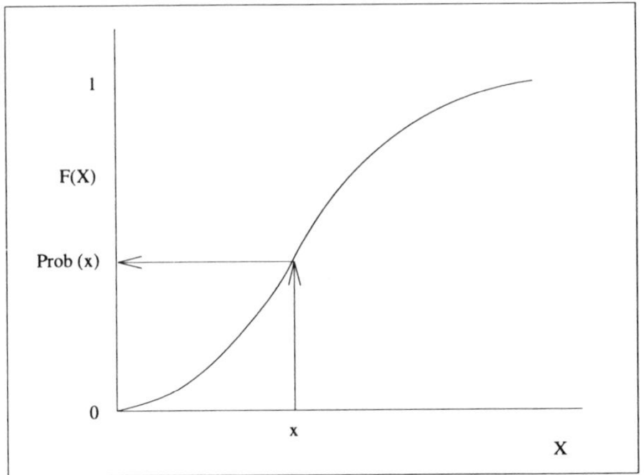

FIGURE 15.5 Sample Normal Distribution CDF

$\mu = 43.5$ and variance $\sigma^2 = 4.8$. To find the probability of completing this project in 45 days or less we note that $x = 45$ days, therefore

$$z = \frac{45 - 43.5}{4.8} = 0.31$$

From Table 15.3 for $z = 0.31$ the probability is 0.62 (i.e., 62% chance of completing the project in 45 days or less).

In some cases probability tables are provided for positive values of z only. In such a case the symmetric properties of the normal distribution are used to find the probability of a negative variate. Simply stated

$$\Phi(-z) = 1 - \Phi(z)$$

[note that $\Phi()$ is the cumulative probability denstiy function of the normal distribution], therefore one would read the value corresponding to $-z$ and calculate the probability by subtracting the value read from the table from 1.0.

For example, the probability of the project in our example being completed in 40 days or less is calculated as follows:

$$z = \frac{40 - 43.5}{4.8} = -0.73$$

From Table 15.3* for $z = +0.73$ the probability is 0.76 and therefore probability of completing the project in 40 days or less would be $1 - 0.76 = 0.24$.

* Note that Table 15.3 provides both plus and minus values for z. This discussion only applies if other tables are used.

TABLE 15.3 Table of Values of the Standard Normal Distribution Function

Z	0	Z	0
0.0	0.5000	−3.0	0.0013
0.1	0.5398	−2.9	0.0019
0.2	0.5793	−2.8	0.0026
0.3	0.6179	−2.7	0.0035
0.4	0.6554	−2.6	0.0047
0.5	0.6915	−2.5	0.0062
0.6	0.7257	−2.4	0.0082
0.7	0.7580	−2.3	0.0107
0.8	0.7881	−2.2	0.0139
0.9	0.8159	−2.1	0.0179
1.0	0.8413	−2.0	0.0228
1.1	0.8643	−1.9	0.0287
1.2	0.8849	−1.8	0.0359
1.3	0.9032	−1.7	0.0446
1.4	0.9192	−1.6	0.0548
1.5	0.9332	−1.5	0.0668
1.6	0.9452	−1.4	0.0808
1.7	0.9554	−1.3	0.0968
1.8	0.9641	−1.2	0.1151
1.9	0.9713	−1.1	0.1357
2.0	0.9772	−1.0	0.1587
2.1	0.9821	−0.9	0.1841
2.2	0.9861	−0.8	0.2119
2.3	0.9893	−0.7	0.2420
2.4	0.9918	−0.6	0.2743
2.5	0.9938	−0.5	0.3085
2.6	0.9953	−0.4	0.3446
2.7	0.9965	−0.3	0.3821
2.8	0.9974	−0.2	0.4207
2.9	0.9981	−0.1	0.4602
3.0	0.9987	−0.0	0.5000

15.5.2 General Notes Regarding the PERT Method

1. The higher the variance on a path the more uncertainty is associate with the estimate at that path.

2. Although it is common in the literature that the distribution categorized in the PERT analysis as discussed in Step 2 above is a beta distribution, this need not be the case. In actual fact PERT only uses the mean, which is a weighted average of the three-time estimates in the analysis. No distribution is actually required in the analysis. It is only needed to facilitate the theoretical proof (see Section 15.6).

3. Notice that the PERT calculations only use the mean in arriving at the

critical path. The variance is only used after the path is determined to assign a level of uncertainty associated with the mean of the determined event. This commonly results in the so-called "merge event bias" leading to an optimistic estimation of the mean of the project time compared to the true mean time.

4. How large should the number of activities on the critical path be so that the Central Limit Theorem holds? This is rather difficult to generalize. The smaller the value of n (e.g., less than 30 activities), the more deviation from normality may be observed. The results would still remain approximate however.

5. How can we assume that the durations of the activities are independent? This assumption is made frequently because it simplifies the analysis, but productivity in construction activities (which determine the duration) provides for strong correlation between the durations. When the concrete formwork on the first section of a building takes longer than expected owing to bad weather, it is highly likely that the same happens on the second section, for example.

15.6 VALIDITY OF THE PERT ASSUMPTIONS*

This section discusses the PERT assumptions, their validity and shortfalls. We will first focus on the assumptions regarding activity times and then discuss the validity of the computation of event times.

15.6.1 Assumptions Regarding the Activity Time

A beta distribution was considered by PERT developers as a possible distribution to characterize the activity time. A beta distribution has four parameters and is therefore represented by $f(x;\alpha,\gamma,L,U)$ where x is a variable, α and γ are shape parameters, and L and U are the end points. Since the extreme points and the mode are available from collected data (an expert's opinion), one additional parameter is required to characterize a beta distribution with its four parameters. The assumption that the standard deviation will be 1/6 the range facilitates this and was therefore adopted.

The beta distribution has certain advantages over other standard distributions (e.g., normal, triangular, gamma, uniform) which makes it more suitable to represent activity times. For example, the probability density function of a beta distribution can attain many shapes, as demonstrated in Figure 15.6. This enables us to represent cases where the most likely value is close to the pessimistic or optimistic values reflecting the envisioned spread of possible durations. The bell-shaped normal distribution does not allow this flexibility

* This section deals with advanced material regarding PERT and may be bypassed in introductory readings.

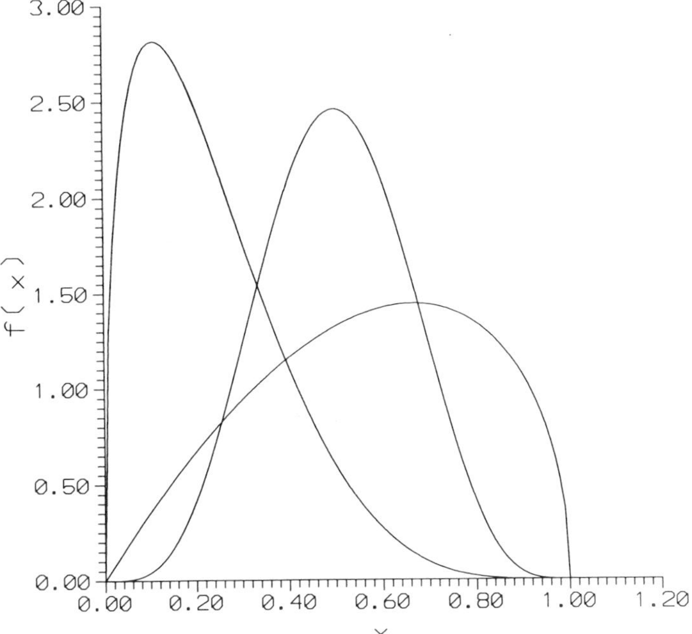

FIGURE 15.6 Beta Distribution Shapes

and requires that the duration spread be symmetric around the mean, for example. A beta distribution is also bound between two points, which makes it more suitable for finite modeling of activity times.

The beta distribution is fully defined by its end points (data limits) and its two shape parameters. Its functional form is given in Equation 15.5:

$$f(x;\alpha,\gamma,L,U) = \frac{\Gamma(\alpha + \delta)\,(x - L)^{\alpha-1}\,(U - x)^{\gamma-1}}{\Gamma(\alpha)\Gamma(\gamma)\,(U - L)^{\alpha+\gamma-1}} \qquad \text{if } L \leq x \leq U \qquad (15.5)$$

otherwise $f(x;\alpha,\gamma,L,U) = 0$. Where $\Gamma() =$ gamma function, $\Gamma(z) \equiv \int_0^{\alpha} t^{z-1}e^{-t}dt$ for all $z > 0$.

Using PERT, we wish to estimate the mean and variance of the beta distribution corresponding to the specified three-time estimates. The mean and variance of the beta distribution are given in Equations 15.6 and 15.7 as functions of the beta parameters α, γ, L, and U.

$$\mu = \frac{\alpha U + \gamma L}{\alpha + \gamma} \qquad (15.6)$$

$$\sigma^2 = \frac{(U - L)^2\,\alpha\gamma}{(\alpha + \gamma)^2\,(\alpha + \gamma + 1)} \qquad (15.7)$$

PERT developers simplified the estimation of the mean and variance to facilitate hand computation procedures by using a weighted average of the optimistic duration D_{opt}, the most likely duration D_{ml}, and pessimistic duration D_{pes} as given in Equation 15.8 and assuming that the standard deviation may be approximated by Equation 15.9.

$$\mu = \frac{(L + 4M + U)}{6} \tag{15.8}$$

$$\sigma^2 = \left(\frac{U - L}{6}\right)^2 \tag{15.9}$$

In fact beta distributions that satisfy these conditions must have the following shape parameters α and γ (see Grubbs, 1962):

$$\alpha = 2 + \sqrt{2} \quad \text{and} \quad \gamma = 2 - \sqrt{2}$$

$$\alpha = 2 - \sqrt{2} \quad \text{and} \quad \gamma = 2 + \sqrt{2}$$

$$\alpha = 3 \quad \text{and} \quad \gamma = 3$$

thereby reducing the possible set of beta distribution that may be used from the set $\alpha > 1$ and $\gamma > 1$ to the three scenarios depicted above.

The approximation is acceptable, however, for many practical situations. A detailed analysis was carried out by MacCrimmon and Rayvac (1964), who studied the validity of the PERT assumptions. Their conclusions and our recommendations are as follows:

1. An error may be induced in the PERT analysis if the activity time has a distribution other than beta (e.g., triangular or normal). An empirical study by AbouRizk (1990) demonstrates that the beta distribution represents construction duration sets very well. The following properties also add to the practicality of using a beta distribution in modeling activity times:

- The beta distribution is continuous
- It has a unique mode between the end points
- Its end points are nonnegative, distinct, and finite

These properties were outlined as requirements for a good distribution that models activity time (see AbouRizk et al. 1991). Therefore, assuming that the beta distribution represents construction duration data is an acceptable and practical one.

2. Assuming that the distribution is beta, an error due to the approximation of the mean and variance by Equations 15.8, and 15.9 rather than through an exact solution of Equations 15.6 and 15.7 occurs. The analysis of this error is discussed below.

Given L, M, U for the duration, estimate the mean and variance μ and σ^2 of a beta distribution (α, γ, L, U). Assuming that we know L and U, two equations are required to solve for α and γ, thus defining a beta distribution. When this is done the values of α and γ are substituted in Equations 15.6 and 15.7 to get μ and σ^2. The solution may be derived as follows. The mode m is given by

$$m = \frac{(\alpha - 1)U + (\gamma - 1)L}{\alpha + \gamma - 2} \tag{15.10}$$

and the variance is given by

$$\sigma^2 = \frac{(U - L)^2 \alpha\gamma}{(\alpha + \gamma)^2(\alpha + \gamma + 1)} \tag{15.11}$$

To facilitate the solution we will use the following standardized values:

$$\psi = \frac{\sigma^2}{(U - L)^2} \tag{15.12}$$

$$\nu = \frac{(M - L)}{(U - L)} \tag{15.13}$$

$$\tau = \frac{\psi}{(1 - \nu)^2} \tag{15.14}$$

The solution of Equations 15.10 and 15.11 in terms of the standardized estimates given in Equations 15.12–15.14 yields the cubic equation given in Equation 15.15.

$$c_3\gamma^3 + c_2\gamma^2 + c_1\gamma + c_0 = 0 \tag{15.15}$$

where

$$c_0 = -12\tau\nu^3 + 20\tau\nu^2 - 11\tau\nu + 2\tau \tag{15.16}$$

$$c_1 = 16\tau\nu^2 + (2 - 18\tau)\nu + 5\tau - 1 \tag{15.17}$$

$$c_2 = -(7\tau + 1)\nu + 4\tau \tag{15.18}$$

$$c_3 = \tau \tag{15.19}$$

Solving the cubic equation (Eq. 15.15) and taking the largest positive real root, an estimate of the shape parameters γ is obtained. Then α is found by substituting the value of γ in Equations 15.10 and 15.11.

Using this estimate of α and γ, the actual value of μ can be obtained using Equation 15.6.

MacCrimmon and Rayvac show that in the worst case scenario the maximum absolute error in the mean estimate could be 33% and in the variance 17%.

3. An error may be induced in the analysis by assuming that the solicited three-time estimates are exact.

The original work on PERT* indicates that those estimates must be made by the technician who understands the operation being considered. In addition Clark (1961) indicates that it can be shown that the best estimate one can infer regarding the time to complete an activity is the most likely values (statistically this corresponds to the mode of a distribution). Beyond that one can probably estimate the extreme end point of a distribution (lowest and highest values encountered). It is very difficult, however, to subjectively estimate the mean and variance of a given distribution.

It has been shown (see Peterson and Miller, 1964) that the mode is the easiest and most accurate statistical subjective estimate that may be elicited from a person familiar with a given population. Estimates of the end points are not as reliable as the mode. In fact researchers argue that estimating the 5th and 95th percentiles of a population (or 2nd and 98th percentiles) are more reliable. For all practical purposes the difference is negligible. Therefore, this assumption is also acceptable, in our opinion, for all practical purposes in construction planning.

More advanced tools may be used to specify a real beta distribution without any assumption being required to avoid the shortcomings discussed above. A recent work by AbouRizk and Sawhney (1993) illustrates how a beta distribution may be precisely specified given certain subjective and linguistic descriptors about the envisioned operation.

15.6.2 Assumptions Regarding the PERT Network Computations

PERT uses the concept that the longest path on the network is the critical path and thus all of its activities are critical. While this might be an acceptable assumption in deterministic analysis such as CPM, it is not quite valid in stochastic analysis. An activity is defined as critical based on whether its delay will extend the project time or not. In a PERT network an activity duration is described by its mean time and a variance to reflect the randomness associated with the time estimate. PERT uses the mean time of an activity in arriving at the critical path without any reference to the variance, thus reducing the analysis to a deterministic one. Consider, for example, the network given in Figure 15.7. The PERT analysis indicates that the project mean time will be 33 days with a variance of 0.33. The critical path is given by 1–2–3–4. The probability that the project will be completed in 34 days or less is then ($z = 3.03$) 99.8%.

* PERT Summary report, Phase I, July 1958 Special Projects Office, Bureau of Naval Weapons, Department of the Navy, USA.

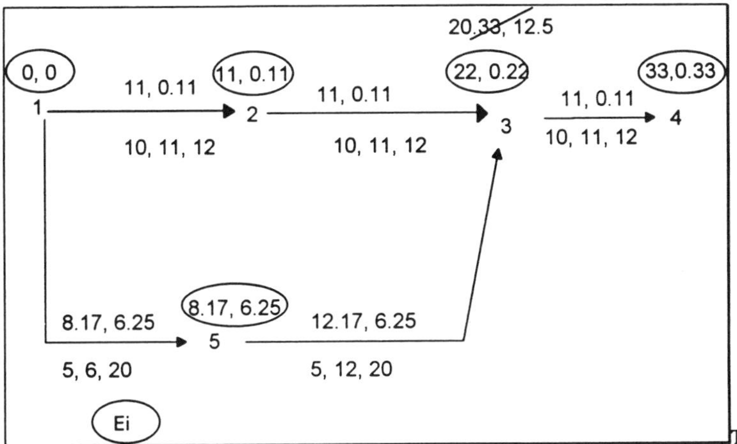

FIGURE 15.7 Example of PERT Network

Assume that the critical path was arbitrarily defined as 1–5–3–4 instead. The mean project time will then be 31.33 with a variance of 12.65. The probability of this project being completed in 34 days or less [z = (34 − 31.33)/12.65 = 0.211] will be close to 59%. In essence the PERT-determined project time distribution with mean 33 and variance 0.33 is overoptimistic because it indicates that the 34-day schedule is attainable with a probability of 99.8% whereas we know, for a fact, that there is one other path in the network with a different estimate of the project time distribution, indicating that the probability of this happening is only 59%! This contradicts the definition of a critical activity, because for this specific project activities 1–5 and 1–3 are more critical, in a sense, than all other activities since by determining the project duration from the path containing them we realize a lesser probability than the PERT one.

It can be proven (see MacCrimmon and Rayvac) that if there exists n distinct paths $P_1, P_2, \ldots P_n$ in a PERT network connecting the origin and the terminal nodes each with a random duration p_i, the PERT calculated mean project time "is always less than, and never greater than, the true project mean." This bias in the mean estimate is termed the merge event bias and is most pronounced when $p_1, p_2, \ldots p_n$ are close to being equal values with PERT neglecting the paths $P_1, P_2, \ldots P_n$ because of the adopted path P_c.

The situation may be viewed in two distinct cases. In the first case the PERT assumption regarding the critical path will hold if there is one predominant path in the network that is much larger in duration than all other competing paths. When there is more than one possible parallel path, problems similar to the one described above start surfacing and PERT estimates the mean to be less than the true mean. Methods for alleviating this problem (such as PNET) have been developed but are not discussed in this book since most of these issues may be better handled through a formal stochastic simulation study as discussed in the next chapter. In such a study the true properties of a distribution (including all its statistical descriptors like the mean and variance) are used and the concept of activity criticality rather than path criticality is introduced to overcome most of the PERT shortfalls.

PROBLEMS

15.1 Devise a schedule for the following network:
What are the probabilities of finishing the job on day (a) 40.0, (b) 46.4, and (c) 55.0?

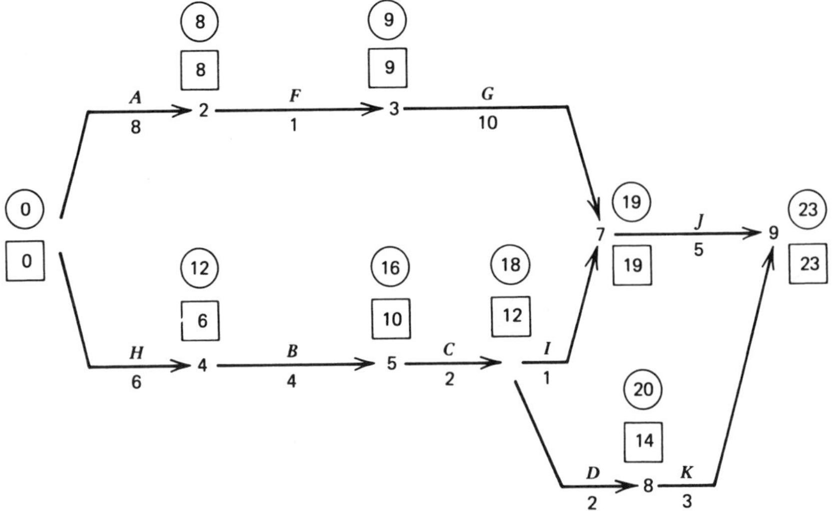

15.2 The optimistic, most likely, and pessimistic durations in days for the activities of HABITAT project are shown in the given table. Determine the project duration.

Activity		Optimistic	Most Likely	Pessimistic
5–10	Obtain tank	9	10	11
10–15	General area of site decided	14	15	21
10–20	Umbilical initial design complete	59	60	80
10–25	Determine biological–environmental requirements	50	60	90
10–30	Determine communication system requirements	40	50	80
10–35	Determine electrical requirements	20	30	40
10–40	Determine required hull modifications	10	15	25
10–45	Determine plumbing requirements	10	15	30
10–50	Determine interior layout requirements	20	30	35
10–105	Testing procedure and apparatus designed	40	50	60
10–110	Submerging procedure designed	70	80	90
15–55	Detailed data collection on area	14	15	16
15–115	Shore station designed	14	15	16
20–60	Umbilical final design complete	50	60	80
20–135	Umbilical materials obtained	50	60	61
25–65	Environmental maintenance system designed	14	15	16
25–70	Sanitary and water facilities decided	14	15	16
25–75	Environmental monitoring and alarm system designed	29	30	31
30–80	System components designed	39	40	41
35–85	Electrical system designed	59	60	61
40–90	Full modifications complete	89	90	91
45–20	Dummy	0	0	0
45–95	System component design finished	9	10	11
50–85	Dummy	0	0	0
50–100	Interior layout designed	19	20	21
55–120	Final site selected	14	15	16
60–135	Umbilical constructed and tested	9	10	11
70–20	Dummy	0	0	0
70–35	Dummy	0	0	0
70–45	Dummy	0	0	0
70–125	System components obtained and installed	29	30	31
80–30	Dummy	0	0	0
80–35	Dummy	0	0	0
80–125	Components obtained and installed	29	30	31
85–20	Dummy	0	0	0
85–125	Wiring fixtures, outlets, etc., installed	25	27	27
90–50	Dummy	0	0	0

Activity		Optimistic	Most Likely	Pessimistic
95–125	System installed	4	5	6
100–125	Interior construction finished	9	10	11
105–125	Testing apparatus constructed	19	20	21
110–130	Cradle and habitat transportation arrangements complete	29	30	31
115–130	Shore station constructed	29	30	31
120–20	Dummy	0	0	0
120–130	Cradle designed and constructed	39	40	41
120–40	Cradle foundation designed and constructed	30	41	60
125–130	Test and evaluation	19	20	21
130–135	Cradle and habitat transported to site	4	5	6
135–140	Habitat lowered and connections made	4	5	6
140–145	Habitat made operational	4	5	6

15.3 An entrepreneur wants to start a recycled paper products plant, for which he has drawn up a list of events. Prepare a network, find the project duration, and draw up a schedule for this project. Assume that all the required funds and resources are available. (a) What is the probability of finishing this project 30 days in advance of project duration? (b) By calculating the expected time

$$t_e = \frac{a + 4m + b}{6}$$

for each activity, determine from your network the expected time of the event "acquisition of production facilities and equipment."

The durations are given in days. The project will start on July 20 and a five-day week will be used.

		Time Estimate		
		Optimistic	Most Likely	Pessimistic
1	Decision to produce and sell	0	0	0
2	Appointment of task force to plan marketing of product	3	5	15
3	Appointment of task force to plan collection of old newspaper	8	15	25
4	Selection of potential communities willing to participate in the project	5	10	30
5	Preparation of collection forecast area	8	16	20
6	Search for potential customers for first year's sales	10	15	20
7	Preparation of total demand forecast	3	5	5

		Time Estimate		
		Optimistic	Most Likely	Pessimistic
8	Preparation of total sales forecast for year 1	1	2	3
9	Preparation of area sales forecast	10	15	20
10	Compilation of operating forecast for year 1	1	3	5
11	Appointment of advertising agency	1	10	20
12	Decisions on promotion program objectives and appropriation	5	10	20
13	Selection of promotion targets	5	5	5
14	Selection of promotion themes	3	5	10
15	Choice of advertising media	5	10	20
16	Planning of advertising copy	5	15	20
17	Planning of direct mail material	5	15	20
18	Signing of advertising contracts	1	5	10
19	Submission of advertising cuts to media	1	1	1
20	Search for collection agents	5	10	25
21	Establishment of collection booths	10	20	25
22	Preparation of direct mail material for mailing	1	1	1
23	Preparation of editorial material for trade journals	3	5	10
24	Preparation of public relations material	3	5	15
25	Choosing of price and discount structure	1	5	10
26	Search for appropriate distributors[a]	20	40	60
27	Mailing of initial information to distributors and salesmen	3	3	3
28	Preparation of distribution plans	5	20	40
29	Instruction of distributors[a]	5	20	60
30	Preparation of an estimate of contribution to profit and overhead by new product	2	3	5
31	Preparation of a demand estimate for second and third years	3	3	3
32	Preparation of estimate of inventory requirements	1	1	1
33	Preparation of estimate of receivables	1	1	1
34	Preparation of estimate of dollar equipment requirement	1	3	5
35	Preparation of estimate of dollar material purchases for the first year	1	2	5
36	Preparation of proposed overall production plans	3	5	10
37	Decisions on an overall production plan	1	2	5
38	Acquisition of production facilities and equipment	10	15	18
39	Installation of production equipment	5	10	20

		Time Estimate		
		Optimistic	Most Likely	Pessimistic
40	Debugging and testing of production equipment	5	10	20
41	Preparation of an estimate of the size of initial production runs	1	2	3
42	Acquisition of materials needed for initial production runs	5	9	15
43	Preparation of schedule for initial production runs	1	1	1
44	Selection of initial foremen	2	2	2
45	Selection of initial work force	5	10	15
46	Training of work force	3	3	3
47	Manufacture of initial production run	3	3	3
48	Commence marketing of product	0	0	0
49	Receiving of samples from initial production run by distributors	5	5	5
50	Decisions on overall marketing strategy	5	10	15
51	Decision on overall strategy for collection of old newspapers	6	12	18

[a] This activity is never actually completed. The estimate refers to the attainment of a sufficient degree of completion for the purpose of launching the new project.

15.4 From the network below, determine the project completion time, the probability of occurrence of which is 0.55.

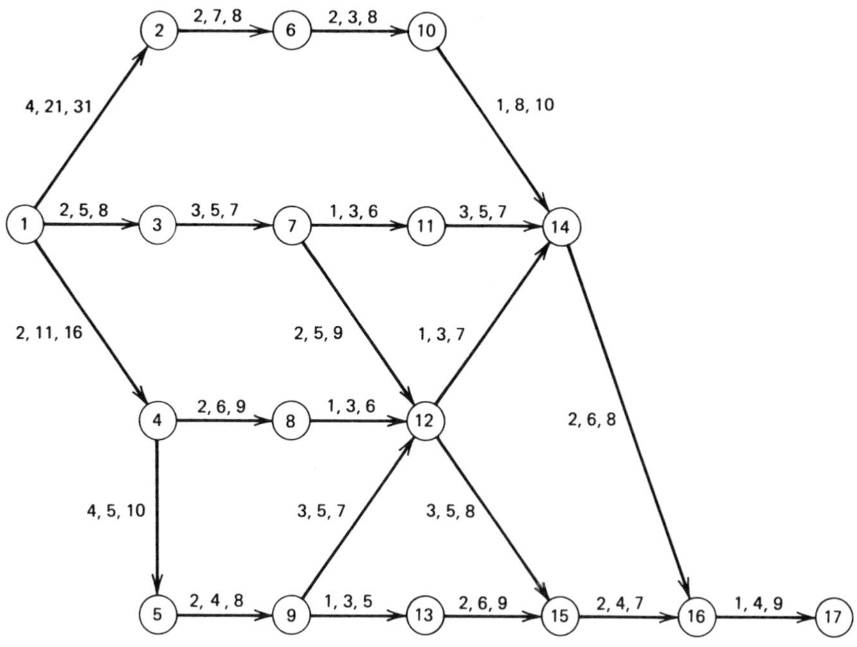

CHAPTER 16

AN INTRODUCTION TO SIMULATION

This chapter introduces simulation and risk analysis using the Monte Carlo simulation technique. Construction projects are often associated with high degrees of uncertainty stemming from the unpredictable nature of construction. One such example is the effect of weather on the productivity and progress of certain activities. The PERT method attempts to estimate the uncertainty in the project schedule as discussed in the previous chapter. Many limitations of the PERT method were discussed. Simulation offers a tool that eliminates many of these limitations. With today's widespread use of microcomputers many of the computational difficulties and time-consuming computations that prohibited the use of simulation in the past have been overcome.

Construction operations are subject to a wide variety of fluctuations and interruptions. Varying weather conditions, learning development on repetitive operations, equipment breakdowns, management interference, and others are external factors that impact the production process in construction. As a result of such interferences, the behavior of construction processes becomes subject to some random variations. It rarely happens on a construction site that the same task consumes the same exact duration on successive occurrences. A truck traveling from one destination to the other on multiple cycles consumes different amounts of time on each cycle, for example. This fact necessitates modeling a construction process as a random (stochastic) process that varies and behaves based upon some prespecified laws of probability.

This chapter is organized as follows: The first section provides an overview of Monte Carlo simulation. The second section provides some insight into the essence of simulation by introducing the reader to random number generation, transformation into appropriate distributions, and statistical analysis of simulation results. This will be followed by two well-known applications of Monte Carlo simulation. The first is a simulation of PERT networks and the second deals with range estimating.

16.1 INTRODUCTION TO MONTE CARLO SIMULATION

The critical path method, precedence diagramming, PERT and even bar charts provide models of the project schedule. Once a model is constructed, the engineer provides input to it and expects to derive some output, based on which more analysis is done or a decision is made. The above mentioned models are fixed in nature since one set of variables is allowed as input and the corresponding set of output is produced.

In simulation analysis the system's model takes input in the form of random variables. The computer then performs experiments with many variations of the input and collects sets of output which are presented to the engineer as statistical distributions. The output may then be statistically analyzed to provide a measure of uncertainty and risk. For a specified model with defined variable parameters, Monte Carlo simulation can be summarized by the following algorithm:

1. Generate a uniform random number on the interval [0–1]
2. Transform the random number into an appropriate statistical distribution (e.g., normal, beta), the resulting number is referred to as a random variate.
3. Substitute the random variates into the appropriate variables in the model.
4. Calculate the desired output parameters within the model.
5. Store the resulting output for further statistical analysis.
6. Repeat steps 1–5 a number of times (note that the generated uniform random numbers must be different in each iteration), when done exit to step 7.
7. Analyze the collected sample of output and perform risk analysis.

This procedure is graphically depicted in Figure 16.1 as it would be applied for simulation of a PERT network.

16.2 GENERATING RANDOM NUMBERS AND TRANSFORMING INTO APPROPRIATE DISTRIBUTION

In the context of simulation, incorporating randomness into a model of a construction operation is attained via a sequence of steps that defines a

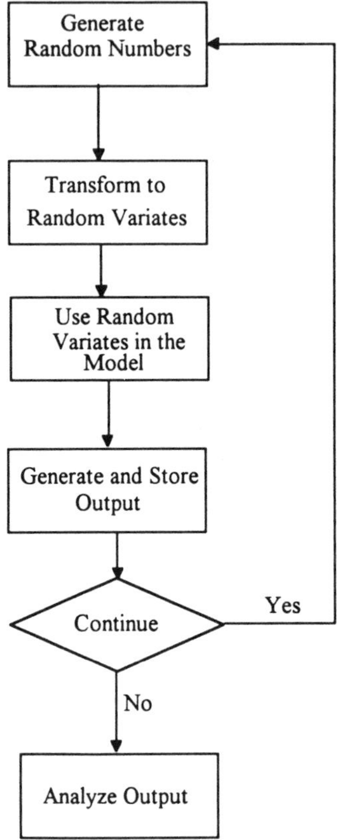

FIGURE 16.1 Flow Diagram for Monte Carlo Simulation

coherent and comprehensive experimentation technique that could be summarized as follows:

1. Generating random numbers that are reproducible (i.e., the simulator should be able to replicate the same procedure when desired).
2. Generating random variates that reflect the true nature of the durations of the work tasks that constitute the process being modeled. For this to properly work an appropriate distribution that is considered to be a suitable model of the input data must be identified and specified (this is commonly referred to as input modeling).
3. Collecting samples of the output parameters of concern efficiently and accurately.
4. Analyzing the samples of the output parameter correctly and developing point and interval estimates for the relevant parameters.

We first introduce some of the popular numerical methods used for generating pseudo random numbers, after which we introduce some basic techniques useful for generating random variates. The reader is then introduced to some of the methods used for modeling input data.

16.2.1 Generating Random Numbers

The computer simulation of a stochastic process is driven by random input processes. The basis for randomness in a simulation experiment lies in the *Random Number Generators*. Consider for example a truck hauling earth from destination A to B. It was observed that it takes the truck anywhere between 10 and 30 minutes to do the traveling on one particular day. In order to model such a phenomenon in simulation we have to have some means of generating random durations on different cycles between the specified limits in an efficient way. The basis for sampling durations between the lower limit of 10 and upper limit of 20 lies in generating a *uniform random number* on the range [0,1] and then *transforming* that number into the appropriate range and according to the appropriate model of the collected data. The transformation of random numbers into an appropriate variate is discussed in the next section. In this section we discuss the generation of uniform random numbers on the range [0,1].

A true random number as defined by mathematicians is very difficult to generate. In today's age of computers, simulators settle for a pseudo random number. Such a number would possess similar attributes to a true random number for functionality purposes (from here on we will use the phrase *random number* to mean *pseudo random numbers*). To get a random number on the range of [0,1] one could throw dice, look in a telephone book, draw numbered balls, use tables like the one produced by the RAND Corporation, or use numerical means. Numerical means are adaptable for computer use and with some care they could be used to generate numbers which appear to be random for all practical purposes. A recursive algorithm that is used to generate random numbers is referred to as a random number generator (RNG). An RNG should produce fairly uniform numbers on the range [0,1] which appear to be independently sampled, dense enough on the interval [0,1], and reproducible. The algorithm should also be efficient and portable for use in simulation programs.

Numerical techniques for random number generation go back to the early 1940s with the midsquare method introduced by Newman and Metropolis. Lehmwer introduced a method in 1951 referred to as the Linear Congruential Scheme (LCS). Today the most widely used version of LCS is the Multiplicative LCS, which could be defined by the following recursive equation on the nth iteration:

$$Z_n = a^*Z_{n\text{-}1 \text{ MOD }} m \tag{16.1}$$

Where Z_0 is a user-defined starting integer value for Z_n-1 (referred to as a *SEED*), m is the modulus usually defined to be a large integer value (ex. 2^{31} - 1), a is the multiplier usually set to the value 7^5 (16,807) (Lewis et al.).
The random number R_n on the ith iteration would be obtained by

$$R_n = \frac{Z_n}{m} \tag{16.2}$$

To generate random numbers one would usually specify a seed number Z_0 as a starting value. The value of Z_1 would then be computed, resulting in R_1 and so on. The values of a and m should be chosen with utmost care or the random numbers will start to regenerate after a certain cycle.

To illustrate the Multiplicative LCS consider the following example. (Note that for illustration purposes we use the small values of a and m. The values previously mentioned would be more adaptable to a computer program.)

$$a = 5 \qquad m = 7 \qquad Z_0 = 9$$

The recursive equation will be:

$$Z_n = 5 \times Z_{n-1} \text{ MOD } 7$$
$$Z_1 = 5 \times Z_0 \text{ MOD } 7$$
$$= 5 \times 9 \text{ MOD } 7$$
$$= 45 \text{ MOD } 7$$
$$= 3 \text{ (3 is the reminder of the division of 45 by 7)}$$

The first generated random number would be $3 \div 7 = 0.4285714$. If one continues on the same line he/she will obtain the following table:

n	Z_n	R_n
0	9	
1	3	0.4285714
2	1	0.1428571
3	5	0.7142857
4	4	0.5714286
5	6	0.8751429
6	2	0.2857143
7	3	0.4285714

Note that on the seventh iteration we obtained the same value of Z as on the first iteration. This was the result of using a modulus equal to 7. Hence it is recommended that you use larger numbers like $2^{31} - 1$.

The numbers generated are fairly uniform on the range [0,1] and appear to be independent, but actually are not. If we used a larger value of m we could have obtained a denser population (in this example we were able to create only 6 random numbers before starting to regenerate).

16.2.2 Transforming a Random Number into a Random Variate

In the previous section we introduced the reader to the basics of generating uniform random numbers on the interval [0,1]. We also mentioned that most random variates are basically transformations of the generated random numbers. There are a number of techniques by which one can transform these numbers into some desired number with a particular distribution. This section introduces some of the more common techniques.

Generating Variates. The most basic and reliable method for generating random variates is the *inverse transform method*. The method is superior to other methods in that there is a one to one correspondence between the random number used and the variate generated. This becomes very crucial when debugging a simulation model or trying to replicate the same experiment for some other purpose. The method would be normally preferred to others when the cumulative density function (CDF) exists and is easy to numerically compute.

In essence the inverse transform method works as follows:

1. Generate a random number R on the range [0,1]
2. Set $X = F^{-1}(R)$
3. Deliver X

Graphically the method simply works by generating a random number R_1 on the interval [0,1]. The number would be mapped into the CDF and down the X axis as shown in Figure 16.2. The value of X is then read off the X axis.

Sample Applications

Generating Uniform Variates on the Range [L, U]. The probability density function of the variable X which has a uniform density is given by

$$f(x) = \begin{cases} \dfrac{1}{U - L} & L \leq X \leq U \\ 0 & \text{otherwise} \end{cases} \tag{16.3}$$

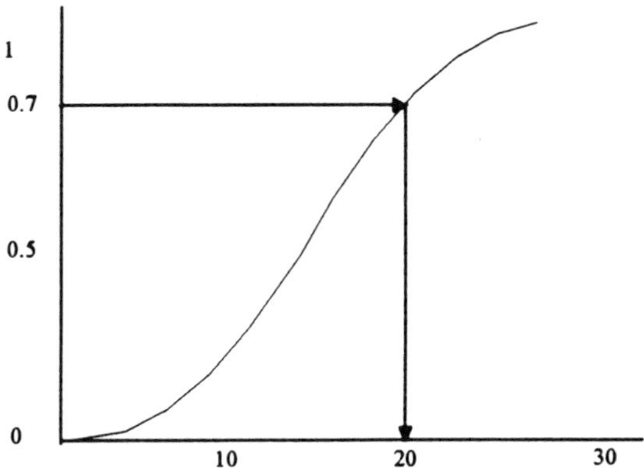

FIGURE 16.2 Cumulative Density Function

The corresponding CDF can be explicitly determined by integrating the PDF; the resulting CDF would be

$$F(X) = {}_{-\infty}{}^{X}\!\int f(x)dx = \begin{cases} \dfrac{X\text{-}L}{U\text{-}L} & L \leq x \leq U \\ 0 & x < L \\ 1 & x > U \end{cases} \tag{16.4}$$

To use the inverse transform method for generating uniform variates one would set $F_x(x) = R$ and solve for X as follows:

$$\frac{X - L}{U - L} = R \tag{16.5}$$

$$\mathrm{X} = \mathrm{F}^{-1}(X) = R(U-L) + L \tag{16.6}$$

Now to generate a uniform random variate on the interval $[L, U]$:

1. Generate R uniform [0,1]
2. Set $X = L + R\,(U - L)$
3. Deliver X which would be UNIF (L,U)

Other transformations for triangular, exponential, normal, and beta distributions are provided in Section 16.7 for the interested reader.

Selecting and Fitting the Appropriate Distribution. We have seen thus far that the computer generates random variates for use in the computer simulation of a given construction model. In order for the procedures described above to successfully proceed, the engineer should specify for the simulation program two basic things regarding generation of random variates, namely, (1) the type of distribution desired for random variate generation and (2) the parameters of the selected distribution.

A simulator can resort to a number of means in the process of sampling random variates in the course of a simulation. For example, an empirical distribution can be constructed from the sample data to model the observations. In such a case it would be necessary to write the code to sample from the empirical CDF. It is readily apparent that such a technique is inconvenient and impractical, particularly when trying to manipulate large sets of data and a multiple number of data sets at the same time.

A classical way of overcoming the disadvantage of the previous technique is by searching for a standard statistical distribution which has a well-defined functional form that closely models the set of observations available. Such distributions are usually supported by simulation software. The problem reduces to finding a distribution, solving for the parameters that fully define the selected distribution, and testing for how well the selected distribution tracks the empirical distribution of the sample data. The simulation literature is extensive on the subject. Section 16.7 provides some insight into the subject. Many commercial software also enables the engineer to fit proper distribution with very little effort (see UNIFIT for example).

16.3 STATISTICAL ANALYSIS OF SIMULATION RESULTS

Given that we have conducted a simulation experiment of n independently seeded runs one can calculate the sample mean $(\overline{X})$ and sample variance S^2 of the parameter X_1 as follows:

$$\overline{X} = \frac{1}{n} \sum_{i=1}^{n} X_i \tag{16.7}$$

$$S^2 = \frac{1}{n} \sum_{i=1}^{n} (X_i - \overline{X})^2 \tag{16.8}$$

If the experiment is repeated with another set of random numbers, the sample mean would likely be different than that obtained previously. This fact requires one to build a confidence interval around the provided estimate. Such an interval would reflect the degree of confidence one has in the results of the simulation (this should not be interpreted as the degree of confidence one has in the results themselves which depend on the model and a multitude of factors).

A 100 (1-α)% confidence interval around a parameter strictly means that if one repeats the simulation experiment (with the same sample size) a large number of times, 100 (1-α)% of the calculated parameters would contain the underlying random parameter we are trying to estimate. So the probability that the underlying parameter of concern lies in the interval would be (1-α).

16.3.1 Developing Confidence Intervals for the Mean and Variance

To construct a confidence interval one has to know the underlying distribution governing the behavior of the random variable of concern. Normal distribution theory can be applied to construct a confidence interval around the estimate when the data to be analyzed is close to being normal. The output from most queuing networks is usually normal or does not deviate much from normality. The development and analysis of output can become fairly complicated if the sample cannot be assumed normal and/or not independent (i.e., correlated). The discussion of both cases is beyond the scope of this text. The reader is however referred to Fishman (1973) and Welch (1983) for a detailed discussion of output analysis under both situations. Herein we limit our discussion to the specific case when the output is from a *transient* simulation and appears to be originating from a *normal* distribution as well as being fairly *independent*. To test for the normality of data the quantile points of the normal distribution can be plotted against the quantile points of data in question (commonly referred to as a q–q plot). If the plot is linear the data is considered to be normal. A simple spreadsheet may be set up to facilitate the procedure. The test is described in Section 16.9.

Having these assumptions in mind (these assumptions are actually valid and account for most construction simulation), if the output parameter of concern X_i $(i = 1 - n) \approx N(\mu,\sigma^2)$ (this would read as X_i are observations originating from a normal distribution with mean μ and variance S^2) and the sample mean and variance are calculated from Equations 16.1 and 16.2 then from the theory of mathematical statistics we know that the statistic $t = (X - \mu)/(S/\sqrt{n})$ follows a t distribution with n—1 degrees of freedom [ususally written as $\approx t\ (n - 1\ \text{df})$]. The t statistic calculated above can be bounded between two critical t values with some cutoff probability α, as shown in Figure 16.3.
In other words,

$$\Pr\{ -t_{\alpha/2}(n - 1) - \frac{\overline{X} - \mu}{S/\sqrt{n}} - t_{1 - \alpha/2}(n -)\} = 1 - \alpha \qquad (16.9)$$

where Pr { } is the probability of occurence.

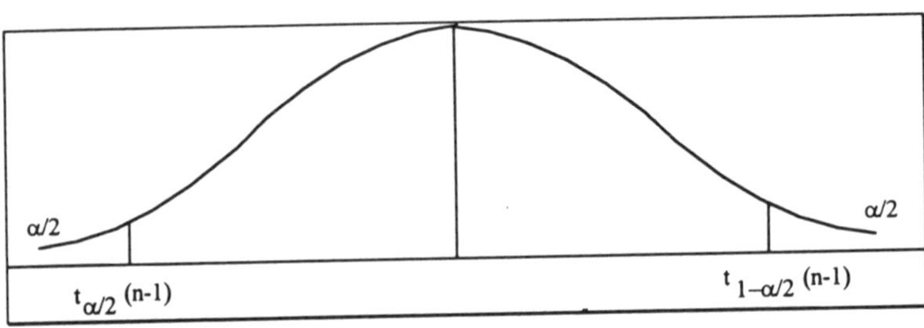

FIGURE 16.3 *T* Distribution with Critical *t* Values

Now the equation above can be rearranged to obtain a simple confidence interval for μ. The $100(1\text{-}\alpha)\%$ confidence interval for the mean μ can be stated as

$$(\overline{X}) - t_{\alpha/2}(n - 1)\frac{S}{\sqrt{n}} \le \mu \le (\overline{X}) + t_{1-\alpha/2}(n - 1)\frac{S}{\sqrt{n}} \qquad (16.10)$$

The probability that μ falls in the interval given above would be $1\text{-}\alpha$.

The confidence interval for the variance is developed in a similar way. In this case however the statistic $(n - 1)S^2 \sigma^2$ follows a chi-square distribution with n-1 degrees of freedom [$x^2(n\text{-}1 \text{ df})$]. The corresponding 100 $(1\text{-}\alpha)\%$ confidence interval (Wilson, 1984) for σ^2 is given by

$$\frac{(n - 1)\,S^2}{x^2_{1-\alpha/2}(n - 1)} \le \sigma^2 \le \frac{(n - 1)\,S^2}{x^2_{1-\alpha/2}(n - 1)} \qquad (16.11)$$

16.3.2 Estimating Quantiles and Probabilities

In the previous sections of this chapter we have seen how one can estimate the mean and variance of an output parameter. In most practical cases though, a simulator might be interested in other statistics about the performance of an operation. Important statistics to a practitioner could be an arbitrary quantile or a probability of exceeding (or not exceeding) a given threshold value of the parameter. In the first case one might be interested in the ninety fifth percentile of project completion time for example. This can be a more interesting parameter as compared to the mean completion time. Simply stated, the ninety fifth percentile has 95% of the possible data below that observation. In the context of the process completion time, 95% of the simulation we conducted would result in a lesser duration than that duration at the ninety fifth quantile. This could be more indicative of completion time of the project than the estimate of the mean completion time. Another interesting and practical question that is often asked is regarding the probability of the completion time of the project not exceeding X units of time? This question is often addressed in PERT-type analysis of scheduling networks. The same theory applies to analysis of simulation output.

16.3.3 Estimating Probabilities

If we are interested in finding the probability that the output parameter of concern X does not exceed a particular fixed value x [Pr $\{X - x\} - F(x)$] we use the estimator $\Phi\,(x - \overline{X}/S)$ where $\Phi(z)$ is the cumulative probability density for a standard normal distribution and $\overline{X}$ and S are the sample mean and variance.

The assumption here is that the output is normal or approximately normal. To illustrate the above assume we are interested in calculating the probability that the time to complete the operation (of 40 truck loads) does not exceed 14 hr (840 min). The parameters are evaluated as follows:

$$\left(\frac{x - \overline{X}}{S}\right) = \frac{840.0 - 759.74}{152.37} = 0.5267$$

$$\Phi\ (0.5267) = 0.7019$$

One can also estimate a confidence interval for the probability just estimated. Although that subject is not within the scope of this book, the reader is referred to Welch (1983) for detailed coverage.

16.3.4 Estimating Arbitrary Quantiles

We define the qth quantile of the random variable X by $X_q \equiv$ minimum value x such that $F(x) \geq q$. This would simply be the inverse CDF evaluated at q [referred to as $F^{-1}()$].

To obtain an estimate for X_q from the collected sample we use the approximation of the binomial distribution by the standard normal distribution for large sample sizes [the proper estimates could be found in Welch (1983)]. The estimator of X_q would be given by

$$X_q = \overline{X} + z_q S \qquad (16.12)$$

where z_q is the critical value from the standard normal distribution at the specified cutoff value q, $\overline{X}$ is the sample mean, and S is the sample variance calculated according to Equations 16.1 and 16.2 (which would be the most likely estimates, not the unbiased estimator of the variance).

A confidence interval around the estimator X_q is given by (e.g., see Wilson 1984)

$$(\overline{X} + z_q S) \pm z_{1-\alpha/2}\left[\sqrt{1 + \frac{z_q^{\,2}}{2}}\right]\frac{S}{\sqrt{n}} \qquad (16.13)$$

16.4 SIMULATION OF PERT NETWORKS

The model to be simulated is the PERT network; its variables are the durations of the activity times which are represented by statistical distributions. The output parameters of concern are the project time as well as all event times. The method of calculating these variables (event times) is the same CPM method discussed earlier in Chapter 5. The method will be to first specify distributions for all activity times, then apply the Monte Carlo algorithm as discussed in the previous section.

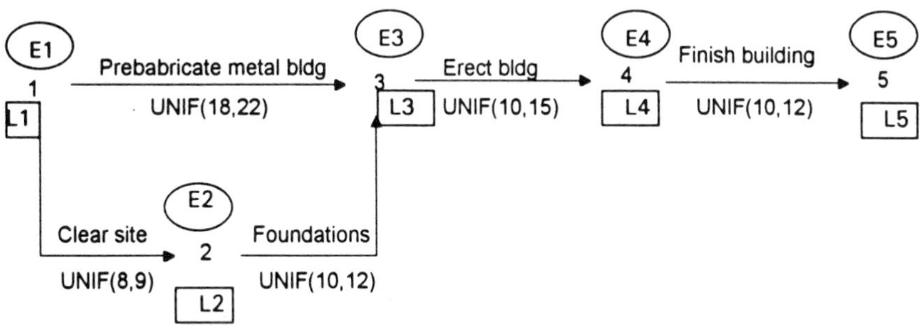

FIGURE 16.4 Convenience Store Problem

Consider the convenience store example discussed in Chapter 15 and given in Figure 16.4. The event times (Ei and Li) are variables now and functions of the durations which are derived from a uniform distribution in this example. Here E2 is calculated as follows:

$$E2 = E1 + \text{UNIF}(8,9)$$

The variable "UNIF(8,9)" is found as discussed in the previous section and will change from one iteration to the other depending on the sampled random number as follows: $\text{UNIF}(L, U) = L + R_i(U\text{-}L)$

To simplify the discussion assume that the random numbers were generated and are as given in Table 16.1.

The calculations are carried as follows.

Iteration 1. Generate duration for activity 1–2 as follows:

$$\text{UNIF}(8,9) = 8 + 0.802773*(9\text{-}8)$$

$$\text{UNIF}(8,9) = 8.802773$$

Compute event e_2.

$$E2 = E1 + 8.802773 = 0 + 8.802773$$

To compute event E3 generate the durations of 2–3 and 1–3 (note that the order of generating random numbers is not important at this point):

$$D23 = \text{UNIF}(10,12) = 10 + 0.5839742*(12\text{-}10) = 19.21306$$

$$D13 = \text{UNIF}(18,22) = 18+0.3032648*(12\text{-}10) = 11.16795$$

TABLE 16.1 Random Numbers for Sample Problem

.8027722	.2311794
.5839742	.7299049
.3032648	.6074464
.8386958	.5411672
.9670481	.9784693
.5330714	.5757428
.7702737	.9305286
.3982053	.2401335
.183295	.1269318
.7588731	.2071984
.2713507	.9150069
.9079139	.7869429
.8653507	.2357711
.329161	.8828779
.1168681	.3982835
0.00165104	.4469764
.8629772	.2588445
.6597677	.0541361
.404589	.4130733
.2020466	.7893257
.629559	.3095249
.906165	.9746047
.2769695	.6841383
0.0000188	.1287186
.0111348	0.0012934

$$E3 = \text{Max}\,\{(\text{E1} + \text{D13}), (\text{E2} + \text{D23})\} = (0 + 19.21306, 8.802773 + 11.16795)$$
$$= 19.970723$$

The remainder of the calculations for this first iteration are shown on the network in Figure 16.5. The backward pass is performed with the durations generated on the forward pass. The results are also shown on the network. Now collect the results of iteration 1 for later use. The results are compiled in Table 16.2

Iteration 2. The computations for iteration 2 are carried in the same manner as iteration 1 except that the random numbers used are different. We start with the random number R6 (from Table 16.1 row 6) and repeat the procedure as before. For brevity the results of the event times only are shown in Table 16.2.

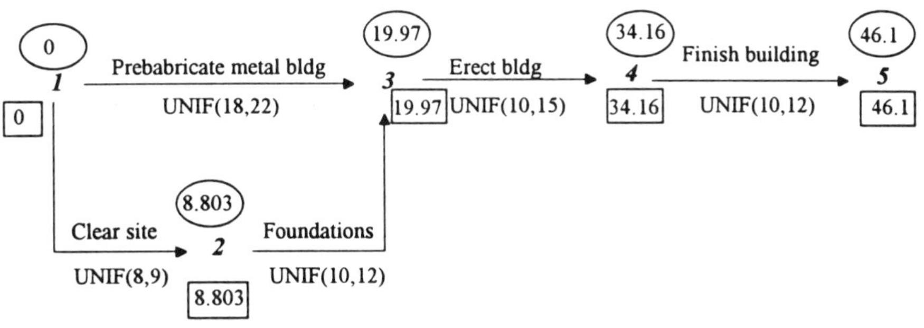

FIGURE 16.5 Forward and Backward Path Calculations for Iteration 1

Iteration 3 to *n*. From this point on we only show the results. The simulation was repeated 10 times with the results tabulated in Table 16.3, which can now be analyzed for risk analysis purposes. Consider for example that the probability of completing the project in 44 days must be computed. This is simply given by

$$\text{Probability } (x < 44) = 66.4\%$$

TABLE 16.2 Simulation Results

Iteration	Event 1	Event 2	Event 3	Event 4	Event 5	Type[a]
1	0	8.802773	19.70723	34.164203	46.048303	A
2	0	8.770274	20.13229	31.04876	42.56651	B
3	0	8.907914	20.638614	32.283824	42.517564	A
4	0	8.862977	20.182517	32.205467	42.609557	A
5	0	8.906165	20.51824	30.51918	40.54145	B
6	0	8.729905	19.944745	32.650635	44.607575	A
7	0	8.930529	20.30297	30.93758	41.35198	B
8	0	8.786942	21.66003	36.07442	46.87099	B
9	0	8.25884	19.7879	31.85327	43.43192	B
10	0	8.974605	20.342885	30.98475	41.012345	B

[a] Path Type A = E1-E2-E3-E4-E5 critical. Path Type B = E1-E3-E4-E5 critical.

TABLE 16.3 Results of the Simulation

Iteration Number	Project Completion Time (Days)
1	46.048303
2	42.56651
3	42.517564
4	42.609557
5	40.54145
6	44.607575
7	41.35198
8	46.87099
9	43.192
10	41.012345
Mean	43.13183
Standard deviation	2.002

Similarly, if we are asked to supply the project completion time corresponding to the 90% chance of completion we do the following:

$$z = 1.29$$

$$x = 2.002*1.29 + 43.1383$$

$$x = 45.72 \text{ days}$$

16.5 RANGE ESTIMATING

Range estimating is a simple simulation exercise carried after an estimate is prepared to reflect the degree of uncertainty associated with the estimate. Range estimating will be presented by covering the method first and then using a commercial simulation add-on to spreadsheets to provide an example application.

16.5.1 Simulation of a Project Estimate

The model to be experimented with in this case is the estimate of the project which is often a summation of individual work items which are extended from quantities and corresponding unit costs. Therefore the only variable to be analyzed is the summation of all costs comprising the project. Since estimates are typically composed of hundreds of individual components the analysis may be simplified by concentrating on those components that are thought to vary by an amount that may effect the bottom line by more than 0.5% (or any other satisfactory percentage).

Each of the uncertain components of the estimates is then estimated by a distribution of the cost rather than a single number. For example, if a given work package is estimated to require 1000 worker-hours with some uncertainty it can be represented by a triangular distribution with the most likely, the pessismistic, and optimistic estimate. The estimate then becomes a most likely use of 900 worker-hours and ranging between 700 and 1200 worker-hours, for example.

To summarize the method the following algorithm is provided:

1. Divide the project into manageable components (e.g., line items or work packages).
2. Identify the uncertain components (those that effect the bottom line).
3. For each of the uncertain components estimate the variability using a statistical distribution (e.g., triangular).
4. Generate random numbers and transform them to the appropriate distribution.
5. Find the project cost on this iteration as the sum of all components (including those that do not vary).
6. Repeat steps 1–5 a number of times.
7. When done construct the cumulative distribution function and calculate all relevant statistics to perform risk analysis.

16.5.2 Sample Application

A highway overpass project is to be estimated. The estimate breakdown and a preliminary estimate was prepared by the estimators. The chief estimator decides to perform a range estimate to derive a value for contingency that will be used at bid time. Table 16.4 shows the line items and the corresponding quantity and unit price. The quantity shown in column 3 of the table indicates that some quantities may vary uniformly within the specified range, reflecting the level of confidence associated with the accuracy in the take-off as well as variability due to what may be encountered in the field. For example, the excavation quantity may vary between 2300 m^3 and 2500 m^3. The quantities were determined to be either deterministic or random following a uniform distribution. The unit costs were considered to be deterministic or random following a triangular distribution. The choice of distribution was arbitrary for illustration purposes (Section 16.7 discusses how a distribution may be selected). The unit price reflects some uncertainty due to attainable levels of productivity in man-hours/unit (or $/unit). Therefore, to place concrete in pier footings the unit cost may range between $320 and $350 per cubic meter with a most likely value of $330. To arrive at such numbers the estimator performs the estimate as for the deterministic case except that when a productivity or quantity may vary he/she will provide the best possible range instead of the one guess previously used.

TABLE 16.4 Line Items from an Estimate with Ranges for Quantities and Unit Cost

Item	Description	Quantity Range	Unit Cost Range
1	Excavation (m^3)	Uniform (2200, 2500)	Triang (10, 11, 13)
2	Backfill (m^3)	Uniform (1700, 2200)	Triang (9, 10, 13)
3	Piling (300 dia m)	160.00	29.00
	Piling (750 dia m)	510.00	Triang (175, 183, 190)
	Bells (1500 dia ea)	42.00	Triang (370, 390, 420)
	Bells (1200 dia ea)	16.00	340.00
4	Cast in place conrete		
	Pier footings (m^3)	73.00	Triang (320, 330, 350)
	Pier columns (m^3)	55.00	Triang (600, 650, 700)
	Abutments (m^3)	635.00	Triang (200, 235, 290)
	Approach slabs (m^3)	55.00	Triang (220, 230, 400)
	Bridge girder (m^3)	1,310.00	Triang (370, 390, 450)
	Parapets incl. finish (m)	171.00	Triang (150, 160, 175)
	Center median (m)	67.00	124.00
5	Concrete slope protection (m^2)	Uniform (1000, 1100)	Triang (42, 45, 50)
6	Hot mix asphaltic concrete paving (m^2)	1,900.00	Triang (17, 18, 19)
7	Reinforcing steel	LS	
8	Expansion joint	LS	
9	Deck waterproofing	Uniform (1800, 2000)	5.70
10	Abutment pier cap	LS	
11	Bearings	LS	
12	Prestressing steel including grouting	LS	
13	Miscellaneous metals	LS	
14	Electrical conduit	LS	
15	Class 5 finish (NIC parapets) (m^2)	565.00	6.00
16	Mobilization	LS	

The estimate is entered into a spreadsheet (e.g., Microsoft EXCEL). An add-on @RISK (Palisade Inc.) was used to conduct the simulation. We have simply specified that the sum of the amounts is the output of interest and fired simulation for 1000 iterations. The results are automatically generated by @RISK and summarized in tabular as well as graphical formats. The simulation results ranged from $1,381,610 to $1,521,289 with a mean of $1,446,842. The results are summarized in the CDF (descending order to

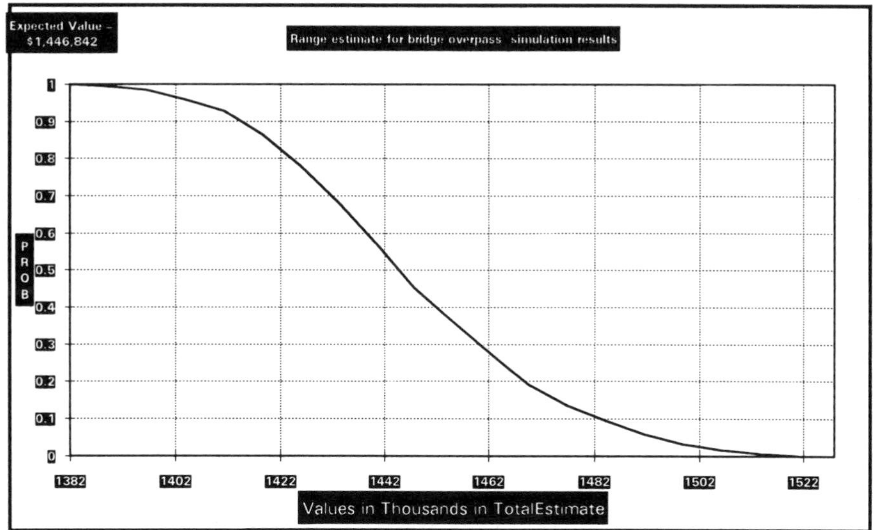

FIGURE 16.6 Range Estimate of Bridge Overpass

facilitate reading the probability of cost overrun rather than cost underrun) given in Figure 16.6.

The interpretation of the chart is straightforward. For any desirable number on the x axis a level of probability is projected at the y axis. For example, if the $1,502,000 was used in the bid on such a project, the chance of a cost overrun is very small (about 2%). For a bid price of $1,422,000 the chance of a cost overrun is projected to be just over 80%. This analysis may now be used to estimate the contingency. For example, a level of 10% chance of cost overrun is acceptable and reflects a bid price of $1,482,000. The difference between this number and the deterministic estimate may be viewed as a contingency sum. Furthermore, during bidding the graph may quickly assist the bidder in knowing how low the bid can be without taking too much risk.

The following sections are advanced topics which may be included in a postgraduate course.

16.6 INPUT MODELING FOR SIMULATION STUDIES

16.6.1 The Inverse Transform Method

Generating Triangular Variates. For a triangular distribution with minimum L, mode M, and maximum U the PDF is given by

$$f(x) = \begin{cases} \dfrac{2(X - L)}{(M - L)(U - L)} & L \le X \le M \\ \dfrac{2(U - x)}{(M - U)(M - L)} & M \le X \le U \end{cases} \tag{16.14}$$

The corresponding CDF would be given by

$$F(x) = \begin{cases} \dfrac{(X - L)^2}{(M - L)(U - L)} & L \le X \le M \\ 1 - \dfrac{(U - X)^2}{(U - L)(U - M)} & M \le X \le U \\ & X < L \\ & X > U \end{cases} \tag{16.15}$$

Now to solve for the inverse CDF we set $R = F_x(x)$ and solve for X in both intervals $[L,M]$ and then $[M,U]$. Solving for that yields the following system:

$$F^{-1}(x) = \begin{cases} L + \sqrt{(M - L)(U - L)R} & 0 \le R \le \dfrac{M - L}{U - L} \\ U - \sqrt{(U - M)(U - L)(1 - R)} & \dfrac{M - L}{U - L} < R \le 1 \end{cases} \tag{16.16}$$

To use the inverse transform method we do the following:

1. Generate R on the interval [0,1].
2. If $R \le \dfrac{(M - L)}{(U - L)}$ then set $X = L + \sqrt{(M - L)(U - L)R}$. Otherwise set $X = U - \sqrt{(U - M)(U - L)(1 - R)}$.
3. Deliver X which would be TRIANGULAR(L, M, U).

Generating Exponential Variates with Mean μ

$$f(x) = \begin{cases} \dfrac{1}{\mu} e^{-x/\mu} 0 \le x \le \infty > 0 \\ 0 \qquad \text{Otherwise} \end{cases} \tag{16.17}$$

$$F_x(x) = \begin{cases} 1 - e^{-x/\mu} 0 \le x \le \infty \\ 0 \qquad \text{Otherwise} \end{cases} \tag{16.18}$$

To use the inverse transform method, we set $F_x(x) = R_0$ (the uniform random number on [0,1]) and solve for X as follows:

$$R_0 = 1 - e^{-x/\mu}$$

$$\Rightarrow e^{-x/\mu} = 1 - R_0$$

$$\Rightarrow \ln(e^{-x/\mu}) = \ln(1 - R_0)$$

$$\Rightarrow -x/\mu = \ln(1 - R_0)$$

$$\Rightarrow x = -\mu \ln(1 - R_0)$$

Also knowing that $1 - R_0$ is only a rescaled uniform random number on interval [0,1] we can replace $(1 - R_0)$ with R, which is a uniform random number on the interval [0,1].

To generate exponential deviates on the range $[0,\infty]$ with a mean μ we do the following:

1. Generate R on the interval [0,1].
2. Set $X = -\mu \ln(R)$.
3. Deliver X, which would be exponential (μ).

In some cases the PDF of a distribution might exist, but the corresponding CDF cannot be defined analytically. Some examples are the beta distribution, normal distribution, and others. In such a case the inverse transform method cannot be used and one would have to resort to some other technique like the *acceptance/rejection* method, the *composition* method, or an analytical approximation of the inverse CDF. Herein we introduce the *acceptance/rejection* method for its simplicity and wide use in simulation packages.

16.6.2 The Acceptance Rejection Method

The acceptance/rejection method is described by Brately et al. after Newman (1951) as follows. Given the probability density function $f(x)$ where x lies on the range $[L, U]$, define the maximum value that $f(x)$ can attain to be c. Now we have $0 \le f(x) \le c$. The method works as follows:

1. Generate X uniform on the interval $[L, U]$.
2. Generate Y uniform on the interval $[0, c]$.
3. If $Y \le f(x)$, deliver X, otherwise go to step 1.

The method is simply trial and error. If one considers the graphical interpretation, the method would become clearer. Basically we are generating a point on the X-Y plane where the PDF is plotted. If the point falls below the PDF curve, it would be of the same distribution as $f(x)$, if not try again.

As an application to the acceptance/rejection method we present a method to generate beta deviates. The method is due to Cheng and could be sumarized as follows:

Knowing the shape parameters a and b of the fitted beta distribution the following is applied:

Set $\alpha = a + b$
If MIN $(a,b) \leq 1$, set $\beta = \text{MAX}(a^{-1}, b^{-1})$
otherwise set $\beta = \sqrt{\{(\alpha - 2)/(2ab - \alpha)\}}$

Set $\gamma = a + \beta^{-1}$

1. All parameters should be larger than zero.
2. Generate two uniform deviates r_1 and R_2 on range [0,1].
3. Set $V = \beta \ln[R_1/(1 - R_1)]$
4. Set $W = ae^{v}$.
5. If $\alpha \ \ln[\alpha/(b + W)] + \gamma V - 1.3862944 \geq \ln(R_1^2 \ R_2)$, set $X = W/(b + W)$ and deliver X (rescale to the appropriate range L, U).
6. Go to step 2.

The generated X would be on the interval [0,1]. The generated deviate can easily be rescaled to the proper range, say $[L, U]$, by

$$X_1 = L + (U - L) * X \tag{16.19}$$

The generated value x_1 would be a beta variate on the range $[L, U]$ and with the shape parameters a and b normally referred to as BETA(L, U, a, b).

16.6.3 Other Techniques

There are other techniques for generating random variates as previously mentioned. The scope however is not within the context of this text. The interested reader should consult other texts on the subject [e.g., see Fishman (1973), Brately et al. (1983), Law and Kelton (1982)]. The last technique we present deals with generating normal deviates. Since the CDF cannot be found analytically, one has to resort to a technique other than the inverse transform. The one we introduce here is due to Box and Muller.

The technique does not fit under the acceptance/rejection method or the composition or the approximation of the CDF; however, it is widely used and accurate. It can be summarized as follows: To generate two normal deviates X and Y from the normal distribution Normal (μ, σ^2), one would generate two random numbers R_1 and R_2 and substitute in the following equations:

$$x = \mu + \cos 2\pi R_1 \sqrt{-2\sigma^2 \log R_2} \tag{16.20}$$

$$Y = \mu + \sin 2\pi R_1 \sqrt{-2\sigma^2 \log R_2} \tag{16.21}$$

The values X and Y would be two independent normal deviates with means μ and variance σ^2. In the context of programming one would call the routine and generate two normal deviates at a time. The first would be used and the second saved for the second call.

16.7 SELECTING AN INPUT MODEL AND TESTING FOR GOODNESS OF FIT

16.7.1 Defining an Empirical Distribution

A simple way of defining an empirical distribution from a sample of observation can be described as follows:

Given a sample of n observations X_i ($i = 1 \ldots n$)

1. Sort the sample observation in increasing order resulting in the new ordered sample $X_{(1)} X_{(2)} \ldots X_{(n)}$.
2. Calculate the probability of sampling a number less or equal to each and every observation.

$$F(x_i) = \Pr(X \leq x_i) = \frac{i}{n}$$

3. This specifies an empirical CDF that would graphically look like the CDF for concrete screeding durations as shown in Figure 16.7.

16.7.2 Selecting a Distribution to Model the Sample Observations

Given that we have collected a sample of observations from a construction site, the first decision to be made would be which statistical distribution to use as a model (e.g., normal, beta, or lognormal). In most cases it is a matter of the physical properties of the random number desired and the experience of the modeler. Modeling durations of work tasks for example requires a bounded distribution since physically any real task will require some amount of time to be accomplished. The time cannot be zero or negative and is usually bounded from the positive side as well, depending on the type of the task. This automatically excludes unbounded distributions like the normal and exponential from being good candidate models (this does not include the truncated versions of these distributions). To better appreciate this phenomenon, consider the case of a normal model for the duration of a work task. There is always a small probability that the duration sampled be negative

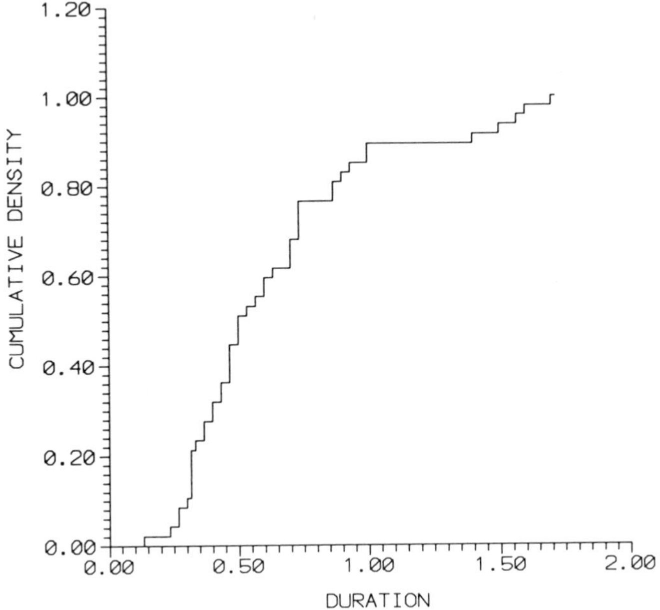

FIGURE 16.7 Histogram of Concrete Screeding Operation

(physically impossible for time) or be very large (infinite) which would be highly unlikely for a duration of a construction work task.

The most basic way of selecting a statistical distribution as a model for a set of data is to relate the sample obtained to the shape (or shapes for families of distributions) of the theoretical distribution. A histogram formed from the sample is analogous to the PDF of the theoretical distribution since both reflect the weight each of the sample intervals (or sample points) should receive in terms of their probability of occurrence (hence sampling). The idea would be to relate the shape of the histogram of the sample to the shape of a known distribution. For example, if the histogram were skewed to the left, a log normal or beta distribution would be a good candidate since both can attain such shapes. The problem with this technique however is the construction of the histogram itself. In the absence of a standard technique to do this, one can easily distort the real shape of the histogram by inappropriately specifying the width of the cells and their locations. A good way to overcome this problem is by using a standard procedure for specifying histograms like Sturge's rule, for example, which could be summarized as follows.

Given N observations X_i to summarize in a frequency distribution (histogram) one would take

$$\text{Number of cells} = 1 + 3.3 \log_{10}(N)$$

$$\text{Width of a cell} = \frac{X_{\text{max}} - X_{\text{min}}}{\text{No. of cells}}$$

$$\text{Lowest cell} = X_{\text{min}}$$

This guideline will usually reveal the general layout of the data provided that the number of cells is in the range $5 \leq$ no. of cells ≤ 15. The most frequently encountered problem with constructing histograms is actually the tendency for specifying more cells than the data can support. Sturge's rule accounts for that in a heuristic manner.

To illustrate the construction of a histogram for a sample data set, we present the data set in Table 16.5 of the time it takes to dump concrete on a concrete pouring operation on a given floor.

To construct a histogram according to Sturge's rule, we need to know the total number of observations and the end points of the data. The calculations would be carried out as follows:

$$\begin{aligned}
\text{Total number of observations} &= 47 \\
\text{Number of cells required} &= 1 + 3.3 \log_{10}(47) \\
&= 7 \text{ cells} \\
\text{Width of each cell} &= \frac{1.700 - 0.133}{7} = \frac{1.567}{7} \\
\text{Lowest cell} &= 0.133
\end{aligned}$$

The histogram is now specified. The second step is to go over the set of observations and count the number of observations that fall into each of the seven cells to construct the histogram which is shown in Figure 16.8.

TABLE 16.5 Cycle Times for Concrete Dumping Task

0.36	0
0.42	0
0.42	0.46
1.20267	0.46
1.20267	0.33
1.98467	0.33
1.98467	0.15
2.76667	0.15
2.76667	0.20
3.54867	0.20
3.54867	0.51
4.33067	0.51
4.33067	0.10
5.11267	0.10
5.11267	0

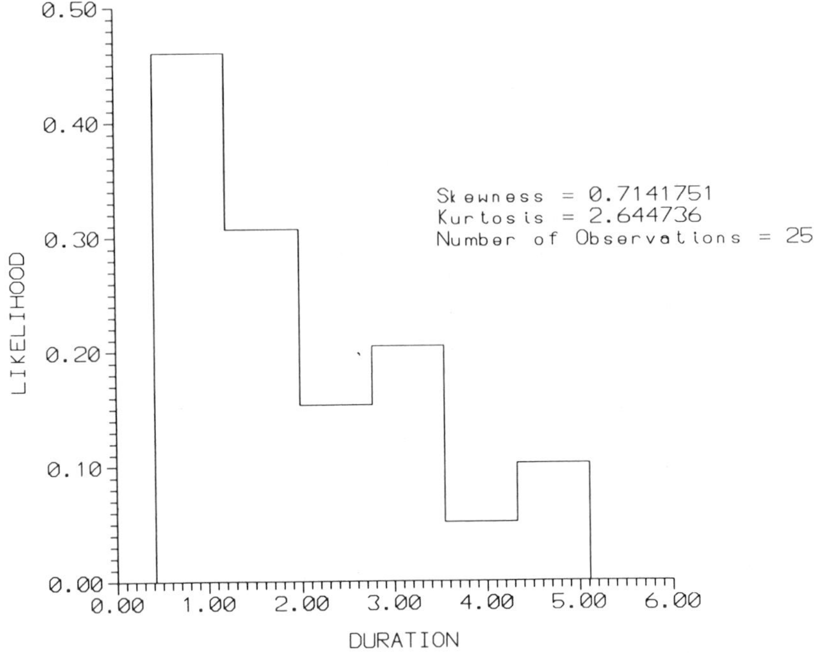

FIGURE 16.8 Flexibility of Beta Distribution

Having constructed an appropriate histogram, the next step in selecting a distribution as an input model is to relate the shape of the histogram to that of a known distribution. A somewhat "bell-shaped" histogram suggests the use of a normal distribution (truncated for duration models) for example.

An even better approach for selecting distribution is to start with a "family" of distributions. Families like the Beta family, the Johnson system, and others give the modeler the flexibility of attaining a wide variety of shapes with the same distribution model. Figure 16.9 shows samples of PDFs from the Beta family attained by varying the shape parameters of the beta distribution.

16.7.3 Estimating the Parameters of a Selected Distribution

Having decided on the type of distribution to use as an input model, we have to solve for the parameters that define that distribution. To give an example, say our model is to be a beta distribution. In this case we have to solve for the end points of the distribution as well as its two shape parameters. To generalize, if we have chosen a distribution with *K* parameters that fully define it, we have to solve for all *K* parameters. In general one could use the maximum likelihood estimates (MLE) of the parameters of concern when

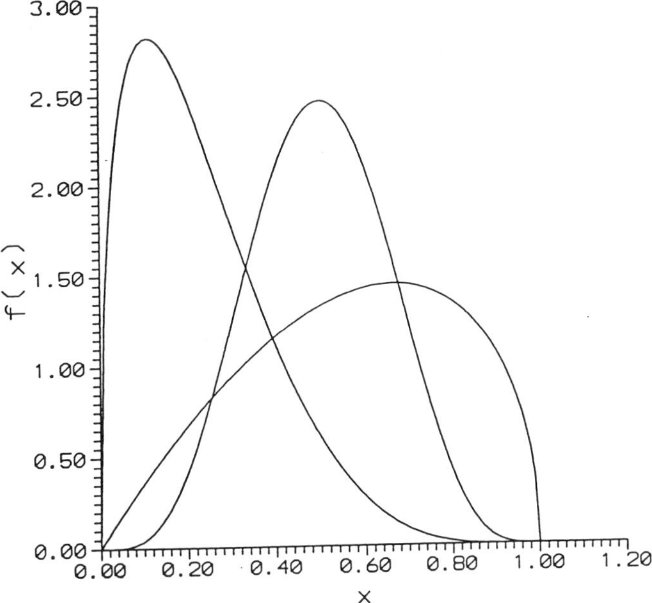

FIGURE 16.9 A *q-q* Plot for Time Required to Complete 40 Truck Loads (Sample of 12 Observations)

the likelihood equations can be numerically solved. To make our point clearer, consider that we are parameterizing a normal distribution. In this case, the parameters that fully define a normal distribution are its mean and its variance. The maximum likelihood estimates of the mean for the normal distribution is given by

$$\mu = \overline{X} = \frac{1}{n}\sum_{i=1}^{n} X_i \tag{16.22}$$

The MLE of the variance is given by

$$\sigma^2 = \frac{n-1}{n} S^2 \tag{16.23}$$

where $\overline{X}$ and S^2 are the mean and variance of the collected sample respectively.

For some distributions like the beta, Weibul, and gamma one would have to solve a system of nonlinear equations for the MLE estimates of the parameters and this would not always be possible analytically so one would have to resort to numeric methods.

When the MLE method is easy to implement it gives the most reasonable fitted distribution, particularly when the set of observations is large.

Another approach to parameterize a distribution is to use the method of moment matching. The method simply states that for a distribution with K parameters, fix K moments of the distribution to the first K moments of the sample and solve K equations for the K unknown parameters. The moment matching procedure is widely used but requires large samples to insure good fits. To illustrate the technique we consider having to fit a beta distribution to a sample of observations X_i. For simplicity we assume that the end points of the distributions are going to be those of the sample and all we need to solve for are the two shape parameters a and b.

The procedure of moment matching would be as follows: set $L = X_{(1)}$ and $U = X_{(n)}$, where L and U are the lower and upper end point of the theoretical beta distribution respectively, and $X_{(1)}$ and $X_{(n)}$ are the lower and upper end points of the sample respectively.

To use moment matching we set the mean and variance of the sample equal to those of the theoretical mean and variance of the beta distribution.

The mean and variance of the theoretical beta distribution are given by

Mean

$$\mu = L + (U - L)\frac{a}{a + b} \tag{16.24}$$

Variance

$$\sigma^2 = (U - L)^2\frac{ab}{(a + b)^2(a + b + 1)} \tag{16.25}$$

Notice that L and U were previously set to the end points of the sample. We set $\mu = \overline{X}$ and $\sigma^2 = S^2$ in Equations (16.24) and (16.25) and solve for the estimates of the parameters a and b to yield

$$\hat{a} = \frac{1 - \overline{X}}{S^2}\,[\overline{X}\,(1 - \overline{X}) - S^2] \tag{16.26}$$

$$\hat{b} = \frac{\overline{X}\hat{a}}{1 - \overline{X}} \tag{16.27}$$

The estimates of the parameters a and b are given as $\hat{a}$ and $\hat{b}$. Now the beta distribution is fully defined and one could proceed to perform goodness of fit tests.

16.7.4 Testing for Goodness of Fit

Having parameterized a distribution one should check for the goodness of fit by comparing the fitted distribution to the empirical distribution and assessing the quality of the fit obtained. Usually one would perform the goodness of fit test by using statistical tests (like the chi-square or the *K-S* tests), or visually assessing the quality of the fit.

The Chi-Square Goodness of Fit Test. The Karl Pearson chi-square test is based on the measurement of the discrepancy between the histogram of the sample and the fitted probability density function. When the discrepancy is large enough, the test rejects the fitted model. The test is derived based on the fact that the parameters of the fitted distribution were found using the MLE approach and could be summarized as follows.

To test that the observations $X_i \{1 \leq i \leq n\}$ follow a particular distribution $f(x)$ where we computed its m parameters using the MLE approach:

1. Select a number of class intervals to group the observations (say c_1, $c_2 \ldots c_n$) and associate the intervals with a weight $p_i = \int f(x)\,dx$. A number of approaches could be used.

Mann–Wald Procedure: Use classes with equal weights p_i so that if you have k classes each would have a weight $1/k$.

$$k = b\left(\frac{\sqrt{2}\,(n-1)}{Z_{(1-\alpha)}}\right)^{2/5} \quad (2 \leq b \leq 4).$$

Set $c_0 = 0$ and $c_n = \infty$ and for $i = 1$ to k we set $F(c_i) = i/k$ for all i yielding $c_i = F^{-1}(i/k)$.

2. Count the number of observations X_i in each cell c_i such that $c_{i-1} < X_i < c_i$ to give N_i.

3. Calculate the chi-square statistic χ^2 as

$$\chi^2 = \sum_{i=1}^{k}\left[\frac{(N_i - np_i)^2}{np_i}\right]$$

where n is the total number of observations, N_i is the respective number of observations in cell i, and $p_i = 1/k$.

4. Compare the value of χ^2 obtained to that from the chi-square tables similar to the one in Table 16.6. Reject the distribution if the $\chi^2_{\text{computed}} > \chi^2_{1-\alpha}(k - L - 1)$ where $k - L - 1$ are the degrees of freedom (note the loss of L degrees of freedom since we estimated the parameters from the sample) for a theoretical chi-square distribution at the α confidence level.

TABLE 16.6 Percentiles of the χ^2 Distribution[a]

Degrees of Freedom (γ)	0.005	0.010	0.025	0.05	0.10	0.20	0.30	0.40	0.50	0.60	0.70	0.80	0.90	0.95	0.975	0.990	0.995	γ
1	0.0^4393	0.0^3157	0.0^3982	0.0^2393	0.0158	0.0642	0.148	0.275	0.455	0.708	1.07	1.64	2.71	3.84	5.02	6.63	7.88	1
2	0.0100	0.0201	0.0506	0.103	0.211	0.446	0.713	1.02	1.39	1.83	2.41	3.22	4.61	5.99	7.38	9.21	10.6	2
3	0.0717	0.115	0.216	0.352	0.584	1.00	1.42	1.87	2.37	2.95	3.67	4.64	6.25	7.81	9.35	11.3	12.8	3
4	0.207	0.297	0.484	0.711	1.06	1.65	2.19	2.75	3.36	4.04	4.88	5.99	7.78	9.49	11.1	13.3	14.9	4
5	0.412	0.554	0.831	1.15	1.61	2.34	3.00	3.66	4.35	5.13	6.06	7.29	9.24	11.1	12.8	15.1	16.7	5
6	0.676	0.872	1.24	1.64	2.20	3.07	3.83	4.57	5.35	6.21	7.23	8.56	10.6	12.6	14.4	16.8	18.5	6
7	0.989	1.24	1.69	2.17	2.83	3.82	4.67	5.49	6.35	7.28	8.38	9.80	12.0	14.1	16.0	18.5	20.3	7
8	1.34	1.65	2.18	2.73	3.49	4.59	5.53	6.42	7.34	8.35	9.52	11.0	13.4	15.5	17.5	20.1	22.0	8
9	1.73	2.09	2.70	3.33	4.17	5.38	6.39	7.36	8.34	9.41	10.7	12.2	14.7	16.9	19.0	21.7	23.6	9
10	2.16	2.56	3.25	3.94	4.87	6.18	7.27	8.30	9.34	10.5	11.8	13.4	16.0	18.3	20.5	23.2	25.2	10
11	2.60	3.05	3.82	4.57	5.58	6.99	8.15	9.24	10.3	11.5	12.9	14.6	17.3	19.7	21.9	24.7	26.8	11
12	3.07	3.57	4.40	5.23	6.30	7.81	9.03	10.2	11.3	12.6	14.0	15.8	18.5	21.0	23.3	26.2	28.3	12
13	3.57	4.11	5.01	5.89	7.04	8.63	9.93	11.1	12.3	13.6	15.1	17.0	19.8	22.4	24.7	27.7	29.8	13
14	4.07	4.66	5.63	6.57	7.79	9.47	10.8	12.1	13.3	14.7	16.2	18.2	21.1	23.7	26.1	29.1	31.3	14
15	4.60	5.23	6.26	7.26	8.55	10.3	11.7	13.0	14.3	15.7	17.3	19.3	22.3	25.0	27.5	30.6	32.8	15

16	5.14	5.81	6.91	7.96	9.31	11.2	12.6	14.0	15.3	16.8	18.4	20.5	23.5	26.3	28.8	32.0	34.3	16
17	5.70	6.41	7.56	8.67	10.1	12.0	13.5	14.9	16.3	17.8	19.5	21.6	24.8	27.6	30.2	33.4	35.7	17
18	6.26	7.01	8.23	9.39	10.9	12.9	14.4	15.9	17.3	18.9	20.6	22.8	26.0	28.9	31.5	34.8	37.2	18
19	6.84	7.63	8.91	10.1	11.7	13.7	15.4	16.9	18.3	19.9	21.7	23.9	27.2	30.1	32.9	36.2	38.6	19
20	7.43	8.26	9.59	10.9	12.4	14.6	16.3	17.8	19.3	21.0	22.8	25.0	28.4	31.4	34.2	37.6	40.0	20
21	8.03	8.90	10.3	11.6	13.2	15.4	17.2	18.8	20.3	22.0	23.9	26.2	29.6	32.7	35.5	38.9	41.4	21
22	8.64	9.54	11.0	12.3	14.0	16.3	18.1	19.7	21.3	23.0	24.9	27.3	30.8	33.9	36.8	40.3	42.8	22
23	9.26	10.2	11.7	13.1	14.8	17.2	19.0	20.7	22.3	24.1	26.0	28.4	32.0	35.2	38.1	41.6	44.2	23
24	9.89	10.9	12.4	13.8	15.7	18.1	19.9	21.7	23.3	25.1	27.1	29.6	33.2	36.4	39.4	43.0	45.6	24
25	10.5	11.5	13.1	14.6	16.5	18.9	20.9	22.6	24.3	26.1	28.2	30.7	34.4	37.7	40.6	44.3	46.9	25
26	11.2	12.2	13.8	15.4	17.3	19.8	21.8	23.6	25.3	27.2	29.2	31.8	35.6	38.9	41.9	45.6	48.3	26
27	11.8	12.9	14.6	16.2	18.1	20.7	22.7	24.5	26.3	28.2	30.3	32.9	36.7	40.1	43.2	47.0	49.6	27
28	12.5	13.6	15.3	16.9	18.9	21.6	23.6	25.5	27.3	29.2	31.4	34.0	37.9	41.3	44.5	48.3	51.0	28
29	13.1	14.3	16.0	17.7	19.8	22.5	24.6	26.5	28.3	30.3	32.5	35.1	39.1	42.6	45.7	49.6	52.3	29
30	13.8	15.0	16.8	18.5	20.6	23.4	25.5	27.4	29.3	31.3	33.5	36.3	40.3	43.8	47.0	50.9	53.7	30
35	17.2	18.5	20.6	22.5	24.8	27.8	30.2	32.3	34.3	36.5	38.9	41.8	46.1	49.8	53.2	57.3	60.3	35
40	20.7	22.2	24.4	26.5	29.1	32.3	34.9	37.1	39.3	41.6	44.2	47.3	51.8	55.8	59.3	63.7	66.8	40
45	24.3	25.9	28.4	30.6	33.4	36.9	39.6	42.0	44.3	46.8	49.5	52.7	57.5	61.7	65.4	70.0	73.2	45
50	28.0	29.7	32.4	34.8	37.7	41.4	44.3	46.9	49.3	51.9	54.7	58.2	63.2	67.5	71.4	76.2	79.5	50
75	47.2	49.5	52.9	56.1	59.8	64.5	68.1	71.3	74.3	77.5	80.9	85.1	91.1	96.2	100.8	106.4	110.3	75
100	67.3	70.1	74.2	77.9	82.4	87.9	92.1	95.8	99.3	102.9	106.9	111.7	118.5	124.3	129.6	135.6	140.2	100

[a] Abridged from Table V of *Statistical Tables and Formulas* by A. Hald, John Wiley and Sons, New York, 1952.

The Kolmogorov–Smirnov Test. The test is based on measuring the maximum discrepancy (gap) between the empirical CDF and the fitted CDF. The maximum positive deviation would be given by

$$D_n^+ = \max \{F(x)_{\text{empirical}} - F(x)_{\text{theoretical}}\}$$

$$= \max \left\{ \frac{i}{n} - F(x_{(i)}) \right\} \qquad \text{for } i = 1 \ldots n$$

The maximum negative deviation would be given by

$$D_n^- = \max \{F(x)_{\text{theoretical}} - F(x)_{\text{empirical}}\}$$

$$= \max \left\{ F(x_i) - \frac{i-1}{n} \right\} \qquad \text{for } i = 1 \ldots n$$

The test statistic would be $D_n = \max \{D_n^+, D_n^-\}$.

The test would be to compute D_n as described above and to compare the value with that from a table of critical $K-S$ values at the appropriate confidence and degrees of freedom. When the computed D is larger than the theoretical one, the distribution is rejected.

Visual Assessment of the Quality of the Fit. Among the techniques most widely used in measuring the goodness of fit of a theoretical distribution to an empirical one is the method of visual inspection of the fit. Although the method does not bare much of statistical or mathematical weight, it proves to be as effective as any other test. The visual assessment of the quality of the fit is usually conducted in conjunction with the statistical tests as an assurance of the test results. Basically the method is simply to plot both the empirical and fitted CDFs on one plot and compare how good the fitted CDF tracks the empirical one. Alternatively one can compare how well the shape of the sample histogram compares to that of the theoretical PDF. When the CDF is available it is always better to compare to it because, as previously mentioned, a histogram can be easily distorted and can practically attain any shape we desire.

16.8 CHECKING FOR NORMALITY OF OUTPUT

We have previously seen that most of the analysis introduced in this chapter was based on the assumption of normality of output data. When the output is not normal the confidence intervals would be only approximate, depending on how deviated the sample is from linearity. An easy way to test whether a given sample has originated from a normal distribution one can do the following (commonly referred to as $q - q$ plot, Hahn & Shapiro, 1967):

1. Order the statistics X_i to form an ordered sample $X_{(1)} < X_{(2)} < X_{(3)} < \ldots X_{(n)}$
2. Plot the coordinates $X_{(i)}$ and $\phi^{-1}\left(\frac{i - 0.375}{n}\right)$.
3. If the points appear to be linear, the sample can be assumed to have originated from a normal distribution.

A sample $q - q$ plot for a concrete screening task is given in Figure 16.10.

Another way to test for normality of a sample is to use the Shapiro–Wilk W statistic. A detailed coverage of the test can be found in Hahn and Shapiro (1967). Royston (1982) extended the test to work with large samples up to 2000. The W statistic lies in the interval [0,1], a value close to 1.0 is a good indication of linearity. For our example the W statistic was computed and found to be 0.9521, which can be considered an indication of a normal sample.

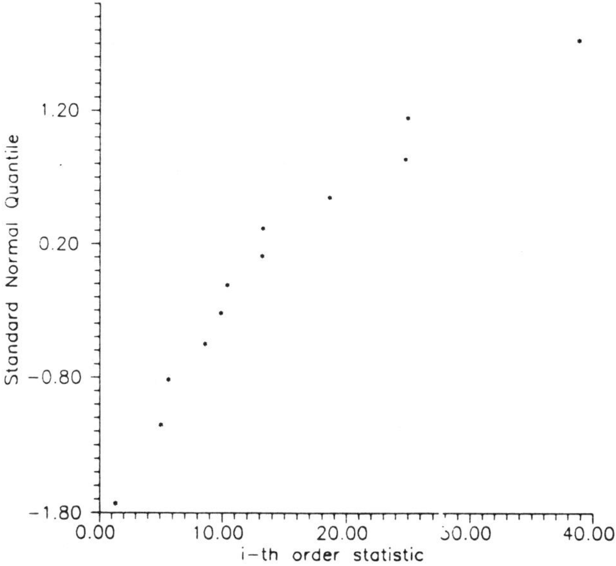

FIGURE 16.10 Sample q — q plot

CHAPTER 17

AN INTRODUCTION TO SYSTEMS SIMULATION

The previous chapter introduced Monte Carlo simulation and some of its applications in construction planning and estimating. The modeling capabilities of the CPM-based planning tools (this includes CPM, precedence diagramming, and PERT) do not allow modeling a dynamic environment such as a construction site. With many recent developments in total quality management, more emphasis is being directed to the production source (i.e., the process level in a construction site). Halpin adds that the fixed CPM-based planning tools are not capable of analyzing and identifying nonvalue-adding tasks and thus eliminating them. For this to occur a model capable of depicting the dynamic flow of resources and their interactions during the construction process must be used. Such tools already exist in the form of system simulation methods. Of these, CYCLONE (Halpin and Riggs, 1992) has received wide attention in construction research. Other commercially available systems include SLAM II (Pritsker, 1987), GPSS, SIMSCRIPT, INSIGHT, SIMAN, and many others.

Both CPM or precedence network planning are best used when the operations of a project have determinable durations. When estimates of activity durations are probabilistic, PERT networks are more useful. Common to these three networking techniques is the requirement that all preceding activities or operations be completed before a node is finally reached or realized. In situations where the realization of a node is dependent on the completion of one or more activities preceding this node (rather than all), and where the performance of these activities is probabilistic, general purpose simulation languages are very useful.

Today's projects are no longer confined to stable environments for which methods are known and durations can be estimated with confidence. Projects involving ocean and space research or exploration often necessitate evaluation of alternative strategies by simulation. If a certain type of hostile environment is encountered, it may become necessary to change the strategy or terminate the project. A prior evaluation of different strategies during the feasibility study stage can be helpful in determining the risk involved and ensuring adequate resources to steer the project to a successful conclusion. Because of the increasing occurrence of such projects, the reader should not limit him or herself to learning such well-known networking techniques as CPM, precedence, and PERT; the reader must also add this new tool to his or her repertory.

17.1 TYPICAL SIMULATION APPLICATION PROBLEM

Suppose a tunnel 3500 m in length, with a diameter of 10 m, is to be built beneath the ocean floor to connect an offshore island to the main land. The geology of the area is not definitely known, and further information will not be available until construction is under way. However, expert opinion gives the probable geologic makeup and associated excavation times, as shown in Table 17.1. Expert opinion also provides the information given in Table 17.2 about the tunnel lining.

Delivery of tunneling equipment and site preparation are carried out concurrently and take 45 and 20 days, respectively. These activities are followed by assembly of the tunneling equipment, which takes 15 days. This is followed by excavation. When some progress has been made on excavation, the tunnel lining is started. The final activity, site clearance and demobilization, takes 10 days.

A CPM network for this project might resemble Figure 17.1. A duration of 20 days is assigned to tunnel excavation I, after which lining can be started. No durations for excavation II and tunnel lining can be determined since various times and probabilities are associated with each, depending on the geology. Hence neither CPM nor precedence scheduling can be used.

With PERT scheduling, probabilistic estimates of activity durations are permitted, but the realization of a node is dependent on the completion of all activities preceding this node. For example, in the network of Figure 17.2, cases I, II, and III are depicted by sets of activities 5–6, 5–7; 5–11, 5–8; and 5–9, 5–10. The network shows that all activities preceding node 11 must be completed. This interferes with the logic since at both nodes only two activities, one for excavation and one for lining, are required, that is, the geology is as shown for either case I, II, or III and correspondingly only one combination of tunnel excavation and lining is required. Node 11 is achieved when any one of the three sets of activities is performed. PERT cannot depict such logic. To overcome these difficulties, simulation may be

TABLE 17.1 Excavation

	Granite		Sandstone		Shale			
	Length	Expected Duration	Length	Expected Duration	Length	Expected Duration	Total Duration	Probability
Case I	1500	400	500	80	1500	50	530	0.2
Case II	2000	500	750	100	750	31	631	0.3
Case III	2100	550	1000	120	400	19	689	0.5

TABLE 17.2 Lining

	Anchor Bolts	Formed Concrete Lining		Shotcrete Lining				
	Length	Expected Duration	Length	Expected Duration	Length	Expected Duration	Total Duration	Probability
Case I	1500	58	500	20	1500	160	238	0.2
Case II	2000	73	750	25	750	100	198	0.3
Case III	2100	80	1000	30	400	60	170	0.5

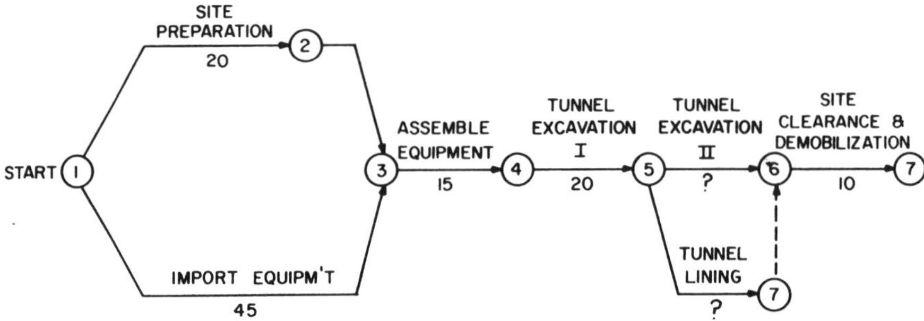

FIGURE 17.1 Network for Tunnel Example

used to construct a network of the system and obtain a project time by simulation for management planning.

The background for the development of simulation methods like CYCLONE lies in the area of stochastic networks. A stochastic network has the following properties:

1. Each network consists of nodes and activities that denote the logic of the project.
2. An activity in the network has associated with it a probability for its performance.

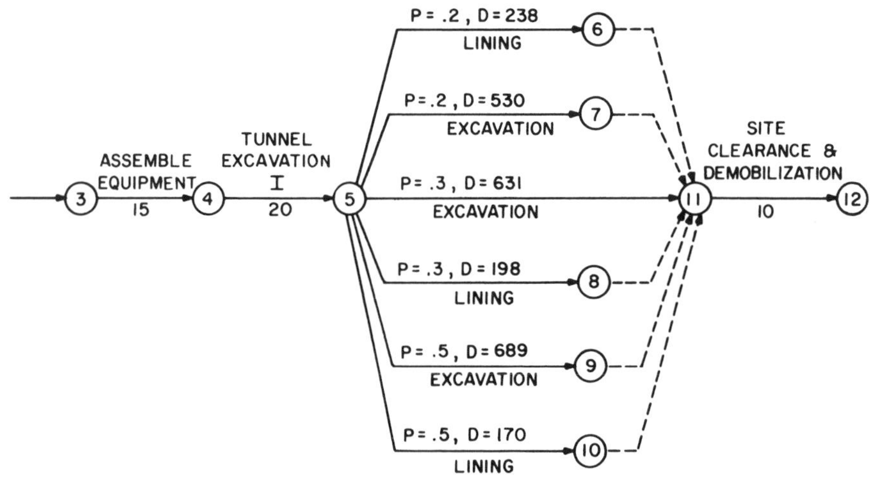

FIGURE 17.2 Probability Network for Tunnel Example

3. Other parameters provide the activity data.
4. A realization of a network is the realization of a particular set of activities and nodes that describes the network for one experiment.
5. If the time associated with an activity is a random variable, then a realization also implies that a fixed time has been selected for each activity.

An attempt is made here to present simulation in a simplified form so that the reader can recognize its usefulness as a network technique.

17.2 CYCLONE MODEL VERSUS OTHER NETWORK MODELS

The objective of any network technique is to model a project by the logical management of its operations. For this purpose special symbols are introduced to each of the three network techniques previously described. CYCLONE modeling, which differs from the other network techniques because it has probabilistic as well as deterministic nodes, has special symbols for node identification. CYCLONE allows not only unidirectional flow, like the other network techniques, but also loops, which may start from any node and be incident to any previous node. This accommodates the iterative nature of operations in research and development projects. CYCLONE therefore places a special emphasis on graphical representation.

An additional purpose of a CYCLONE model is actually to simulate the project by repeatedly incurring the network logic. This means that each activity is released (performed through simulation) for a large, prespecified number of times, so that statistical data can be collected at the desired nodes for analysis of the project. To do this, the given problem must be analyzed and separated from irrelevant factors. Thus a network model is required. CYCLONE notation will therefore be discussed in the next section and later used in the development of a CYCLONE model for a project.

17.3 CONSTRUCTION PROCESS SIMULATION

Most simulation methods offer modeling paradigms and experimentation environments. CYCLONE provides the modeling elements and methods that a modeler can use to represent a construction operation in much the same way as a scheduler would build a CPM network for a construction project (i.e., by specifying activities, their logical relationships, durations, and resource requirements). To model an operation using CYCLONE, the modeler focuses on the involved resources and their interactions. A resource can be in one of two states, active or idle. An active state of a resource is represented by a square element while the idle state with a circle element. The resource

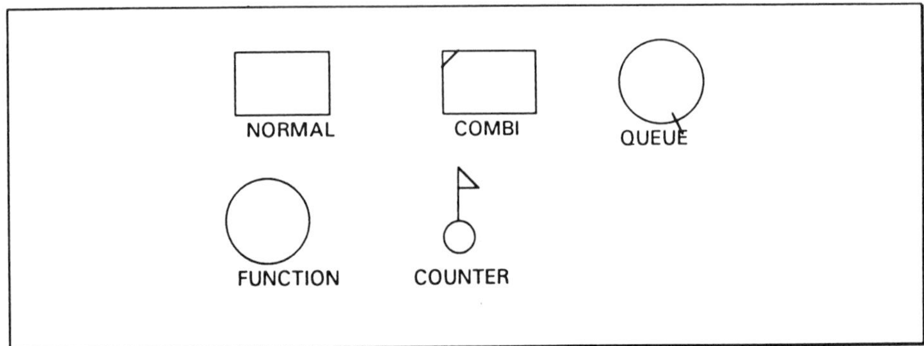

FIGURE 17.3 CYCLONE Elements (Halpin, 1973)

move between the two states and thus from one activity to the other in the model. It is essential that the reader makes the distinction between this method and a static system like CPM.

17.4 BUILDING A CYCLONE MODEL

A CYCLONE model is constructed using the CYCLONE elements shown in Figure 17.3.

The rules for structuring CYCLONE network models using these elements may be summarized as follows:

CYCLONE Element	Description and Rules for Model Building
NORMAL	The NORMAL is not a constraint task. Any resources that arrives to a NORMAL is given access and is immediately processed. It is like a serving station with infinite number of servers. Can be preceded by all other CYCLONE elements except for a queue node. Can be followed by all other elements except for a COMBI.
COMBI	Is a task that is constraint by the availability of more than one type of resource. A resource arriving to a COMBI will have to wait until all other required resources are available before it is given access to the task. Can be preceded by Queue nodes only.

CYCLONE Element	Description and Rules for Model Building
	Can be followed by all other elements except COMBIs.
QUEue	A QUEue node is a waiting area for a resource. Therefore it makes sense to use it only when a task is restraint. A resource arriving at a QUEue node will stay in the node until a COMBI is ready to process it.
	A QUEue node has one other function in the MicroCYCLONE (Halpin and Riggs, 1992) implementation namely to multiply resources when specified. In other words a modelor can specify that once a resource enters a specified QUEue node it will multiply into a finite number of duplicate resources.
	Can be preceded by any element except a QUEue node.
	Can be followed by COMBIs only.
FUNCTION	The FUNCTION element was devised so it would provide some flexibility. Different computer implementations of CYCLONE have somewhat different functions. In MicroCYCLONE one type of function is allowed, namely the consolidate function. Its job is to take units and consolidate them into a specified number. Any unit arriving at this function will accumulate until a threshold value is reached at which point only one unit is released from the function (all others are destroyed).
	Can be preceded by all elements except QUEue nodes.
	Can be followed by all elements except COMBIs.
COUNTER	The counter keeps track of the number of times units pass it. It does not alter any of the resources or their properties; it just increments and keeps track of cycles and few other statistics.
	Can be preceded by all elements except QUEue nodes.
	Can be followed by all elements except COMBIs.

Preparing a CYCLONE model requires the modeler to focus on the resources, their active and idle states and make use of the building blocks discussed above to properly represent an operation. All key resources involved in the operation to be modeled must be first identified. Once this has been accomplished all work tasks (active states of a resource) are defined. These would be represented using the rectangular CYCLONE elements. If the task requires a combination of resources to be achieved, a COMBI element is used to model it. If it is nonconstraint by such combinations of resources, then a NORMAL element is used.

The next step is to define the resource involvement in the tasks and decide on where they are initialized (resources may be initialized only in front of COMBI tasks) and where they should wait when a constraint task is not available for service (i.e., it is waiting for other resources before it can proceed). These waiting areas are modeled with the circle element known as QUEue nodes in CYCLONE terminology. Now that all tasks and waiting areas have been defined the logical relationships between work tasks (i.e., precedence and sequencing of the tasks) are established by connecting the COMBIs, NORMALS, and QUEue nodes with directional flow arrows indicating where the resource would be moving from and to upon completion of a task. This makes up the CYCLONE network.

To illustrate how a CYCLONE model may be built, a simple model of an excavation process is presented. Consider that a certain amount of earth is to be moved from location A, loaded on trucks and hauled to location B where it is dumped as shown in the schematic diagram of Figure 17.4. Once the truck dumps its load, it returns for another cycle until the project is complete.

The modeling part of CYCLONE simulation can be summarized as follows:

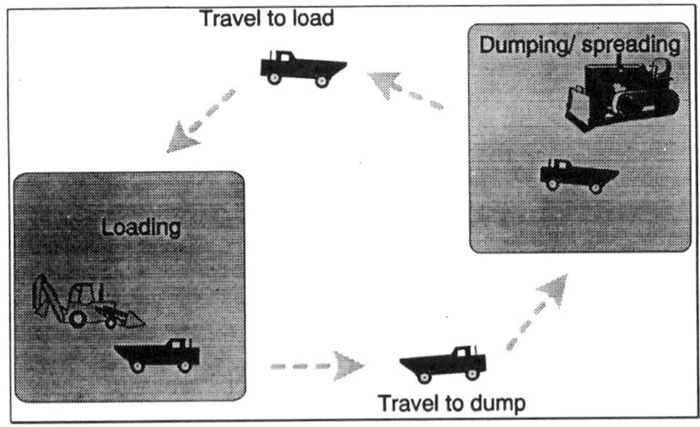

FIGURE 17.4 A Schematic Illustration of an Earth-Moving Operation

1. Resources are trucks, loaders, and dump spotter. Each of these resources has a well-defined cycle associated with it. The cycles are composed of individual well-defined tasks.
2. The tasks with which the loader is involved are

   Loading. This task requires that both the truck and a loader be available before it may commence, hence it is defined as a COMBI.

   The tasks that the truck is involved with include:

   Loading defined as a COMBI (same reason as above).

   Hauling to dump site. This only requires that the truck be available and does not require the loader, therefore it is defined as a NORMAL.

   Dumping. Requires a dump spotter and the truck, therefore it is constrained and is defined with a COMBI.

   Returning to loading site. NORMAL (same reason as for hauling).

   The dump spotter is only involved with the dumping task (COMBI).
3. Now the model can be graphically represented by visualizing the flow of resources in the project and using the defined elements to represent that flow.

Consider first the truck cycle. It gets loaded, it travels to the dump site, dumps it load, and returns. This can be represented using the CYCLONE elements as shown in Figure 17.5.

This model is missing few elements before it becomes correct. Since "Load" and "dump" are COMBI tasks they must be preceded by queue nodes as shown in Figure 17.6.

Now the trucks wait in the prescribed queue nodes awaiting the other resource before the respective tasks may begin. The cycles of the other two resources are given in Figure 17.7.

In order to form the complete model, the three cycles are combined. Since the "load" task is the same for both the loader and truck cycles, only the

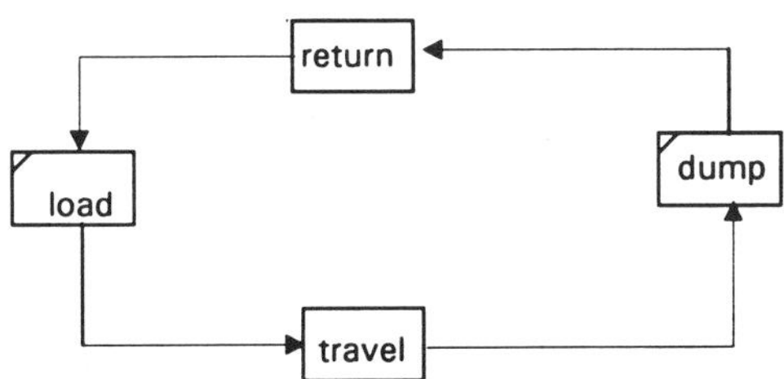

FIGURE 17.5 CYCLONE Model

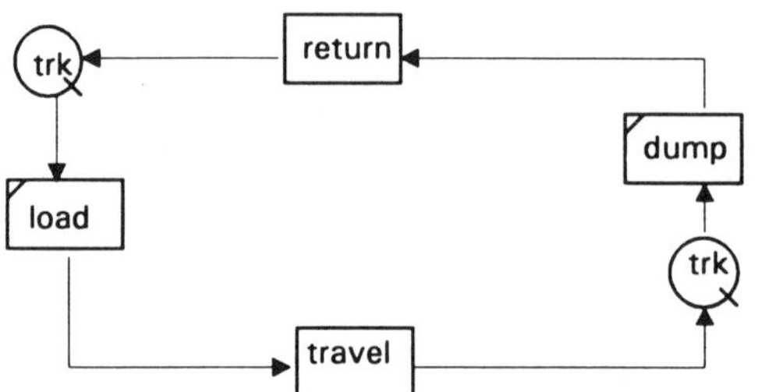

FIGURE 17.6 CYCLONE Model with Queues

loader queue must be added. The same applies to the spotter queue. The compete model is given in Figure 17.8.

Notice that the counter has been added so production may be measured during the simulation. The choice of placing the counter after the dump task signifies that production is realized only after one truck load is dumped. Also added to the model are the duration of the various tasks, which in this case were either beta or uniformly distributed and the initial number and location of the resources in the system which is denoted by an asterisk.

To review the model consider how entities will flow through it. The loader will simply engage in loading; once complete it returns to its waiting location (queue), the truck engages in loading, then proceeds to traveling, waits in a queue before dumping its load, and finally triggers the production counter once the dump is complete, and travels back to the loading queue. The spotter only engages with the dumping tasks and waits in its queue for another truck to arrive. The simulation actually involves doing just this (i.e., moving the resource when permitted and keeping track of time). This is referred to as discrete event simulation since the computer maintains a clock that will advance only when an important event takes place (e.g., travel compete).

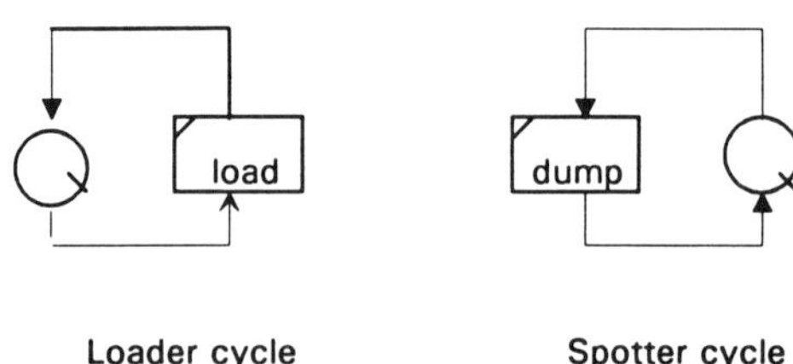

FIGURE 17.7 CYCLONE Models for the Load and Dump Subprocesses

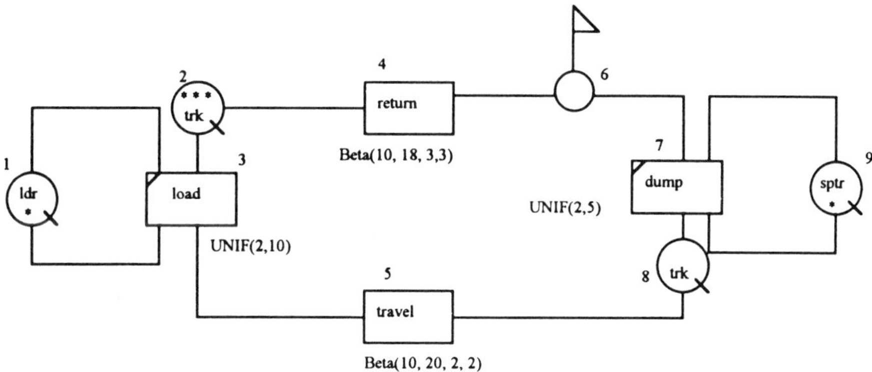

FIGURE 17.8 Final CYCLONE Model for the Earth-Moving Operation

17.5 WHAT HAPPENS DURING SIMULATION?

After a CYCLONE model like the one described in the previous section is built, a computer program like MicroCYCLONE (Halpin, 1992) is used to perform the simulation experiment of the model just as the Primavera Project Planner is used to process a CPM network plan, for example. A simplified description of what happens during simulation follows:

- All resources are initialized at the queues where they will start (e.g., trucks are queuing at the "trk" QUE waiting for loading at the beginning of simulation).
- Work tasks require time to be accomplished. In simulation the modeling element holds the resource progress for the period of time required for processing to simulate this.
- Entities (resources) flow through the network until they reach a task. If the node is a normal, they are delayed by the amount of time required to compete the tasks. If it is a COMBI they wait in a QUEue node until other required resources are available at which point they are delayed as in the case of a normal task.
- If an entity passes a QUEue with a generate effect it will multiply before exiting the QUEue. If it passes through a consolidate function it is accumulated until the desired number of accumulations is reached, at which point one unit exits the node. When an entity passes by a COUNTER it registers a production of one unit.
- All of the above takes place within the MicroCYCLONE program which does all the time keeping for event scheduling, bookkeeping for statistics collection, and control to insure that the model is properly simulated.

17.6 EXPERIMENTING, ANALYZING, AND SIMULATING

Once a model is built, it can be entered into a computer program like MicroCYCLONE to process it and perform the simulation study. Control information must be also specified to assist the system in terminating the simulation experiment. In CYCLONE this may be specified as the total simulation time or total number of cycles on the cycle counter. From the simulation study we first get an estimate of the production of the operation, the hourly production rate, and other measures of equipment utilization.

To illustrate the experimental aspect of simulation, the earth-moving model described above is entered into MicroCYCLONE. The model is simply transformed into an input file as shown in Figure 17.9. (Note that many CYCLONE implement graphical model building on the computer screen and bypass this process such as DISCO and COOPS).

The simulation is carried and the production graph produced during simulation is given in Figure 17.10. The graph shows the change in hourly production with time progressing from 0 to 22 hours.

Since the model has certain random components, a number of runs need to be performed and proper statistical analysis carried, as described in the previous chapter. The simulation experiment was repeated 30 times with different seed numbers each time. The results indicate that the mean time required to complete the 100 truck loads of earth is 1315.8 time units with a standard deviation of 16.4 time units. The 95% confidence interval on the mean was also found to be 1315.8 ± 5.85 time units. It should also be noted that resource statistics help in analyzing the process. The average waiting

```
NAME 'TRUCK EXAMPLE' LENGHT 100000 CYCLE 100
NETWORK INPUT
1 QUE 'LOADER'
2 QUE 'TRUCK WAIT'
3 COM 'LOAD' SET 1 PRE 1 2 FOL 1 5
4 NOR 'RETURN' SET 2 FOL 2
5 NOR 'TRAVEL' SET 3 FOL 8
6 FUN COU QUANTITY 1 FOL 4
7 COM 'DUMP' SET 4 PRE 8 9 FOL 6 9
8 QUE 'TRK WAIT DUMP'
9 QUE 'SPREADER'
RESOURCE INPUT
1 AT 1
3 AT 2
1 AT 9
DURATION INPUT
SET 1 UNI 2 10
SET 2 BETA 10 18 3 3
SET 3 BETA 10 20 2 2
SET 4 UNI 2 5
ENDDATA
```

FIGURE 17.9 MicroCYCLONE Input File

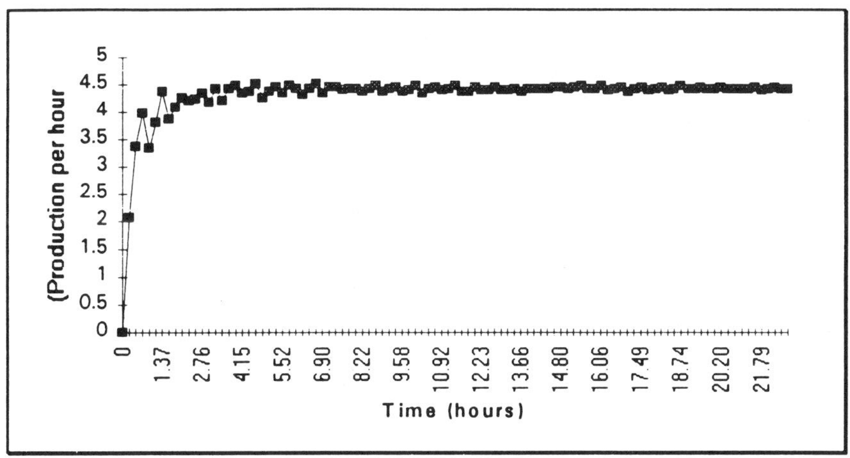

FIGURE 17.10 Production Results from Simulation Run

time for trucks at the location of the loader was reported to be 0.89 time units, whereas the average waiting time for the loader was 7.10 time units implying that the operation may be improved by manipulating the resources (in this case increasing the loaders, for example).

MicroCYCLONE provides many reports regarding the simulation experiment. Figure 17.11 represents a process simulation summary report. It shows that the run length was 1351.1 minutes for 100 truck cycles (i.e., 100 trucks passed the production counter) with a total of 4.44 truck loads produced per hour.

A sensitivity analysis can be conducted and resource allocation adjusted to improve productivity. The report shown in Figure 17.12 shows 19 different combinations of trucks and loaders with the resulting value of productivity and time required to completed 100 truck loads. Such a report may be used to manipulate resources, thus achieving a particular objective.

Upon completing this analysis the model may be enhanced to better reflect the actual construction operation. For example, the additional set of tasks shown added to the original model given in Figure 17.13 models the truck breakdown during hauling. The probabilistic branching feature allows the modeler to include the 5% chance of truck breakdown.

Run length	= 1351.1
Number of cycles	= 100
Units per cycle	= 1
Total production	= 100
Units produced per hour	= 4.44

FIGURE 17.11 Sample Process Simulation Summary Report

Run #	TOTAL TIME	PRODUCTIVITY	TOTAL COST	UNIT COST	DESCRIPTION
1	1351.07	4.44	0.00	0.00	3
2	1036.67	5.79	0.00	0.00	4
3	847.52	7.08	0.00	0.00	5
4	741.04	8.10	0.00	0.00	6
5	669.45	8.96	0.00	0.00	7
6	646.96	9.27	0.00	0.00	8
7	639.48	9.38	0.00	0.00	9
8	639.48	9.38	0.00	0.00	10
9	639.48	9.38	0.00	0.00	11
10	639.48	9.38	0.00	0.00	12
11	639.48	9.38	0.00	0.00	13
12	639.48	9.38	0.00	0.00	14
13	639.48	9.38	0.00	0.00	15
14	639.48	9.38	0.00	0.00	16
15	639.48	9.38	0.00	0.00	17
16	639.48	9.38	0.00	0.00	18
17	639.48	9.38	0.00	0.00	19
18	639.48	9.38	0.00	0.00	20
19	639.48	9.38	0.00	0.00	21

FIGURE 17.12 Sensitivity Analysis Sample Report

17.7 AN OVERVIEW OF SLAM II—A COMBINED DISCRETE EVENT AND PROCESS INTERACTION SIMULATION LANGUAGE

General purpose simulation languages are becoming more and more utilized in different fields. A general purpose language allows the modeler to prepare a model using one of three simulation views: Process interaction (same as CYCLONE, i.e., using nodes and elements to graphically represent a model), discrete event using written code (much like writing a program in C or FORTRAN), and continuous (the simulation is driven by uniform increments of time rather than discrete jumps as events take place). Many implementations like SLAM II allows a combined model of all three views.

17.7.1 SLAM II—Process Interaction Models

SLAM II provides many modeling features. The very basic ones are discussed in this chapter. First the most-used modeling elements are presented

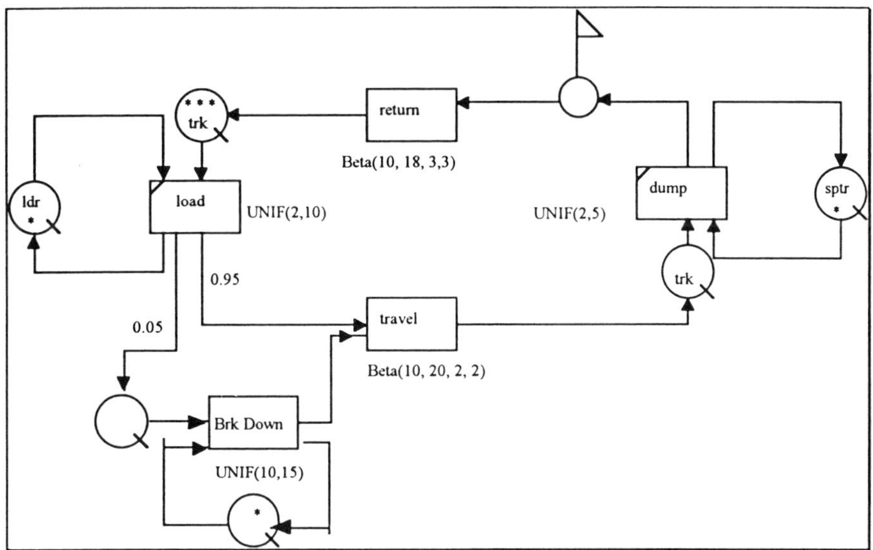

FIGURE 17.13 New CYCLONE Model

and a simple example is provided to illustrate model building with SLAM II. Then an example illustration is given to highlight the added flexibility of SLAM II compared to CYCLONE.

Modeling Elements. SLAM II provides many modeling elements, the description of which is beyond the scope of this book. The reader is referred to Pritsker (1987) for a detailed description of SLAM II. In this section we focus on 10 elements which are very useful in model building. These are the create, terminate, await, goon, activity, select, free, resource, and assign nodes shown in Figure 17.14.

The first difference between SLAM II and CYCLONE is that SLAM II enables the modeler to distinguish between different types of resources. A resource may have attributes associated with it, thus enabling the system to distinguish between say a 30-ton truck and a 70-ton one. To get the flow units into the system, SLAM II uses the create node. Here random arrivals can be generated according to many distributions. Once a resource is entered into the model it may be identified by passing it through an ASSIGN node using the ATRIB( ) variable. The same assign node may be used to set other variable values in the model. The flow unit will now be processed in the system in the same manner as with CYCLONE except that the variety of nodes offers more flexibility. The unit passing through an activity node is delayed by the duration of the activity. If it passes through an await node it waits until a resource of the required type is available in the resource block. Once a resource is available, the unit holds on to it until it is no longer needed, at which point it should release it by passing through a FREE node.

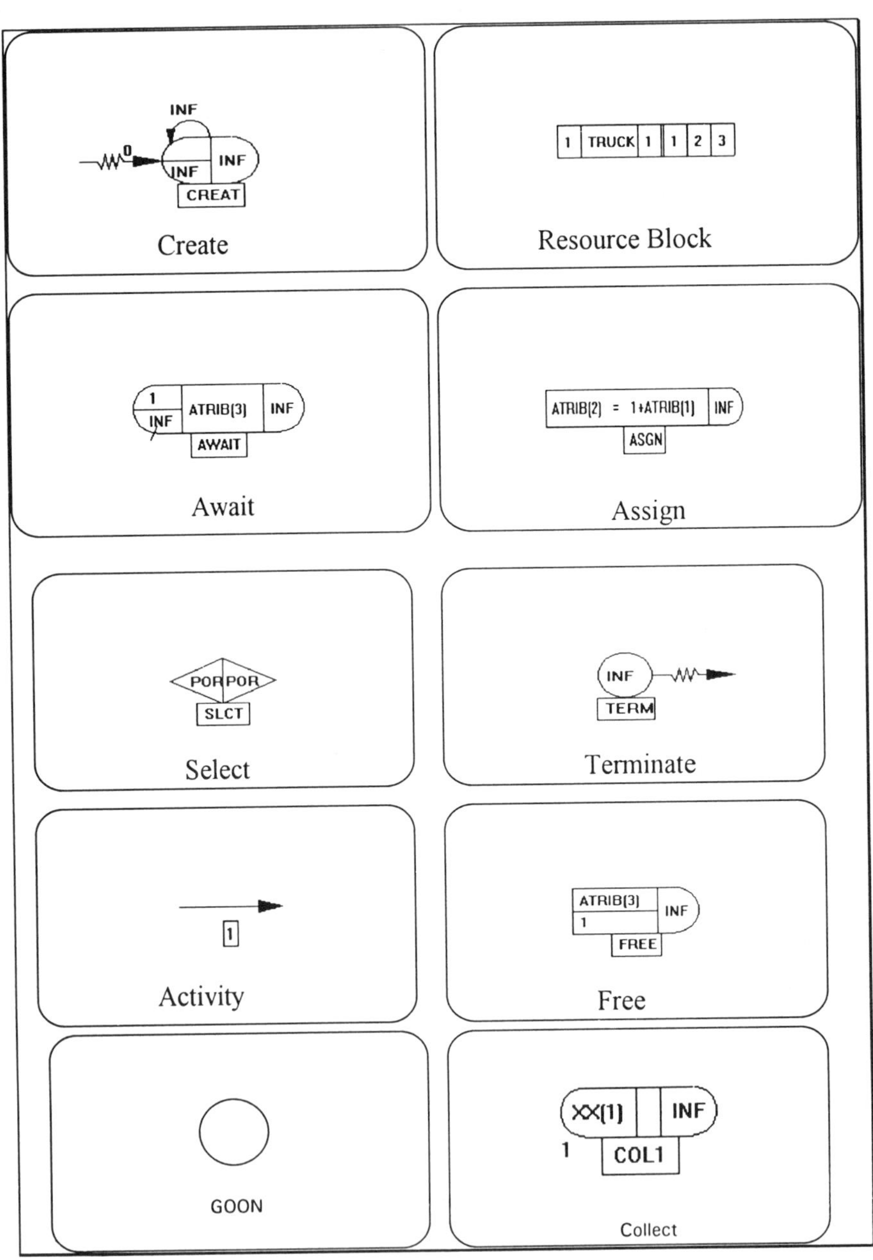

FIGURE 17.14 SLAM II Nodes

The unit may also pass through a select node, allowing it to make a decision as to which activity it should go to if more than one pass is available. Finally once the unit has completed its work it may be terminated in a terminate node.

The functions of the various SLAM II elements given in Figure 17.14 can be summarized as shown in Table 17.3.

TABLE 17.3 SLAM Elements

SLAM II Element	Description
Create	Generates flow units into the system model. The time between generated entities and a maximum number may be specified.
ASSIGN	An entity passing through this node triggers assignment of variables specified within the node. Variables include ATTRIBUTES of a unit, SLAM II variables and so on.
Activity	The activity is not a node, rather an arrow connecting two nodes (similar to activity on arrow description). May be a service activity or a regular one. The service activity is constrained by availability of specified servers and an entity must wait at a queue node before a server is ready (similar to a COMBI). A flow unit arriving at an activity is delayed by the specified duration of that activity. Once the duration expires the activity is released to the connected node.
AWAIT	A unit arriving at this node must wait until the resources specified for that node are available at the resource pool.
FREE node	A unit passing through the free node triggers the return of the resource to the pool, signifying that the unit no longer requires the resource.
RESOURCE BLOCK	Acts like a resource pool. Each resource will have its own resource block (pool).
SELECT node	Allows the passing entity to be routed to one of many choices; for example, to different activities or queues depending on conditions that may specified at the select node and the properties (attributes) of the passing entity.
COLLECT node	Collects statistics wherever it is inserted in the mode.
QUEUE node	Allows the arriving entity to wait until a service activity has an available server ready to process the entity. This is different than an Await node as it does not require any specific resources only the activity server which is tied to the activity itself.
TERMINATE	Destroys the entity. Can also keep count of number of entities processes and may be used to control simulation length.

One of the advantages of SLAM II over CYCLONE and similar languages is its flexibility. In this regard the language allows the user to incorporate his own code into the program if the modeling elements are not enough. For example, the activity time on a SLAM II network may be a user-defined function, written in FORTRAN or C, that will manipulate data and check for various conditions before it decides what duration will be used. Event nodes may be used in SLAM II to write any amount of FORTRAN code that manipulates most model features and accesses most system variables. SLAM II also allows the user to build many networks that use the same resource and communicate with each other, which increases the model's compactness. In addition, SLAM II allows combined discrete event and continuous simulation modeling, which may prove very useful for systems that move continuously in time (e.g., weather conditions) rather than discretely as events take place.

17.7.2 An Example of a SLAM II Application

This project models the tunneling operation at the WMATA Metro line in Washington, D.C. (Fig. 17.15) The operation was documented by A. Hijazi during the year 1986. Owing to the lack of some data, certain assumptions have been made where necessary.

General Description. The tunneling procedure used was the Earth Pressure Balance (EPB) method. The method was invented by the Japanese in 1974 for excavating underground in soft soil. The fundamental idea of the EPB shield is that the ground at the face can be controlled to avoid settlement by balancing the muck discharge from the soil chamber against the muck intake from the cutter head, and by preventing ground water from running in. The ground water and the earth pressure is encountered by a plastic resistance pressure of the muck mass in the cutter pressure chamber.

The main features of the EPB are:

1. A screw discharger for discharging the excavated muck. Excavated muck is compacted into a low permeability sand plug via a sand plug formation zone near the discharge outlet.
2. The sand plug formation zone, adjustable in length to suit the varying soil conditions, holds the earth and ground water pressure.

The process of tunneling with the EPB machine starts with the digging and preparation of a pit to form a shaft. When the shaft is prepared the machine is assembled at the bottom of the shaft. When the machine advances muck has to be carried out of the tunnel, and precast concrete units have to be brought in to the excavation face. The liners are then placed to support the tunnel, the supported face is grouted, a jacking system used by the EPB machine to advance is placed on the liners, the machine is adjusted, and the

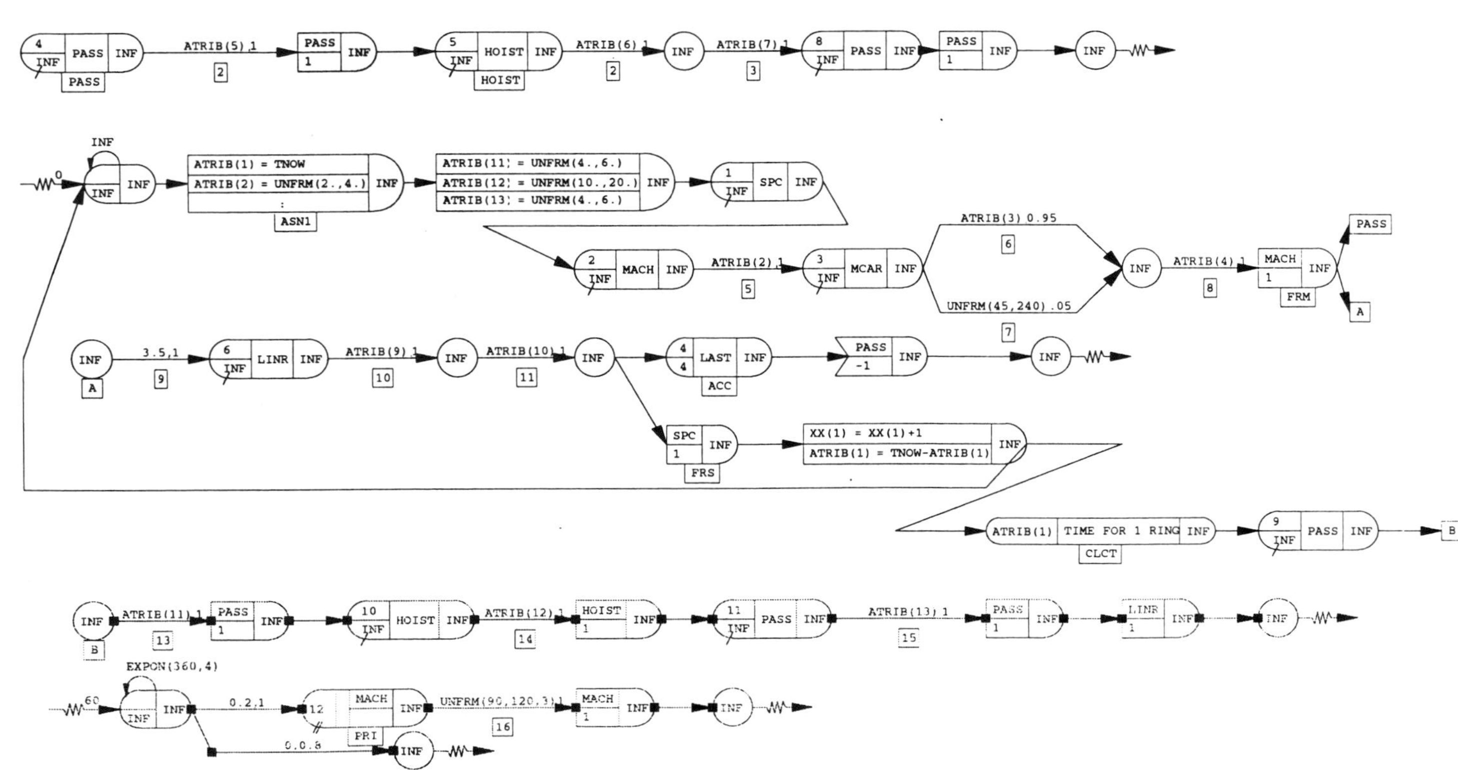

FIGURE 17.15 SLAM Model for Metro Line, Washington, DC

process is repeated. As the machine advances a track used by the train that carries the liners and the muck cars that carry the muck is extended after four rings have been installed. The machine advances in increments of 4 ft, the amount necessary to install one ring of liners. In addition, when the machine is down and work is stopped until it is fixed, the probability of hitting boulders exists and has to be taken into account.

Objectives of the Simulation Experiment. The objective of this experiment is to model the operation described above, figure out the attainable levels of daily production (to enable estimating the completion date of the project), and finally experiment with two different versions of the operation (one that uses muck cars to dispose of the muck and another that uses a conveyor system) and comment on the benefits of using one or the other. The last step requires designing an experiment whereby all sampling is reproducible to insure fair comparisons of the alternatives.

Model Description. The whole operation is modeled as a process interaction SLAM II model. The model is shown in Figure 17.15. The model basically consists of two networks—one to model the operation as described above, and one to model the machine breakdown.

In the first network an entity is created and cycled through the system until 10 h of continuous work is completed. Three different interacting cycles can be noticed in the model. The cycle of the machine as it excavates and advances, the cycle of the muck cars as they travel from the shaft to the machine and back to the shaft, and the cycle of the liners traveling from the shaft to the excavation face and back to the shaft. Each cycle interacts with the other cycles. Note that the muck cars and the cars that carry the liners share the use of the same track and the same hoist to get out of the shaft area.

Transactions. The flow unit in the system is the excavation unit of 4 ft. It enters the system and waits for the necessary resources required for its processing (machine, muck cars to receive during excavation). It then duplicates into two units, one to travel to the shaft area where the muck is carried outside and another to signal the removing of the jacking system and consequently place the liners when all resources are available. The first unit exists as the muck cars empty and return to the excavation face. The second entity again duplicates into two after grouting is completed—one to model the accumulation of four cycles necessary to extend the track after four advances of the machine, and the other signal that excavation can commence by freeing the excavation space since liners are installed and the area is cleared.

Only one unit is created and cycled throughout the system, but each time with changing attributes for service durations.

The breakdown is modeled by creating a breakdown flow unit that causes the preemption of the machine, delays it, and then frees the machine and exists.

Thirteen attributes were used with the flow unit modeling the service durations of each activity to allow use of common random numbers.

ATRIB(1)	Collects the time the cycle starts
ATRIB(2)	Duration for placing the jacking system
ATRIB(3)	Duration for excavating 4 ft
ATRIB(4)	Duration for maintaining the machine after each advance
ATRIB(5)	Duration for traveling to the shaft area (muck car)
ATRIB(6)	Duration for transporting muck from shaft to outside
ATRIB(7)	Duration for returning from outside to shaft
ATRIB(8)	Duration for traveling to excavation face empty
ATRIB(9)	Duration for placing liners
ATRIB(10)	Duration for grouting
ATRIB(11)	Duration for liner cars returning to shaft
ATRIB(12)	Duration for getting new set of liners
ATRIB(13)	Duration of transporting liners to face of excavation

All durations were sampled from stream 9.

Other activities like breakdown of machine, extending track, and duration for retrieving and reexcavating upon hitting a boulder were sampled from different streams.

Definition of Files

File No.	Type	Content	Ranking
1	Await	Excavate space	FIFO
2	Await	EPB machine	FIFO
3	Await	Muck cars	FIFO
4	Await	Passage	FIFO
5	Await	Hoist	FIFO
6	Await	Liners	FIFO
8	Await	Pass	FIFO
9	Await	Pass	FIFO
10	Await	Hoist	FIFO
11	Await	Pass	FIFO
12	Preempt	Machine	FIFO

Network Variables. Two global variables were used:

1. XX(1) captures the total number of cycles (i.e., number of liners installed during the simulation period of 10 hours).

2. XX(2) captures breakdown time in any given shift. XX( ) are user defined variables in SLAM

Both variables are initially set to zero.

Statistics Collection. The statistic of interest in this experiment is the total time it takes to produce one ring, which constitutes a production unit and is a measure of the construction performance. The total time required to produce one ring was collected in ATRIB(1). As the entity starts a cycle ATRIB(1) is set to TNOW, as the entity flows out of the cycle signaling the completion of one ring ATRIB(1) is set equal to TNOW–ATRIB(1), capturing the time that unit spent in the system.

Another statistic of interest is the total time the machine is down. That is captured in XX(2). Both statistics were collected via a COLCT node.

The total number of rings installed in any given day can be obtained from the number of observations recorded in the standard summary.

Analysis of Results. The objectives of the simulation experiment were two-fold. On one hand there is an interest in determining the attainable levels of daily production with the EPB method for the condition set (i.e., the assumed breakdown, time required to excavate 4 ft, etc.), and on the other hand two alternatives of the method (using a muck care train to carry the muck out of the tunnel vs. using a conveyor system) were to be evaluated.

To determine the daily production, a production unit was defined as one set of rings installed in position (4 ft). Twenty simulation days were run. Each day was assumed to be 10-hr continuous work shifts. The total number of liners installed during the shift was counted as the flow unit reaches the collect node. Two assumptions were made in this case:

1. The work on the following day does not resume from where it was stopped but rather starts anew.
2. The 10 hr of work were continuous with no lunch breaks.

The two alternatives were to be compared as fairly as possible. Since the system does not reach steady state it would be incorrect to compare alternatives without employing the idea of common random numbers. Since the activities sample their durations randomly, it was necessary to:

1. Dedicate different streams for different categories of activities.
2. Sample all durations before the unit interacts with the system components.

The seed numbers were fixed at the beginning of any simulation day, and the same seeds used for both alternatives for a given day.

Summary of Results. The two versions of the model were executed for 20 one-day runs with each day equivalent to 600 min of time. The results are summarized in Table 17.4.

A 90% confidence interval on the mean difference in average time to produce one ring is calculated to be [3.119, 5.731].

Although a difference between the two systems exists ($t = 5.858 > 2.861$), the reduction in time required to produce one ring is not large enough to prefer one alternative to the other. Such a decision will have to be completed based on issues other than production levels achievable.

Attainable levels of productions with the muck car system:

Mean = 5.25 rings per day

STD = 1.01 rings per day

95% confidence interval on mean is given as [4.803, 5.697]

For the conditions specified (time required for various activities, levels of breakdown in machine, probability of hitting boulders, etc.) the attainable level of production is approximately 5 rings per day or an advance rate of about 20 ft.

With varying site conditions, like softer ground, less boulders, and so on the model can be modified accordingly by sampling lesser duration and using different probabilities.

There is no added advantage to using a conveyor system unless the cost justifies it, however, in terms of attainable levels of production; both systems are likely to produce about 5 rings per day for the conditions sets.

17.8 APPLICATIONS OF SIMULATION IN CONSTRUCTION

Although it is difficult to represent a construction project for simulation purposes, a number of successful simulation applications have been recorded. One of the prime applications of simulation is on the process level. In this regard the focus is lowered to the process level where a project is envisioned as a collection of processes.

17.8.1 Process Modeling and Simulation

Construction process modeling has matured over the years with numerous examples given in Halpin and Riggs (1992). These include earth moving, pavement construction, concrete placement on high-rise buildings, cladding, tunneling and underground pipe-jacking, and a wealth of others. The objective of process simulation ranges from productivity measurement and risk analysis to resource allocation and site planning. With an accurate representation of the activities constituting the process, the modelor can estimate the

TABLE 17.4 Summary of Runs—Metro Line, Washington, D.C.

	Muck Car System				Conveyor System				
Run No.	Time (1 ring)	Breakdown	No. of Rings	No. of Breadowns	Run No.	Time (1 ring)	Breakdown	No. of Rings	Difference
1	99.2	98.5	6	1	93.3	98.5	6	1	5.9
2	95.3	0	6	0	89.0	0	6	0	6.3
3	111	102.0	5	1	103.0	102.0	5	1	8.0
4.	97.4	0	6	0	89.6	0	6	0	7.8
5	109	119.0	5	1	102.0	119.0	5	1	7.0
6	92	0	6	0	85	0	7	0	7.0
7	93.3	0	6	0	87.1	0	6	0	6.2
8	114	0	3	0	114	0	3	0	0
9	91.0	0	6	0	85.0	0	6	0	6.0
10	94.5	0	6	0	88.4	0	6	0	6.1
11	105	100	4	1	105	100	4	1	0
12	90.6	0	7	0	83.2	0	7	0	7.4
13	141	107.0	4	1	141	107	4	1	0
14	109	0	5	0	109	0	5	0	0
15	109	0	4	0	109	0	4	0	0
16	107	116	4	1	107	116	4	1	0
17	94.9	0	6	0	88.5	0	6	0	6.4
18	107	110	5	1	107	110	5	1	0
19	116	114	5	1	109	114	5	1	7
20	91	0	6	0	83.6	0	7	0	7.4

$X_1 = 103.35$ $X_2 = 98.94$ Mean difference = 4.425
$S_1 = 12.24$ $S_2 = 14.04$ Standard deviation = 3.378

production of the process and the probabilities of meeting a given schedule. As resources in construction include heavy and expensive equipment, various allocations can be examined and the most suitable ones adopted.

17.8.2 Claims Analysis and Dispute Resolution

Simulation has been found to be an attractive tool in mitigating construction disputes. The most common cases which can benefit from simulation analysis include productivity loss due to weather, interruptions due to changes, owner or trade-contractor interferences, or unexpected conditions. The contractor often seeks compensation based on loss of productivity or change impact for nonrecoverable costs or delays due to conditions beyond his control. Simulation models have been successfully used to provide an accurate representation of the original condition as expected at time of bid and the condition that would have been encountered after the new facts arise. Since productivity loss is derived from factors such as weather, labor skill, and site conditions, the simulator can build the simulation model and introduce the new facts to study and analyze their impact on the productivity, cost, or time of the project. A recent example of this was reported by AbouRizk and Dozzi (1992). In their work the authors developed a simulation model for the jacking operation of a bridge as part of a mediation between the public owner and a steel contractor. The original anticipated condition based on drawings provided by the owner reflected lower and less complex jacking operations. The design was proven nonworkable and the parties agreed to mediate a settlement for the total compensation and time extension. The CYCLONE model built for the process accurately predicted the total number of man-hours lost because of the added levels of complexity which when combined with the new required quantities of steel provided the total added cost.

17.8.3 Project Planning and Control

Most of the work in project planning and control dealing with simulation has been of a hybrid nature. In general, CPM-based methods are used and a form of Monte Carlo simulation allows evaluation of the network. Dedicated systems have been in use and are available from commercial software vendors.

The major problem with simulating a construction project is its size and complexity. This, in general, does not provide a positive return on the effort invested in building a simulation model. It is not uncommon to have a project with over 2000 activities in a construction project. Much research is still needed to provide a simple, efficient, workable, and accurate method for construction project simulation. The CPM network techniques are still the leading control method used at the project level. This, however, does not provide the analyzer with all benefits of simulation and is therefore of limited use to practitioners.

17.8.4 Simulation of PERT Networks

The previous chapter demonstrated how Monte Carlo simulation of a PERT network may be performed. The main problem with such an approach is having to write the code that performs the simulation (some commercial software actually evolved an automating simulation of such networks, such as P3 by Primavera Systems Inc.). General purpose simulation systems can be used to accomplish this task. Any PERT network may be converted to an equivalent CYCLONE, SLAM II, or other network. To illustrate, a PERT network will be converted to the equivalent CYCLONE network. Halpin and Riggs demonstrated the feasibility of the method as follows:

Activities on the PERT network become CYCLONE NORMAL's or COMBI's much in the same way as converting a CPM network into a precedence diagramming network. Activities in series on a PERT network are straightforward to convert, as shown in Figure 17.16.

Activity 1–2 becomes a NORMAL which is followed by 2–3, another NORMAL.

Modeling a merge node requires the use of a COMBI preceded by QUEUE nodes, as shown in Figure 17.17.

The main reason for the use of a COMBI is to ensure that activity 2–3 does not start until 4–2, 1–2, and 5–2 have been completed. A COMBI in CYCLONE is constraint and as such requires that one flow unit reach each of the three preceding QUEUE nodes before it may commence. Had we used a NORMAL, the logic of the original network would not have been preserved.

Converting the PERT network into a CYCLONE model may present some advantages. For example, once a network has been converted, CYCLONE permits cycling within any portion of the network. This may help in reducing the total network model structure. Besides the probabilistic study that can be performed, CYCLONE also allows probabilistic branching to model various decisions within the network.

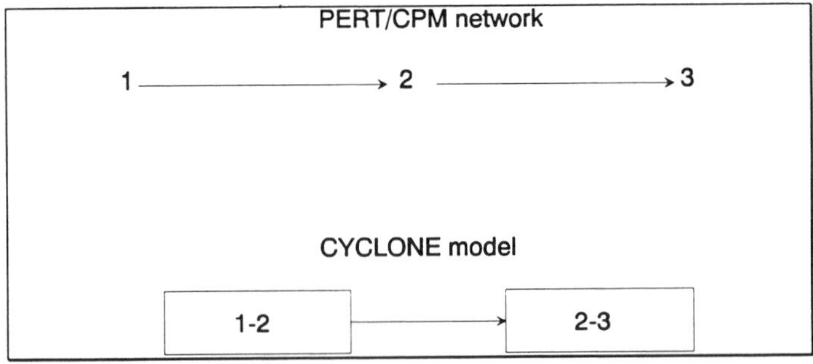

FIGURE 17.16 PERT To CYCLONE Conversion

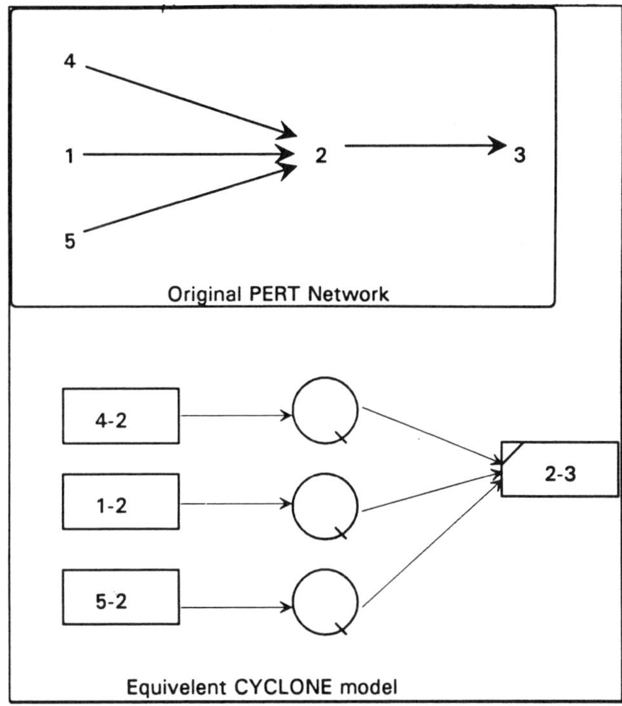

FIGURE 17.17 Modeling a Merge Node in CYCLONE

A sample CPM network is shown in Figure 17.18 with the corresponding CYCLONE model in Figure 17.19. The CPM activities are defined by their node numbers, whereas the CYCLONE tasks use the same designation to facilitate comparison. For example, activity 1–3 is represented with NORMAL 1–3, activity 34 by COMBI 3–4, and so on. The reader should notice that since event 3 was a merge node collecting two activities, the corresponding CYCLONE task 3–4 is a COMBI preceded by two nodes. This will insure that the task will not commence unless one unit reaches Q1 and one

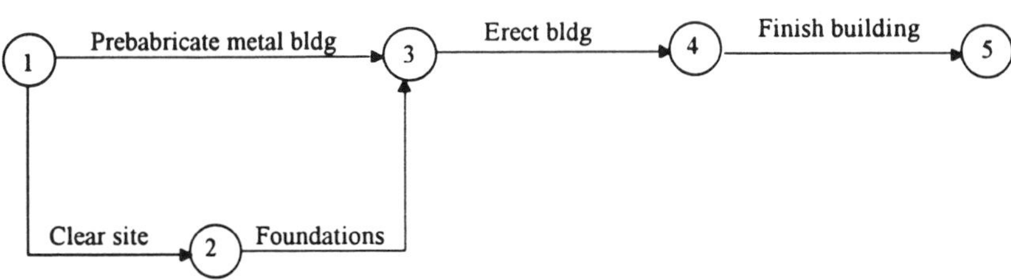

FIGURE 17.18 Sample PERT

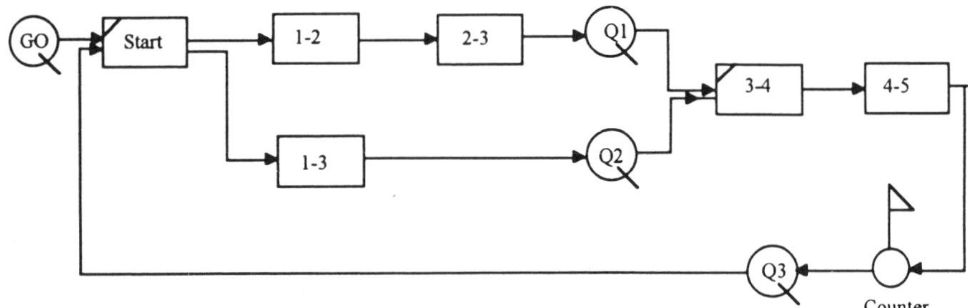

FIGURE 17.19 Equivalent CYCLONE Model

unit reaches Q2. The only way this could happen is if tasks 2–3 and 1–3 have been completed, which preserves the original CPM network logic. For the simulation to work, only one unit needs to be initialized at QUE node GO and one at QUE node Q3. These unit will trigger the task to start once its requirements are satisfied. The other QUE nodes will receive those same units once the proper tasks have been completed. The simulation and comparison of the results is left for the reader as an exercise.

17.8.5 Sample Application of PERT Network Simulation

A multistory building project is divided into the activities shown in Table 17.5. The duration and precedence relationships of each activity is provided in the same table. Such information may be entered into a computer program like P3 to perform the scheduling computations and estimate the project completion time.

The network is not shown, but can easily be produced by a scheduling program. The network was converted to a CYCLONE model, the input file of which is given in Figure 17.20.

Activities with uncertain duration components were estimated using three-time estimates as shown in Table 17.6. A beta distribution was then fit using the PERT method and three experiments were carried out. In the first experiment the mean project completion time using the PERT method previously discussed was calculated. The second experiment was a simulation using the equivalent CYCLONE model with beta distributions fit to the same PERT specifications. The third experiment used a triangular distribution, the mode of which was calculated to match the mean and variance used in the PERT analysis.

The results are given in Table 17.7. It can be seen that the PERT estimate was optimistic compared to the triangular or beta simulation models. It

TABLE 17.5 Activity details for Dabbas example

ID	Name	Duration	Successors
1	Move in and start office	2	2, 3, 4
2	Survey site for bldg.	1	5
3	Rough grade grubbing etc.	2	5
4	Suvery site for sewer line	1	10
5	Excavate half site	14	6, 7
6	Pour foundations A-C	1	8
7	Excavate other half	14	9, 10
8	Rebar basement A-C1	2	11
9	Rebar basement C1	4	12
10	Excavate and lay sewer	11	43
11	Pour basement A-C	1	13
12	Pour basement C1	2	14
13	Cure basement A-C	2	15
14	Cure basement C1	2	15
15	Form and pour columns floor 1	2	16, 17
16	Remove basement forms	1	19
17	Rebar floor 1	4	16
18	Concrete floor 1	3	19
19	Form and pour columns floor 2	2	20, 21
20	Rebar floor 2	6	22
21	Walls in basement	4	23, 28
22	Concrete floor 2	3	24
23	Mechanical and electrical basement	4	25, 31
24	Walls and columns floor 3	2	26, 27, 28
25	Finish basement	4	33
26	Brick walls floor 1	4	29
27	Rebar floor 3	4	30
28	Walls floor 1	4	31, 36
29	Brick walls floor 2	2	37
30	Concrete floor 3	3	32
31	Mechanical and electrical floor 1	4	33, 39
32	Forms for roof	2	34
33	Finish floor 1	4	41
34	Pour concrete roof	3	35, 36
35	Cure and remove forms	2	37, 38
36	Walls floor 2	2	39
37	Brick walls floor 3	3	43
38	Walls floor 3	2	40
39	Mechanical and electrical floor 2	2	40, 41
40	Mechanical and electrical floor 3	2	42
41	Finish floor 2	3	42
42	Finish floor 3	2	43
43	Clean up and test	1	—

```
NAME 'DABBAS EXAMPLE' LENGTH 10000 CYCLE 1
NETWORK INPUT
1 COM 'MOVEIN & START OFFIC' SET 1 PRE 74 75 FOL 2 3 4 75
2 NOR 'SURVEY SITE FOR BLDG' SET 2 FOL 44
3 NOR 'ROUGH GRADE GRUBBING' SET 3 FOL 45
4 NOR 'SURVEY SITE SEWER' SET 4 FOL 47
5 COM 'EXCAVATE HALF SITE' SET 5 PRE 44 45 FOL 6 7
6 NOR 'POUR FOUNDATIONS A-C' SET 6 FOL 8
7 NOR 'EXCAVATE OTHER HALF' SET 7 FOL 9 46
8 NOR 'REBAR BASEMNT A-C1' SET 8 FOL 11
9 NOR 'REBAR BASEMNT C1' SET 9 FOL 12
10 COM 'EXCAVATE & LAY SEWER' SET 10 PRE 46 47 FOL 72
11 NOR 'POUR BASEMNT A-C' SET 11 FOL 13
12 NOR 'POUR BASEMNT C1' SET 12 FOL 14
13 NOR 'CURE BASEMNT A-C' SET 13 FOL 48
14 NOR 'CURE BASEMNT C1' SET 14 FOL 49
15 COM 'FORMPOUR COLS 1' SET 15 PRE 48 49 FOL 16 17
16 NOR 'REMOVE BASE FORM' SET 16 FOL 50
17 NOR 'REBAR FLOOR 1' SET 17 FOL 16
16 NOR 'CONCRETE FLOOR 1' SET 16 FOL 51
19 COM 'COLS FLOOR 2' SET 19 PRE 50 51 FOL 20 21
20 NOR 'REBAR FLOOR 2' SET 20 FOL 22
21 NOR 'WALLS IN BASEMNT' SET 21 FOL 23 53
22 NOR 'CONCRETE FLOOR 2' SET 22 FOL 24
23 NOR 'MECH & ELEC BASEMNT' SET 23 FOL 25 55
24 NOR 'WALLS & COLS FLOOR 3' SET 24 FOL  26 27 52
25 NOR 'FINISH BASEMNT' SET 25 FOL 57
26 NOR 'BRK WALLS FLOOR 1' SET 26 FOL 29
27 NOR 'REBAR FLOOR 3' SET 27 FOL 30
28 COM 'WALLS FLOOR 1' SET 28 PRE 52 53 FOL 54 59
29 NOR 'BRK WALLS FLOOR 2' SET 29 FOL 60
30 NOR 'CONCRETE FLOOR 3' SET 30 FOL 32
31 COM 'MECH & ELECC FLOOR 1' SET 31 PRE 54 55 FOL 56 63
32 NOR 'FORMS FOR ROOF' SET 32 FOL 34
33 COM 'FINISH FLOOR 1' SET 33 PRE 56 57 FOL 67
34 NOR 'POUR CONCRETE ROOF' SET 34 FOL 35 58 59
35 NOR 'CURE & REMOVE FORMS' SET 35 FOL 38 61
36 COM 'WALLS FLOOR 2' SET 36 PRE 58 59 FOL 62
37 COM 'BRK WALLS FLOOR 3' SET 37 PRE 60 61 FOL 70
38 NOR 'WALLS FLOOR 3' SET 38 FOL 64
39 COM 'MECH & ELEC FLOOR 2' SET 39 PRE 62 63 FOL 65 66
40 COM 'MECH & ELEC FLOOR 3' SET 40 PRE 64 65 FOL 68
41 COM 'FINISH FLOOR 2' SET 41 PRE 66 67 FOL 69
42 COM 'FINISH FLOOR 3' SET 42 PRE 68 69 FOL 71
43 COM 'CLEAN UP & TEST' SET 43 PRE 70 71 72 FOL 73
44 QUE, 45 QUE, 46 QUE, 47 QUE, 48 QUE, 49 QUE, 50 QUE, 51 QUE, 52 QUE, 53 QUE,
54 QUE, 55 QUE
56 QUE, 57 QUE, 58 QUE, 59 QUE, 60 QUE, 61 QUE, 62 QUE, 63 QUE, 64 QUE
65 QUE, 66 QUE, 67 QUE, 68 QUE, 69 QUE, 70 QUE, 71 QUE, 72 QUE,
73 FUN COUNTER QUA 1 FOL 74
74 QUE, 75 QUE
DURATION INPUT
```

FIGURE 17.20 Equivalent CYCLONE File for Example

```
SET 1 2
SET 2 1
SET 3 2
SET 4 1
SET 5 BETA 12 23 1.837 4.543
SET 6 1
SET 7 BETA 7 20 3.294 4.492
SET 8 BETA 1 10 2.938 4.617
SET 9 BETA 6 17 1.837 4.617
SET 10 BETA 7 15 4.667 2.333
SET 11 1
SET 12 2
SET 13 2
SET 14 2
SET 15 2
SET 16 1
SET 17 BETA 6 15 2.130 4.636
SET 16 BETA 4 13 2.130 4.636
SET 19 2
SET 20 BETA 4 14 4.443 3.397
SET 21 BETA 5 10 1.968 4.592
SET 22 3
SET 23 BETA 8 14 4.617 2.938
SET 24 2
SET 25 BETA 10 16 2.333 4.667
SET 26 4
SET 27 4
SET 28 4
SET 29 2
SET 30 3
SET 31 4
SET 32 2
SET 33 BETA 6 19 2.754 4.654
SET 34 3
SET 35 2
SET 36 2
SET 37 3
SET 38 2
SET 39 BETA 7 11 4.667 2.333
SET 40 BETA 5 14 4.270 3.681
SET 41 BETA 10 16 4.494 1.728
SET 42 BETA 7 22 2.456 4.673
SET 43 1
RESOURCE INPUT
1 AT 74
1 AT 75
ENDDATA
```

FIGURE 17.20 (*Continued*)

TABLE 17.6

Act ID	L	M	U	Mean	SD	Shape Parameters of Equivalent Beta Distribution	Mode of Equivalent Triangular Distribution
5	12.00	14.00	23.00	15.17	1.83	1.837, 4.543	10.5
7	7.00	12.00	20.00	12.50	2.17	3.294, 4.492	10.5
8	1.00	4.00	10.00	4.50	1.50	2.938, 4.617	2.5
9	6.00	8.00	17.00	9.17	1.83	1.837, 4.543	4.5
10	7.00	13.00	15.00	12.33	1.33	4.667, 2.333	15
17	6.00	8.00	15.00	8.83	1.50	2.130, 4.636	5.5
16	4.00	6.00	13.00	6.83	1.50	2.130, 4.636	3.5
20	4.00	10.00	14.00	9.67	1.67	4.443, 3.397	11
21	5.00	6.00	10.00	6.50	0.83	1.968, 4.592	4.5
23	8.00	12.00	14.00	11.67	1.00	4.617, 2.938	13.0
25	10.00	12.00	16.00	12.67	1.33	2.333, 4.667	10.0
33	6.00	10.00	19.00	10.83	2.17	2.754, 4,654	7.5
39	7.00	10.00	11.00	9.67	0.67	4.667, 2.333	11
40	5.00	10.00	14.00	9.83	1.50	4.270, 3.681	10.5
41	10.00	15.00	16.00	14.33	1.00	4.494, 1.728	17.0
42	7.00	11.00	22.00	12.17	2.50	2.456, 4.673	7.5

should also be noted that the results are not general but rather specific to this example. The objective was to provide the reader with an example that would enable him/her to perform similar experiments. Another observation is that simulation provides confidence intervals for the mean estimate; this is not available from the PERT analysis. It should also be mentioned that in this experiment the results are not significantly different.

TABLE 17.7

		Simulation	
	PERT	Triangular	Beta
Project total time	133.67, 5.01	135.87, 6.17	134.83, 5.12
Confidence interval 95% on mean	Not applicable	135.87 ± 2.21	134.83 ± 1.83

CHAPTER 18

MATERIAL MANAGEMENT

Successful completion of projects requires all resources to be effectively managed. The management of materials is a major part of the resources that we manage. This chapter considers material management as a means to better productivity, which should translate into reduced costs.

In 1983 the Business Roundtable commented on the role of materials management and said: "The construction industry lags far behind the manufacturing industry in applying the concepts of materials management."

More recently the construction industry has become more cognizant of the importance of the management of project materials and equipment, which can amount to 50% or more of project costs. Labor productivity gains of up to 6% (CII, 1988) have been estimated as possible through improved materials management. Traditionally labor productivity receives most attention because the productivity of direct work can be measured at a reasonable cost. Within the construction industry there is a wide divergence in the degree of materials management applied to construction projects. There is no common methodology used to measure the effectiveness of the materials management functions.

Some parts of the construction industry have been very aware of the importance of management of materials. For example, the large Engineering Procurement Construction (EPC) contractors have established programs which evolved from manual to computerized materials tracking systems. These larger EPC contractors realized the need for better materials management, especially for large complex projects which utilized thousands of components. The negative cost impact of shipping delays and poor procurement procedures became increasingly more important to the project and, therefore, by necessity these companies led the way in developing good materials management systems.

The general building and construction industries are beginning to realize the tremendous potential that material management has for increasing productivity and safety on construction projects. Smaller construction projects do not warrant elaborate material management systems, but regardless of size, most companies are realizing the need for some system, whether manual or computerized. Studies have shown that material handling is a large percentage of site labor. In a series of 22 productivity studies carried out in Ontario (O'Brien, 1989), it was found that mechanical and electrical tradesmen were spending only 32% of their day on direct installation work; they spent 20% on material handling, 15% on indirect work, and the remaining 33% on ineffective and miscellaneous operations. Although many areas required improvement, material handling was significant and especially noteworthy. Direct installation was increased to 52% and material handling was reduced to 12% by means of a productivity improvement program.

Other studies show similar ratios of direct to indirect work with material handling and waiting for materials amounting to a significant percentage of the man-hours. These macro studies provide a focus for actions and opportunities for productivity improvement. Traditionally most effort has devoted to the analysis of direct operations such as cutting, assembly, and joining of components. On-site productivity can be improved by reducing the man-hours spent on indirect work, waiting time (including waiting for materials), and material handling. Further, the worker should have the right materials at the right time, which requires more than good material handling; it requires good material management.

Material handling and movement can be a hazardous activity, and most trades persons are not trained in the lifting and transportation of materials; the worker is highly trained for specific trade tasks. Good material management will, through planning and control, improve the efficiency of this operation and thereby reduce risk and improve productivity. Productivity must be considered with the associated level of safety; productivity and safety are closely related.

18.1 MATERIAL MANAGEMENT STEPS

There are several functions or steps within the scope of material management and each of these steps can give rise to potential problems. The more the responsibility is divided, the more potential problems that exist.

Figure 18.1 shows the steps in material management and the pertinent actions related to these steps. Some actions are described in terms of the documentation produced, such as receiving report and vendor data.

Sequence	Contributing Action/Documents
1. RFQ (Requistition) ↓	Drawings, specifications Material bills Terms and conditions
2. Bids ↓	Approved bidders list Pre qualification of bidders Bid evaluations
3. P.O. (Purchase Orders) ↓	Bid clarification Notice of award
4. Expediting ↓	Vendor data Manufacturer inspection Delivery Routings
5. Transport ↓	Carrier and route Ownership en route Customs
6. Receiving ↓	Inspection and acceptance Receiving report Storage
7. Inventory ↓	Dispersal (i.e. material handling) Inventory level Surplus disposal

FIGURE 18.1 Material Management Steps

18.2 SCOPE OF MATERIALS MANAGEMENT

This section deals mainly with the attributes of material management and the responsibilities of those involved in carrying out the material management functions. A detailed understanding of each contributing function is required in order to comprehend the interfaces between material management functions. A materials management system includes the major functions of identifying, acquiring, distributing, and disposing of materials on a construction site (CII, 1988).

By definition, material management is the management system for planning and controlling all of the efforts necessary to ensure that the correct quality and quantity of materials are properly specified in a timely manner, are obtained at a reasonable cost, and are available at the point of use when required (Roundtable, 1982).

Each firm has its particular materials management system and usually the responsibility for the various activities has been spread between engineering, purchasing, and construction. Some assign full responsibility and accountability to a material manager, but for most firms the responsibility is divided and therefore prone to problems.

Figure 18.1 represents the logical steps in the process from identifying matrial needs to delivering the materials when required at the point of use. Each step is a link in the chain of events and the strength of the chain is only as good as the weakest link. From a review of these steps it is apparent that the system can break down in numerous places as the result of misdirected effort or lack of effort by the many individuals involved. These steps and their interfaces are where breakdowns in the process can occur with resulting delays in delivery, which then require additional effort to correct any wrongs. It always costs more if things are not done right the first time. Delays and additional effort are costly, reduce efficiency of a construction operation because of reallocation of resources, and therefore negatively impact productivity.

The steps shown are only the key elements and are part of the whole material management process which by definition includes the planning and controlling of all supporting efforts.

18.3 RESPONSIBILITIES

The purpose for clearly establishing the responsibilities and authority of the participants is not for attaching blame should something go wrong in the process, but to communicate clearly what is expected and avoid misunderstandings as to who does what and when. The scope of each participant's involvement must be clearly defined. If not, increased effort will be expended to rectify missed expectations in quantity, quality, or cost. Unexpected effort reduces productivity of the operation. A quality effort is required in all parts of the project, otherwise poor quality in the material management process becomes apparent immediately at the point of use. By comparison, poor quality of engineering, for example, may not become apparent at all.

Several participants contribute to the material management process and the scope of their involvement should be clearly stipulated in the contractual documents. Figure 18.2 shows the contractual relationships (shown with double-ended arrows) and the key documents that are used to establish the scope of material management of each participant. An efficient material management system leads to improved productivity and must necessarily include all participants. The alternative is an inefficient, incomplete plan which will prove counterproductive.

If an owner purchases a long-lead item and later assigns the purchase order to the contractor, a clear understanding of the purchase order is required, as well as full knowledge of any relevant correspondence, to ensure that nothing is overlooked.

18.4 LEGAL ASPECTS OF DOCUMENTS

A purchase order (PO) is a contract which specifies technical, delivery, warranty, and cost details for the goods and services to be provided. Besides

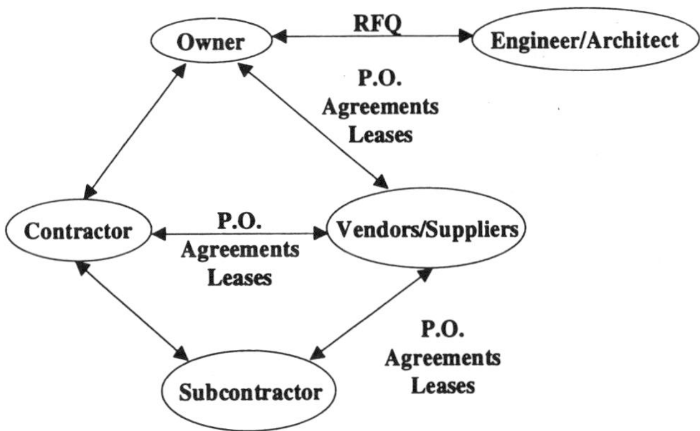

FIGURE 18.2 Relationships and Key Documents

providing technical specifications, it is important to clearly state the required delivery date and whether the goods are FOB (free on board) at the supply source or the delivery destination. Vendor information such as engineering data sheets and drawings are often required in order to finalize the engineers drawings, specifications, or both. The PO must clearly define what drawings are required and when. Two examples are the final anchor bolt layout for a piece of machinery or electrical grounding locations. This information is required in order to complete foundation design and construction, and late delivery of vendor information could result in delays and additional costs to overcome the effects of delay.

Leases should be read carefully to understand the limitations respecting insurance coverage. The wording on leases leaves little doubt that the lessee assumes responsibility for the piece of equipment or vehicle that is being leased from the lessor.

The legality of contracts is readily ascertained, but it is not apparent to many involved in procurement of materials that some of the otherwise bothersome paperwork has legal implications. A bill of lading transfers responsibility for the goods from the supplier to the transporter of the goods. Upon delivery to the receiver of the goods, at both ends of the shipping phase, the bill of lading is signed as being correct by the receiver of the goods.

Signed releases by inspectors are another group of documents that allow fabrication or processing to proceed during the manufacture of items. Pressure vessels and piping are examples of fabricated items that have mandatory hold points for inspection and further fabrication is not allowed until signed approval is received from an authorized inspector.

It is commonly required to produce mill test reports for materials used in the shop fabrications or field construction. Also, reports for specific testing of materials are required to fulfill the contractual requirements for supply of some goods. In fact, materials should not be unloaded without the proper and complete documentation. The absence or delay of the required documentation can delay fabrication progress and potentially delay the project. Besides, there are serious longer term legal implications for use of material that has not been properly certified as acceptable for use in a particular application. A dramatic example of such an occurrence is the major fire that resulted from the use of improperly identified material at the Syncrude Canada Ltd. Tar Sand facility at Mildred Lake, Alberta. Cost to repair the fire damage exceeded 100 million dollars and loss of production is estimated to be several hundreds of millions of dollars.

It is usually worthwhile for the buyer to take advantage of payment terms and discounts that are offered by the seller. The invoice for payment will stipulate these terms, such as 2/10 NET 30. These terms mean 2% discount will be allowed by the vendor if the invoice is paid within 10 days of the delivery date with the full amount due within 30 days, with interest being charged at a predetermined rate on accounts outstanding thereafter. Although such terms and notations are commonly used, they are not standard or uniform. The terms and conditions of each purchase transaction must be carefully reviewed to the mutual satisfaction of both the buyer and seller.

18.5 INTERFACES AND IMPLICATIONS ON PRODUCTIVITY

Figure 18.3 illustrates the numerous logical components in a material management system. As with any system, most problems arise at the interface between functions. At these interfaces misunderstandings or lack of understanding occurs and management attention is therefore required. A system consists of documentation, procedures, and trained personnel to execute the functions.

All major departments in a company are involved in a material management system. The key departments are Engineering, Procurement, and Construction. Project needs are identified usually by Engineering, which also determines the specifications and quantities. Engineering generates a Request for Quotation (RFQ) which is completed by the Procurement department.

Procurement develops a bidders list, solicits quotations or bids, evaluates bids with engineering assistance for technical aspects, and issues a purchase order. Other departments within Procurement expedite, inspect, and look after transportation.

The Construction Department receives materials, inspects materials upon receipt, and stores and issues materials to the work stations.

Computerized systems are increasingly being used, especially for large projects. These systems are effective if they provide a communication tool and save time in the execution of certain functions such as quantity takeoffs and material lists. Any system that satisfies these needs will improve productivity by minimizing the cost to acquire materials and improving efficiency

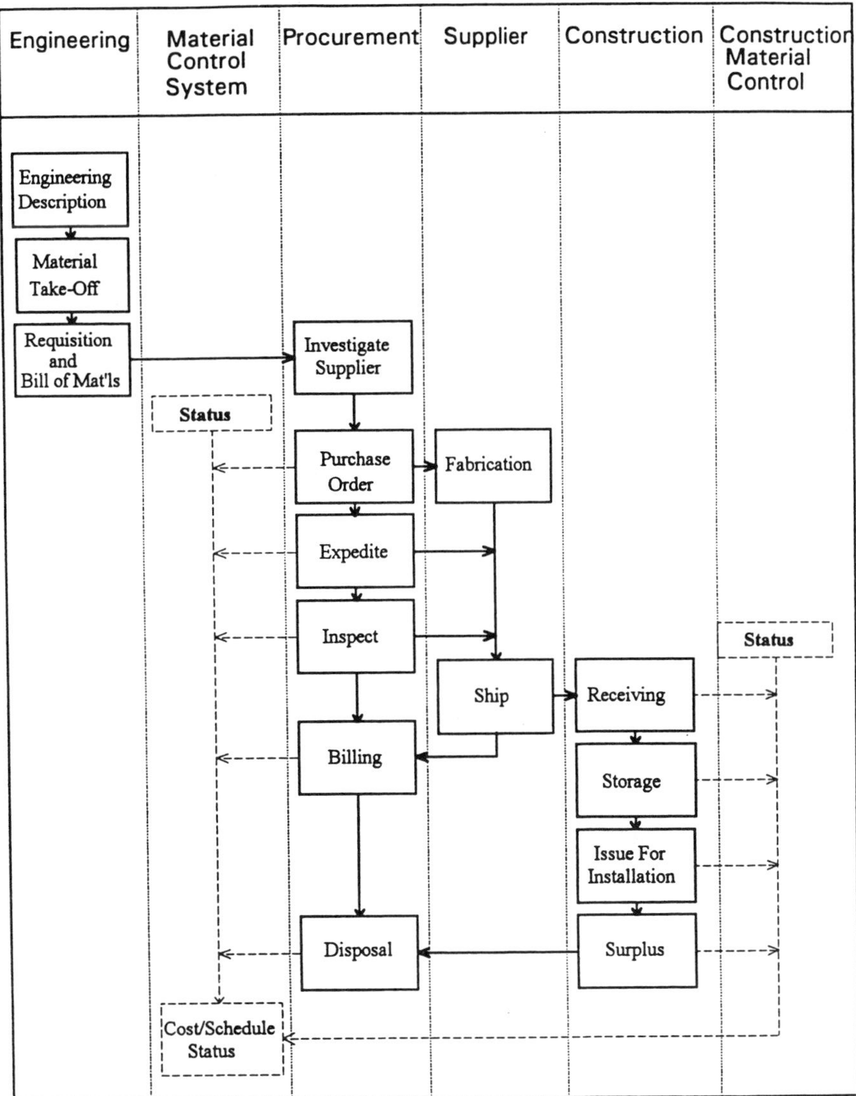

FIGURE 18.3 Construction Materials Management

at the site, especially by the timely delivery of the required materials. The level of sophistication will depend on several circumstances such as company size and project size and complexity. An organized program has a positive effect on staff and clients, creating a perception of an organized, well-planned project and an organization dedicated to improving productivity. Creating the right attitude is often as important as the function itself.

Quantity errors result in shortages or surpluses. Shortages disrupt the work pattern and require replanning around the shortages. If the system cannot detect shortages far enough in advance of material needs, the result is last minute shuffling of work crews.

Substandard materials are cause for rejection and usually require additional man-hours for installation. Inferior quality products such as a lower grade of wood may require additional man-hours during installation to overcome excessive warpage. Rejected material must be removed and replaced, both operations requiring unanticipated man-hours. Beware that a dollar saved in purchasing is not necessarily a dollar saved on the project.

The most noticeable effect of poor materials management is that of delays in delivery which disrupt the work flow and require replanning. Disruptions cause lost time and necessitate nonproductive work to remedy the situation.

Those who have experienced projects where everything happens as planned will recall that the right material was delivered at the right time, and morale rises on well-managed projects because everything happens as planned.

On a well-planned project, the right material arrives at the right time at the right place. Conversely, poorly managed materials programs give the impression that management (or engineering) does not care, and the workers' productivity declines.

With an integrated material management system, materials are more likely to be available when needed and craft supervision can plan the work around material availability. Returning to a work area to replace a previously missing item wastes man-hours. It has been reported that craft foremen spend as much as 20% of their time hunting for materials and another 10% of their time tracking purchase orders and expediting (Bell et al., 1987). Numerous actions are required to deliver the right material to the right place at the right time and the potential for delays is great. A material management system has a payback commensurate with the amount and quality of the input effort.

18.6 PREPLANNING

Front end planning is probably the single most important determinant of a successful materials management effort on a project (CII, 1985). An integrated activity is required rather than material management performed on a fragmented basis with minimal communication and no clearly established responsibilities assigned to the owner, engineer, or contractor. Decisions made during the early planning stage are critical for overall success of the project.

Early planning means early communication. Important cost-related decisions are made early, such as site access and lay-down areas, schedule compression, cash flow restrictions, expenditure approvals, and audit requirements. All of these have an impact on costs and productivity.

18.7 MATERIAL CONTROL

The material control function includes determination of quantities, material acquisition, and distribution. The objective is to purchase materials in a timely manner to avoid costly labor delays.

Bills of materials establish quantities and materials specifications define quality for ordering. A milestone schedule should be established for major items so that a complete plan is available (see Fig. 18.4). This plan considers the required at site date and takes into account final issue of drawings and data for vendors, vendor data, manufacturer's schedule, and delivery time. All items in the chain are required in order to avoid delivery delays.

Field material control is required to plan storage and issue of materials. A material management system should provide an alert for potential shortages. Control of inventory is required to prevent theft, unauthorized issue, and warranty protection for environmentally susceptible items. For those who have had to find material under the snow, the benefits of well-planned storage are readily evident. Some have experienced the frustration of sloppy warehousing when, for example, only six out of eight low-priced gaskets are available and the schedule for this task must be revised, which requires time to move and set up crews and equipment for the substitute task.

There are various techniques which can be adopted. A "just in time" plan requires careful planning and a good system. Purchases are made just in time to assure timely deliveries, and this technique yields additional benefits in the area of cash flow. Expenditures are made only when required and no sooner. "Just in time" is better suited to large purchases but also applies to bulk materials such as ready-mix concrete and asphalt paving.

Another technique is the "inventory buffer" approach which can be costly from the standpoint of cash flow and losses due to theft. In a more extreme case, increased storage and double handling may result in increased costs and lower productivity. However, it is unrealistic to expect every piece to arrive at exactly the right time. Some buffer is required depending on the material item and complexity of the job. Most foremen will want assurance that all materials are available for a particular operation to avoid false starts because of shortages.

Tradeoffs are necessary between just in time and inventory buffers. The inventory buffers are a form of assurance to provide continuous and uninterrupted work operations. However, the more material that must be stored at the site, the more double handling that is required.

Computer systems serve two basic functions—they provide a network for communications and a source of readily accessible data. For material control a database is used to provide commodity descriptions and prices. As part of the overall materials management, some database systems track the status of major pieces of equipment and critical items. Spreadsheets are a convenient tool for tracking.

Legend	
1 Data shts complete	7 Prelim vendor data
2 RFQ issued	8 Certified data
3 Receive bids	9 RAS date
4 Bid tab complete	10 ETA jobsite
5 Eng'g release PR	
6 PO award	

COMPANY XYZ
RE-VAMP PROJECT

MECHANICAL PROCUREMENT SCHEDULE

CONTRACTOR CO.
Status to 90 08 20

		June	July	August	Sept.	Oct.	Nov.	Dec.	Forecast Del.	Vendor Promise ETA	Req'd At Site RAS	Bidders (R=Received) (N=No bid)
Week beginning Monday		1 8 15 22 29	6 13 S 20 27	3 10 17 24	31 7 14 21 28	5 12 19 26	2 9 16 23 30	7 14 21 28				
Week Number		1 2 3 4 5	6 7 S 8 9	10 11 12 13	14 15 16 17 18	19 20 21 22	23 24 25 26 27	28 29 30 31				(S=Successful Bidder)
P-3306A/B M.F.O.H. Liquid Pumps	P		S									New Impellor
Sulphur guard bed	A/F											
X-632 SGB Steam pre heater	P		2 S	3 45 6	7 8		10		14 wks			Shop (Dim and Cor checks)
Sulphur guard bed	A/F	1	S 2	3 45 6	7	8	10					
V-687 & V688 Reactor and clay filter	P		S 2	3 45	6	10			12 wks			
Jet A Merox	A/F	1	S 2	3 45	6	10						
V-659/660/666/667/672 Misc. vessel work	P		S 1	2 3	45 6	10			6 wks			
Jet a merox (Supply mat'l)	A/F		S	2 3	45 6	10						
Catalyst & Acetic Acid Eductor	P		S									Hold
Jet a Merox	A/F		S									
B.S.-1/2/3/4/5/6/7/8 Basket strainers	P		S 1	2 3	45 6	10			8 wks			
Jet a Merox	A/F		S									
Catalyst & acetic acid addition drums	P		S									Hold
Jet a Merox	A/F		S									
P-262 Condensate pump	P	1	S 2	3 45	6 7	8	10		12 wks			New
Jet a Merox	A/F		1 S	2 3	45 6 7	8		10				
P-266 Caustic pump	P	1	S 2	3	45 6		10		10 wks			
Jet a merox	A/F		1 S	2 3	45 6 7	8	10					

FIGURE 18.4 Sample Procurement Milestone Schedule

More comprehensive, integrated systems address all materials management functions for both engineered equipment and bulk material. There are different costs associated with the system chosen. Equipment, software, and training costs will depend on the size and complexity of the system. Care must be exercised in selecting the system because of costs and staffing required and to avoid costs disproportionate with the size of operation or company. For large projects with thousands of material items, computerization is necessary; for smaller projects, manual methods or spreadsheets will suffice.

A technique for cost and technical control is the use of item codes for each different piece within each group of items. For example, for a valve, the size, type, metallurgy, manufacturer, applicable standard, type of service, pressure rating, whether it is a screwed fitting or welded, and any other special feature can be captured in an alphanumeric item code. This information can be transferred to a bar code on each individual piece.

Bar coding similar to that used for household consumer goods utilizing scanners is used for inventory identification. This is an emerging technology which is currently under development but has the potential to improve material control with standardization of item codes and improved scanning equipment.

18.8 PROCUREMENT

Procurement includes purchasing of materials, equipment, supplies, labor, and services for a project. Associated with purchasing are the related activities of tracking and expediting, routing and shipping, inspection and acceptance, handling and storage, and disposal of surplus. Procurement can be grouped under three categories—the procurement of materials, labor, and subcontracts.

Four cost categories (Barrie and Paulson 1978) must be considered to optimize the procurement of materials for minimum cost, and to some extent these same considerations apply to the procurement of labor and subcontracts. These four cost categories are purchasing, shipping, holding, and shortage costs. Tradeoffs between the categories must be optimized to achieve minimum costs. In these discussions the maximization of productivity is equated to the minimization of total project cost.

On large projects it is common practice to produce a procurement schedule for major pieces of equipment and a subcontract schedule. For example, Figure 18.4 is a portion of a schedule taken fron an actual project. A scrutiny of the equipment plan reveals milestone dates for the completion of key steps such as issue of the RFQ, purchase orders, vendor drawings, bid requests, award dates, and required at site dates.

18.9 MATERIAL HANDLING

A large percentage of site labor activity involves material handling. As stated previously, in the Ontario studies (O'Brien, 1989) 20% of the labor content was initially for material handling until a concerted effort was made to reduce this to about 17%. Productivity and safety can be improved by reducing material handling.

Material handling consists of operations such as unloading, storing, sorting, loading, moving materials and tools to the installation area, and hoisting. Most material handling is performed by tradesmen who have had little formal training in material movement. Tradesmen are highly trained and qualified to perform specific trade-related tasks. The exceptions are ironworkers and boilermakers who are trained also in lifting materials. Most tradesmen dislike material handling because it can be tiring, dirty, and dangerous. On large sites special material handling crews are used to improve job efficiency as well as safety.

There are several techniques for improving material handling at sites. Good housekeeping benefits materials handling as well as safety. Accumulated waste and reusable material combined with poor storage planning impede the flow of material on a project.

Material handling can be categorized as follows:

1. Packaging, containerization.
2. Movement to site.
3. Off loading at site and storing.
4. Hoisting and vertical handling.
5. Horizontal movement.

Containerization and packaging require careful planning and organization. Various types of pallets, containers, and protection are available. The sequence of packaging or loading is important also. For example, trucks delivering steel to a congested urban site, with little storage for hi-rise steelwork construction, must be loaded such that the top-most steel pieces on the truck are the first required from the truck load. Shaking out the steel in these situations can drastically reduce site efficiency.

The use of skid-mounted equipment and equipment modules reduces on-site construction labor. Modules require considerable preplanning and because of the up-front engineering and manufacturing effort a considerable amount of man-hours is transferred from the field to a shop environment. Overall the result is to improve labor productivity on the site and lower project costs.

Movement to site usually involves trucking but can include rail, ship, or air transport. Planning the arrival of shipments at the site is important so that crews and equipment can be available when required. Unplanned shipments result in waiting time for the trucker or a deployment of manpower to handle unplanned shipments. The most efficient methods of material movement could, for example, require winter roads, night shipments, wide-load permits, or consideration of partial load restrictions on roads during spring thaw.

For special size loads, route planning is necessary to avoid stalled shipments. It is costly to halt movement of a large load while low overhead electrical wires are raised.

Unloading at sites requires trained material handling crews with the proper handling equipment. As in the hi-rise building example, material should go directly to its final destination. Avoid handling material several times; do it once if at all possible.

Vertical movement and hoisting require material and personnel hoists, cranes, or other lifting devices. Several decisions are required in planning the equipment for a job. The required capacity, most suitable type (i.e., mobile, crawler, or fixed hoisting equipment), and best location on the site are examples of decisions that have a direct impact on project productivity. For placing concrete, which method is best? The choice could be a concrete pump or a tower crane, considering that the tower crane will be required to handle formwork. Horizontal movement methods depend on the material being handled. Trucks and trailers are the usual conveyances but conveyors and cranes are also common. Insufficient pieces of equipment, capacity, or size have the obvious negative effects which lower productivity.

Space requirements are usually at a premium because several trades may require the same space. At different times many key decisions relating to space, egress and access are made that affect productivity. Considerations are traffic movement at the site, proximity of building and obstructions, types of roads, turning space, and parking just to name a few.

The concepts of materials management are essentially the same for large or small projects, the differences are a matter of degree in the areas of organization and staffing, documentation, vendor relations, and computerization.

A key action item to improve material management and therefore productivity is to assign the responsibility for the materials management function. This up-front action provides a proper focus and starts the preplanning process.

Materials management requires a corporate strategy which commits the entire organization toward this function. A combination of a suitable organization and systems is required to ensure that the right material arrives at the right time at a reasonable cost.

CHAPTER 19

PROJECT MANAGEMENT INFORMATION SYSTEM

The purpose of a project management information system (PMIS) is for gathering, recording, filtering, and disseminating pertinent information.

The information is both verbal and documented and we must manage both types.

Projects are run with communications and good communications improve productivity. Also we must be prepared for the information explosion that occurs on projects. Approximately 70% of the information on a project is initiated during the **Develop** stage (see Chapter 1) of the project life cycle. This is the stage after funds are authorized and before construction starts. This explosion is accelerated even more so during fast-track projects.

The information system requires a plan which is devised for the benefit of the users and which will provide the information required for control of the work. Ideally a system should be easy to learn and easy to use, neither too complex nor too minimal to be of much use. Most individuals inherit an existing system within the company structure. Project managers are not so inhibited and usually can modify an existing system to suit their project or in some instances they can institute a special information system for that particular project.

Whether an information system is existing or is produced for a special purpose, management needs to ask the question "Who needs to know what and why?" This sounds simple, and it is for small organizations. However, as the size of the organization grows, a proactive effort is required to avoid excessive complexity and costs.

In order to answer the previous questions (Who needs to know what and why?) we should first ask "Who is involved?" (i.e., who are the players?)

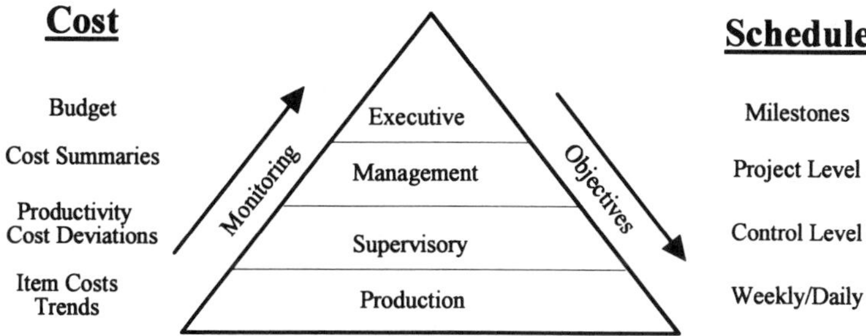

FIGURE 19.1 Information Flow: Roll-Up Technique

and "Who does what?" (i.e., what are the functions of the members of the organization?). The following discussion considers these questions in the control of a project, and can be extrapolated to a company organization. Information flows both up and down the organization. As the flow occurs, information is filtered. By necessity some repackaging of the information is needed. As scope objectives flow downward from the executive level through management and supervision levels to production, an increasing level of detail is required. Conversely, in reporting upward the information must be presented in a summary form for each level of management. Figure 19.1. shows this flow and illustrates the roll-up technique for cost and schedule reporting.

19.1 WHO IS INVOLVED? WHAT DO THEY DO?

In smaller projects the organizational functions are often combined until, ultimately, for the one-person company, all functions are performed by that single person. Nevertheless, the same main functions must be performed, but to a degree commensurate with the size of project and organization.

Consider the case of an owner and his appointed contractor; further consider a large project. A review of the organization chart will reveal what groups are involved and who are the key players.

The client is a key party because he has originated the project, and his main functions are to esetablish the requirements, provide financing, approve expenditures and changes, and monitor schedule, cost, quality and progress. Within a contractor's organization, groups perform functions which are functional and project related, if we accept that most projects are executed with a matrix form of organization. For the smaller company these functions are those related directly to the project and those pertaining to a specific project. Regardless of the size of the organization, the functions performed are similar and vary only in degree.

The following descriptions list a number of functions which require or generate specific communication documents.

Executive management monitors cost, and revenues, and progress and provides support and direction for project management.

Project Management reports project status to the owner and executive management, directs the project, approves schedules, estimates, expenditures, and contracts, plans and organizes the projects, and controls costs and progress.

The procurement department prepares requests for quotations and bidders lists, obtains prices, evaluates quotations, awards purchase orders, expedites, inspects, and arranges transportation, awards contracts, and approves payment of invoices.

Project controls (i.e., the costing and scheduling groups) generates most of the documents used by management for control and status reporting. Their functions are to prepare estimates and schedules, gather and report costs, prepare change orders, and report cash flow position.

As construction begins, the center of gravity of project activity shifts to the field. Construction management organizes the site and executes the work. Also, the construction group acquires manpower, tools, and equipment obtains permits, and administers subcontractors.

Engineering, whether internal to the organization or in the form of a consultant, generate considerable documentation but also require much information. Engineering produces drawings and specifications, initiates requests for purchase or quotation, reviews vendor data, and reports on engineering progress.

The finance and accounting department needs a constant stream of documents in order to accumulate the actual revenues and expenditures. For each project this department pays the bills and payroll, and keeps a commitment register of costs committed but not yet processed through the accounting process.

Much information is needed and a considerable amount of documentation is produced within an organization. As many note, it is a "paper" jungle. Each group produces certain deliverables. For example, the procurement department produces purchase orders, contracts, inspection reports, and material receiving reports. Project controls produce schedules and change orders. Engineering produces drawings and specifications, and so on.

19.2 WHO NEEDS WHAT FROM WHOM?

The next question to ask is "Who needs what from whom?" or "What do they need to do their work?" Unnecessary distribution is costly. For example, in a large EPC firm processing a letter is estimated to cost between $50 and $100. A useful technique to plan the distribution of information is a document distribution list, a partial example of which is shown in Figure 19.2. Many large companies have controlled the flow of information within their organiza-

NOTES:

1. Show number of copies in appropriate square spaces
2. *Responsible for project file

	Code	Description	Total		Client				Construction		Engineering Manager			Mechanical Engineer	Instrument Engineer		Piping Supervisor	Piping Material Engineer	Piping Material Control	Insulation and Coating	Pipe Stress	Model Shop		Vessel Design Supervisor	Structural Design Supervisor	Architectural		Electrical Design Supervisor	Electrical MTO	Vendor Data	Project Procurement	Contracts Supervisor	Cost Engineer	Scheduling	Traffic
PROJECT CORRESPONDENCE (LETTERS, TELEXES, TELEGRAMS)	C-1	CONTRACTURAL (confidential)																																	
	C-2	TO CLIENT																																	
	C-3	FROM CLIENT																																	
	C-4	TO FIELD																																	
	C-5	FROM FIELD																																	
	C-6	TO VENDORS																																	
	C-7	FROM VENDORS																																	
	C-8	LICENSORS																																	
	C-9	SUBCONTRACTORS																																	
	C-10	OTHERS																																	
	C-11																																		
PROJECT DIRECTIVES	D-1	MEETING NOTES																																	
	D-2	JOB MEMOS																																	
	D-3	JOB PROCEDURE MANUAL																																	
	D-4	INTEROFFICE																																	
	D-5																																		
PURCHASING DOCUMENTS	P-1	REQUEST FOR QUOTATION																																	
	P-2	QUOTES																																	
	P-3	BID SUMMARY																																	
	P-4	PURCHASE ORDER																																	
	P-5	PURCHASE ORDER REGISTER																																	
	P-6	EXPEDITING REPORTS																																	
	P-7	INSPECTION REPORTS																																	
	P-8																																		

FIGURE 19.2 Partial Distribution Chart

tion by the use of such charts. Electronic mail, if carefully controlled, will provide another level of security, but can also create a proliferation of data.

In summary, the information plan needs to answer the question "Who needs to know what and why?" and this is answered by asking:

- Who is involved?
- Who does what?
- Who needs what information from whom?
- What information do they need to do this work?

19.3 INTEGRATED MANAGEMENT INFORMATION SYSTEM

Similar to a policy and procedures manual for a company, the overall controlling document for a project is the Project Procedure Manual, which is management's communication tool. This manual contains such items as code of accounts, work breakdown structure, expenditure approval procedures, overall coding of documents and correspondence, project scope definition, and any other project-specific information which must be disseminated uniformly.

All companies have an integrated information system which includes estimating, job costing, accounting, payroll, and scheduling; these are manual systems for the most part. Some organizations have integrated computerized systems, but commonly the job costing, payroll, and accounting are integrated but estimating and scheduling are separate.

A management information system is integrated by means of a cost code of accounts. The following is a list of components for a computerized project control system with reports available for each.

Estimating	Payroll and personnel
Performance analysis	Small tools control
Progress control and reporting	Subcontract administration
Cost control	Document control
Scheduling	General accounting
Material management	Commitment register

The features of a Progress Reporting System are:

Work breakdown structure
Physical percent complete
Planned and actual hours and costs
Productivity analysis and productivity database
Progress reporting at detail or summary level

Project accounting includes:

Accounts payable	Commitment register
Journal vouchers	General ledger processing
Purchase and expenditure register	Open purchase order
Contract costs	Project cost detail or summary

A project cost system must interface with accounting systems and report costs of materials, labor, equipment, materials, labor hours, and commitments. This system should keep track of change orders also. The objective of a cost system is to track and forecast costs for comparison against budgets.

A material management system tracks materials from the requisition stage through to surplus disposal. A complete system includes requisitioning, purchasing, shipping, tracking, inspections, receiving, warehousing, issuing, and inventory control.

Several levels of schedules as discussed previously are required to meet the needs of the management hierarchy and therefore should be capable of the "roll-up" techniques, as illustrated in Figure 8.11. A scheduling system must schedule the activities, identify critical activities, level resources, and be produced graphically.

19.4 PROGRESS REPORTING

Reporting and feedback must be accurate and timely if it is to be effective for control purposes. A myriad of reports can be generated, but caution must be exercised to avoid unnecessary paperwork, especially with computer-based systems.

Feedback must occur to the workers as well as to supervisory and management levels. Good communication will improve productivity. People need to know how they are performing against expectations.

Management-level reporting, that is, for owner and contractor executive levels and for the project management team, must provide a straightforward statement of actual accomplishments versus planned cost and schedule objectives, forecast final costs, and complete schedules. It should also review current and potential problems and indicate project management actions taken to overcome the effects of the problems. These requirements are similar on any size project and relatively independent of the sophistication of the techniques that are used. Sophistication evolves with project size and complexity.

A "Monthly Progress Report" provides the essential information and contains the following:

1. Summary of project status
2. Financial summary (Fig. 19.3)

(All Figures in $ x 1000)	CONTROL BASE	CURRENT BUDGET	ITC	EXPENDED MONTH	EXPENDED TO DATE	COMMITTED TO DATE	ITC CHANGE
DIRECT FIELD MATERIALS	$2,094	$2,088	$2,088	$0	$14	$415	$0
DIRECT FIELD SUBCONTRACTORS	$7,239	$7,407	$7,478	$24	$24	$688	$71
TOTAL DIRECT FIELD COSTS	$9,333	$9,495	$9,566	$24	$38	$1,103	$71
INDIRECTS	$50	$50	$50	$0	$0	$16	$0
TOTAL FIELD COSTS	$9,383	$9,545	$9,616	$24	$38	$1,119	$71
HOME OFFICE LABOUR	$1,136	$1,115	$1,242	$83	$347	$463	$127
HOME OFFICE EXPENSE	$1,296	$1,275	$1,304	$80	$328	$454	$29
TOTAL	$11,815	$11,935	$12,162	$187	$713	$2,036	$227
CONTINGENCY	$1,622	$1,718	$1,491	$0	$0	$0	($228)
TOTAL CONTRACT VALUE	$13,437	$13,653	$13,653	$187	$713	$2,036	($1)
HOME OFFICE MANHOURS	49	49	50	3	14	20	1

ITC=INDICATED TOTAL COST

FIGURE 19.3 Financial Summary

3. Milestone schedules (Fig. 19.4)
4. Home office progress (Fig. 19.5)
5. Schedule status
6. Resource utilization, especially for manpower and any staffing or labor problems (Fig 19.6)
7. Procurement status (Fig. 19.7)

Each of these items is described briefly in the sections below. A sample set of reports for a medium size project are included (see Figs. 19.4–19.7).

19.4.1 Summary of Project Status

This report is a short, overall summary of project status. It contains a brief narrative description of the status of each major phase, provides quantitative information such as the physical percentage complete compared with scheduled completion and forecast "at completion" costs against budget. Major accomplishments, changes, and problems are outlined here.

19.4.2 Procurement Status

This item reviews contracts and purchase orders awarded during the period and those currently out for bid. Status of the manufacture and delivery of purchased items including expediting information is significant. A schedule showing actual procurement status and contract awards compared with the original plan is useful.

19.4.3 Construction Status

This unit of the Progress Report should provide a description of work accomplished during the period, significant work to be accomplished in the next period, and a discussion of major problems with solutions or proposed solu-

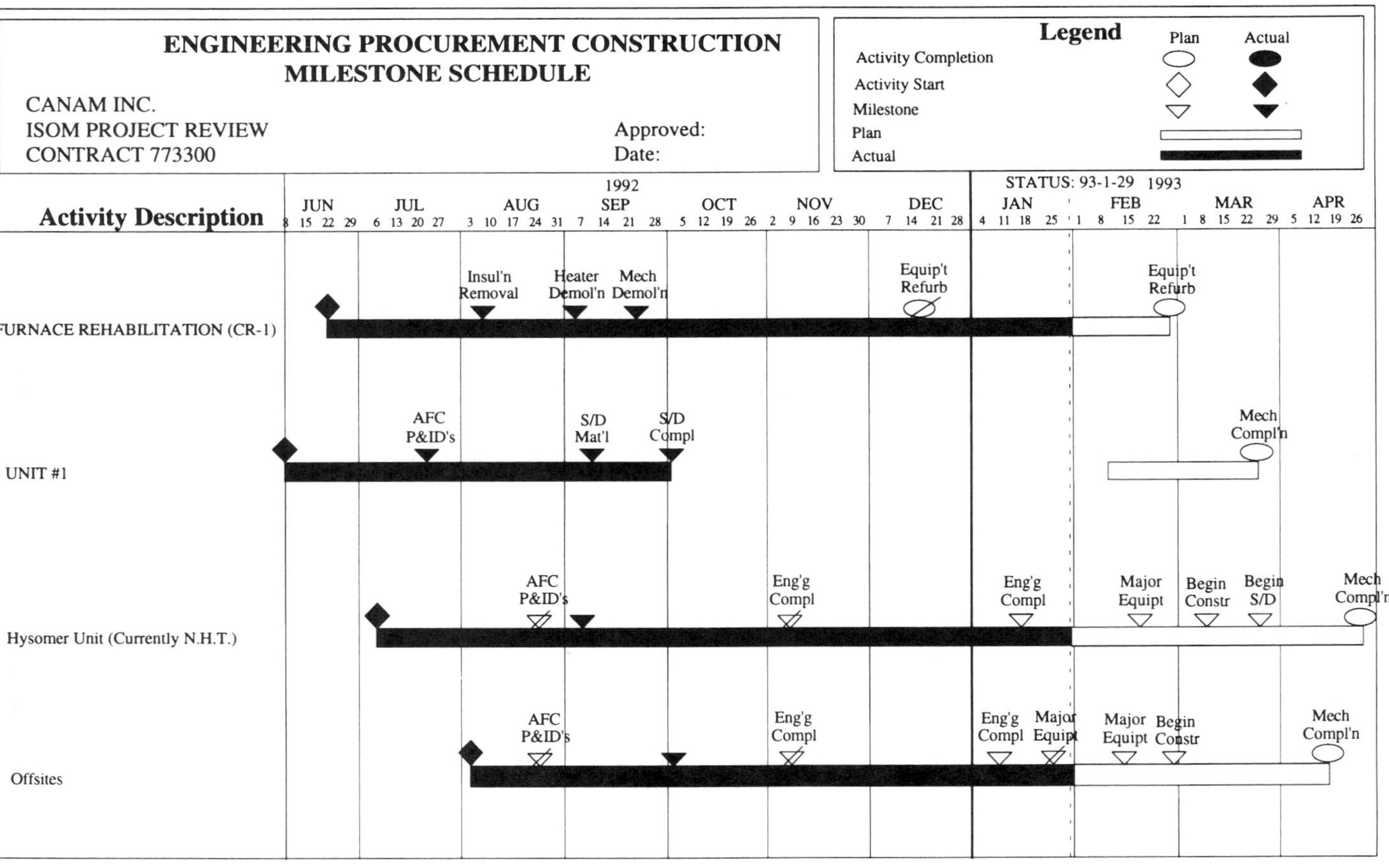

FIGURE 19.4 Milestone Schedules

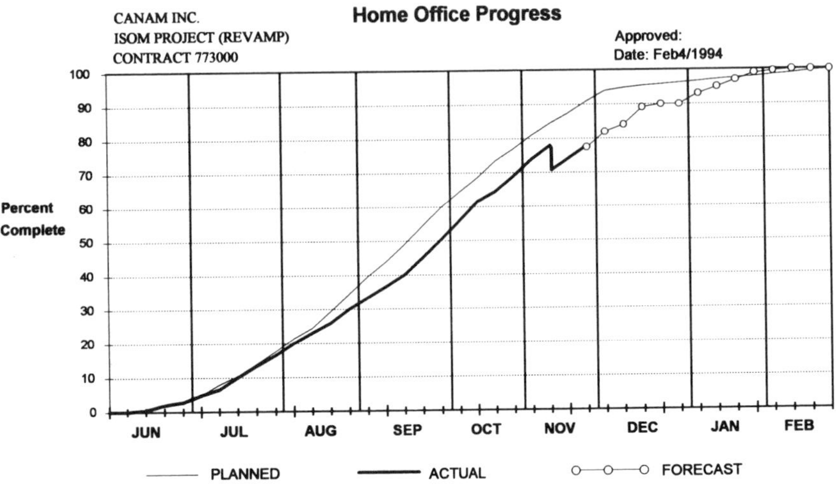

FIGURE 19.5 Home Office Progress

tions. Quantitative information such as productivity for major cost accounts not only provides control information but enhances team building.

19.4.4 Schedule Status

This item should contain the summary level schedules, comparing actual progress to the plan with an explanation of the problems and the indicated solution or measures being adopted to solve the problems.

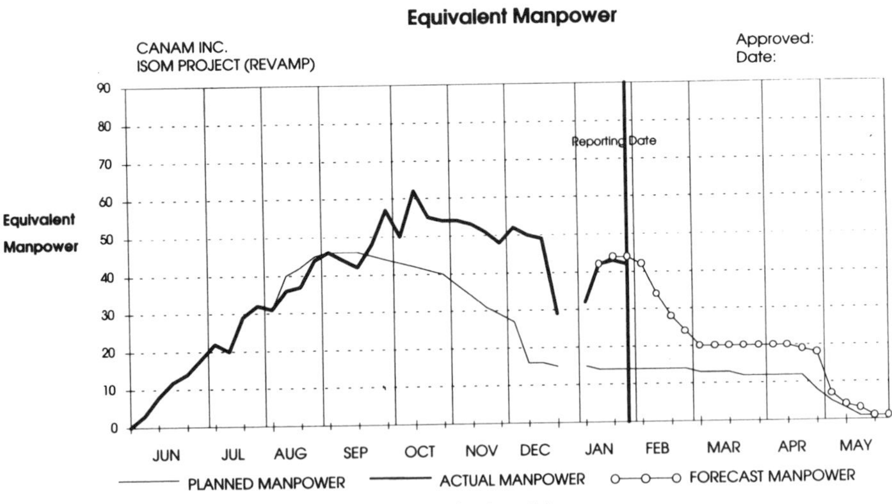

FIGURE 19.6 Equivalent Manpower

Procurement Status

	Required	R.F.Q.'s		P.O.'s		Comments
		Plan	Actual	Plan	Actual	
Vessel P.O.'s (4 Vessels)	3	3	3	3	2	
Compressor P.O.	0	0	0	0	0	
Exchanger P.O.'s	4	3	2	3	2	
Heater P.O. (1 Heater)	1	1	1	1	1	
Pump P.O.'s (8 Pumps)	6	6	5	6	4	
Misc. Equip't Revamp P.O.'s	11	11	10	8	6	
Piping Bulk P.O.'s	30	12	12	12	12	
Instrument P.O.'s	30	16	15	13	12	
Electrical P.O.'s	6	3	3	1	1	
Total P.O.'s	91	55	51	47	40	

FIGURE 19.7 Procurement Status

19.4.5 Financial Status

This summary should show actual recorded costs, committed costs, and estimated costs to complete. It should compare "at completion" costs with project budgets and identify and explain changes from the previous report. An evaluated contingency should be included so that an overall estimate of actual costs at completion is provided.

The format of these reports varies from company to company and the reports shown are part of one example set.

19.5 RECORDS RETENTION

Records must be maintained for monitoring work, future estimating, claims support, and legal requirements. Retention of records carries a cost. As a starting point five questions should be considered in order to determine the extent of records:

1. What does the law require me to keep?
2. What does the contract require me to keep?
3. What is required to control the ongoing work?
4. What historical data is required?
5. What do I need to protect my rights?

PROBLEMS

19.1 For a project of your choice (hospital, industrial plant, etc.) develop distribution charts for all the required procurement documents such as request for quotations, purchase orders, contracts, reports for expediting, material receiving, damage, backcharges, surplus, shortages, test certificates, and so on.

Also include engineering documents such as drawings, specifications, drawings and specification schedule report, transmittal letters, vendor prints, vendor data, correspondence, etc.

An interesting extension to this exercise is to arrange a meeting of the interested parties (role players for this hypothetical project). Have each of the role players discuss and later defend their needs for the documents that they requested. Note the "turf" issues that arise.

CHAPTER 20

CLAIMS

Claims are a common occurrence in the construction industry. Most claims are legitimate and do not give rise to disputes or confrontation between the owner and the contractor. Two situations can arise. The owner and contractor can come to an agreement on the claim and then the owner issues a change order. If the parties disagree a dispute arises which must be resolved. It is important to remember that most contracts stipulate a claims procedure where claims must be submitted within a prescribed period of time.

In recent years, the construction industry has experienced an increase in claims and disputes that require dispute resolution when the parties cannot resolve their differences by negotiation.

This chapter reviews the causes of claims, types of claims, contracting methods, and contract clauses that apply to claims, resolution of disputes, quantification of claims, and comments on the prevention of claims.

20.1 CAUSES OF CLAIMS

Claims occur when one of the parties to a contract seeks consideration, change, or both from an expressed or implied contract provision. Claims result from:

1. Contract documents
2. Actions of parties to the contract, including owners, designers, contractors, and suppliers
3. Force majeure considerations
4. Project characteristics

20.1.1 Claims Arising from the Contract

There are several types of contracts which provide different contracting strategies. The two main types are fixed price and cost reimbursable, and there are several variations on these two types. The construction contract sets forth the intentions of the owner and the method of execution of the project. An owner chooses the method of contracting and the type of contract, which should meet his primary objectives.

An owner's contract objectives are all or some of the following:

1. Schedule goals
2. Requirement for an advanced price guarantee
3. Flexibility to make changes within reason
4. Assessment and assignment of risk in order to transfer risk to those best able to carry it

These objectives also apply to contractors and their relationship with the owner, other contractors, and suppliers.

Contracts may contain conflicting provisions in technical specifications and codes, or may include "no damages for delay" clauses which may be contradictory to the requirements of the funding authority.

In construction contracts, the parties have an a priori agreement for modifying the contract while the work is being performed. Modifications to the contract are facilitated by "changes" and "equitable adjustment" clauses which provide for time extensions, differing site conditions, suspension of work, and termination for convenience.

It may be argued that these clauses only invite claims, but the absence of such clauses will probably lead to litigation. If relief is denied under the contract, the contractor may seek relief in court.

Conversely, exculpatory clauses seek to exonerate the owner and transfer to others the risk for problems that arise. Disclaimers for differing site conditions, subsurface problems, and underground utility interference are typical exculpatory clauses that create disputes. Notwithstanding the contract language, the courts lean toward an equitable resolution of a dispute.

Often a combination of the two basic contract types are used with various contracting strategies. Some of the more common strategies are:

1. Unit price contracts
2. Guaranteed maximum price
3. Construction management
4. Design-build
5. Partnering

The type of contract and strategies create the environment for claims. Conventional wisdom states that the contractor assumes a larger portion of the risk in fixed-price contracts, whereas the owner assumes a greater share in cost-reimbursable contracts. Fixed-price contracts require more documentation and more effort to control the quality. In cost-reimbursable contracts, the owner must spend more time in administrative control. However, these types of contracts are more flexible regarding scope and schedule changes. Although legitimate claims are made in both types, in cost-reimbursable contracts the claims usually result only in a budget adjustment, whereas in fixed-price contracts the claim results in a contract price adjustment.

Contract language is often the source of problems that lead to claims. It is difficult to write an agreement which precisely states the requirements of the project, and is therefore subject to interpretation. Ambiguities arise because of the use of vague clauses such as "reasonable period of time," "or equal," and "in accordance with trade practices."

20.1.2 Claims due to the Action of Participants

Architects and engineers (A/E) are the cause of many claims. Incomplete information on drawings and design errors are the leading causes of claims. Another group of causes is due to A/E failure to perform, and in a timely manner, such services as shop drawing review, change order approval, inspections, clarification of drawings and specifications, and correction of design errors.

A lack of design coordination and inadequate design review also are manifested in errors, omissions, schedule conflicts, and interferences between the various system designers.

Claims arise because of unrealistic contract schedules, attempts to fast-tract the schedule, performance specifications, and underestimated project costs.

Contractors often bear the financial burden of the project's problems and must aggressively seek relief through claims. However, contractors are often the cause of claims.

A common cause of distress to contractors is mistakes in estimating the cost of the work. Underbidding leads to a claims mentality because the contractor will attempt to mitigate the forecasted loss if the bid price is below the anticipated cost. The contractor may attempt to cut corners which will result in substandard quality.

Poor construction quality is a common source of claim. Correction of defective materials and workmanship increases cost and may cause schedule delays.

Inadequate performance by a contractor usually results in cost overruns which likely will lead the contractor to recover the cost overruns through claims. Labor and equipment, cash flow, and a multitude of potential management problems can contribute to poor contractor performance. The comments that apply to contractors similarly apply to subcontractors and suppliers. The lack of coordination of contractors and suppliers is often a problem which creates conflicts and claims.

Owners have contractual responsibilities such as the need to obtain permits and licenses, timely award of the contract, and adequate financial resources to meet progress billings.

Changes result in claims. Owners must control changes and process change orders in a fair and equitable manner.

20.1.3 Force Majeure Causes of Claims

Force majeure contract clauses refers to occurrences which are beyond the reasonable control of any party to the construction contract. These are stated as "Acts of God" or "Unavoidable Casualty." Claims for a time extension are usually permitted. These claims are due to severe weather, floods, fire, sabotage, and so on.

20.1.4 Project Causes of Claims

Projects that are complex, large, remotely located, in congested areas, or require technology at the cutting edge are subject to construction claims. Some examples are nuclear power plants, process plants, unique structures, underground construction, earthworks, and renovation projects.

20.2 DELAY CLAIMS

Delays are a major cause of claims. A review of the many causes of claims will immediately identify which causes give rise to delays. Standard construction contracts recognize the following delays:

1. Excusable delay—entitles a contractor to a time extension. These arise because of owner-initiated actions or changes, severe weather and other force majeure considerations, and design problems. Some contracts contain a "no damages for delay" clause which attempts to prevent a contractor from recovering any cost of delay for any reason.
2. Inexcusable delay—caused by events which should have been reasonably expected or generally of the contractors own making.
3. Compensable delay—entitles a contractor to both an extension of time and additional compensation. Some examples include scope changes, late supply of owner materials or information, impeded site access, out of sequence work requested by owner, and differing site conditions.

Delay should not be confused with the concepts of suspension or disruption of the work. Suspension is a temporary work stoppage that may or may not delay the project. Disruption is the interruption of the contractor's planned work flow, but may not involve any delay. Claims due to these concepts are based on other factors, in addition to delays caused by productivity related issues.

20.3 EVALUATING DELAY CLAIMS

Critical path methods are admirably suited to "after the fact" analyses of schedule delays and impacts. They are superior to bar charts because they depict the sequence and interrelationship between tasks and component parts of the project. Network schedules permit a better evaluation and comparison of work plans and methods, and the float for each activity. The CPM software packages readily produce bar charts from precedence networks which facilitate the comprehension of the overall project plan.

Some difficult aspects of delay claims involve the analysis of concurrent delays, and contribution of each concurrent delay to the total impact on the schedule. Another aspect is the analysis of the impact of a delay on other indirectly affected activities. One technique that may prove useful in analyzing these difficult aspects is to use a "snap-shot" technique (Tardif, 1988). The idea is to obtain a "snap shot" of the project at critical periods (i.e., before and after a major impact to the schedule).

The net impact of actions should be evaluated; acceleration or schedule compression, must be considered also. The topic of float and who owns it is a subject of many claims.

Before acceptance of a change order, the contractor should revise the project schedule. This revision may show that the new activities have used up float but have not extended project duration. The change order represents the cost of any additional labor or equipment used in these new activities plus a provision for the contractor's profit. One may argue that if project duration has not increased, the contractor has no legitimate right to additional overhead costs. This argument is far from true. If the contractor has gone through physical, economic, and financial feasibility analysis and optimization there may not be much float available. In a resource-adjusted schedule, all activities drawing on a scarce resource are critical and important for maximum efficiency. Even when there is float on the activities affected by the changes, change orders may result in less effective utilization of resources or may involve additional funding cost, even though project duration remains unchanged. The contractor should therefore study the impact of the change on the remaining activities.

In the following example a change order has been issued by the owner for extra work. The owner feels that no payment other than that for extra work is justified since the changes did not affect the critical path. Figure 20.1 shows the original network plan.

Activities *A*, *B*, and *D* require a bulldozer. The contractor has only one machine. When activity *A* finishes at day 8, there are still four days before activity *B*'s latest start time. Activity *B* is started at day 8 and finished at day 12. Activity *C* takes two days, so activity *D* starts on day 14 and ends on day 16 (see Fig. 20.2). The contractor then moves the machine to another project.

In Figure 20.3 activity *E* is added by the owner to the activity network

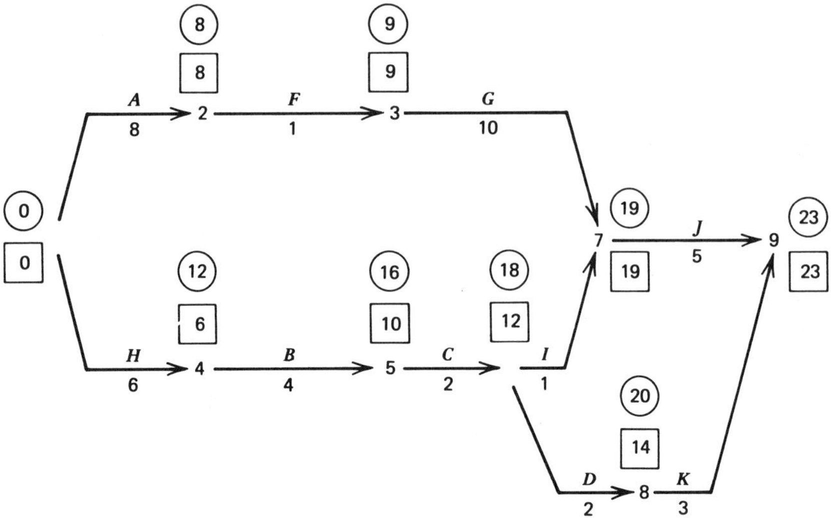

FIGURE 20.1 Claim—Original Schedule

of Figure 20.1. The critical path is not affected; hence there is no change in project duration.

Should the contractor be paid only for the extra work involved in performing activity *E*? This question can be answered by studying the impact of the change, as shown in Figure 20.4. The use of a bulldozer on activities *A*, *B*, and *D* is plotted. Though *B* is finished at day 12, *C* cannot start until day 14. Thus *D* will not start until day 16 and will not be completed until day 18. Since the addition of activity *E* causes the bulldozer to remain idle for two extra days, the contractor should request compensation for this owner-imposed idle time. The claim also includes work that is required beyond the scope of the contract.

If the contractor does not agree with the time or price mentioned in the change order, he must appeal it within the prescribed period of time stated in the changes clause of the contract. Any disputes arising may therefore be settled in accordance with the disputes clause of the agreement.

The claim should be short, concise, and comprehensive. It must include the where, when, why, how, and how much facts and details of the claim. It should consist of (1) an introduction, (2) an explanation, (3) labor, equipment, and material costs, (4) schedule disruption cost, (5) overhead, and (6) markup.

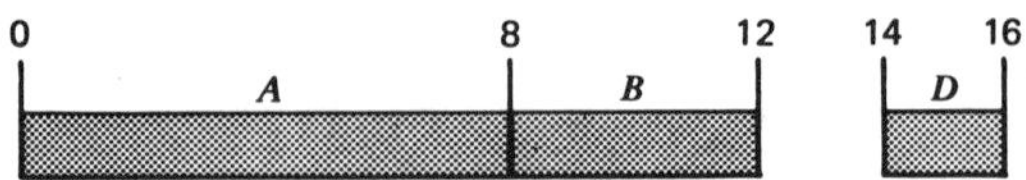

FIGURE 20.2 Claim—Equipment Utilization

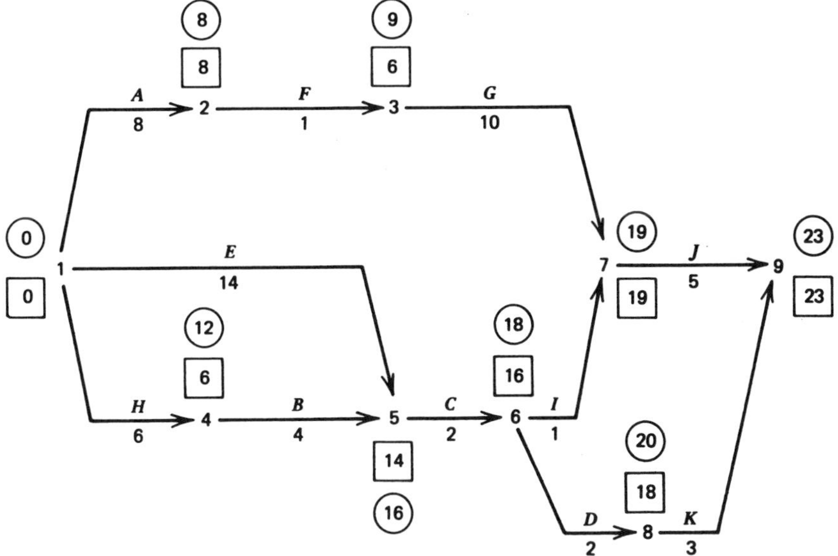

FIGURE 20.3 Claim—Revised Schedule

20.4 EVALUATING COST OF CLAIMS

Direct costs are priced in accordance with the contract change clauses. Typically the method used for pricing a claim are lump sum, unit prices which are specified by additions and deletion clauses in the contract, or force account change orders which allow the contractor to recover reasonable time and costs based on an itemized accounting review.

Impact costs are the increased cost on activities due to separate items of work or actions. For example, an increase in the scope of work of an item may require overtime to complete it on time. This may dilute the supervision and the labor resources that are available. These costs are often referred to as the "ripple effect" because they affect other activites in a project like ripples spreading across a pond. An example of this type of impact is the increase in the amount of reinforcing steel that must be installed with the formwork which has not changed. The resulting slowdown in reinforcing steel installation will also slow down the formwork unless some other action is taken to accelerate the reinforcing steel installation work. Impact cost are

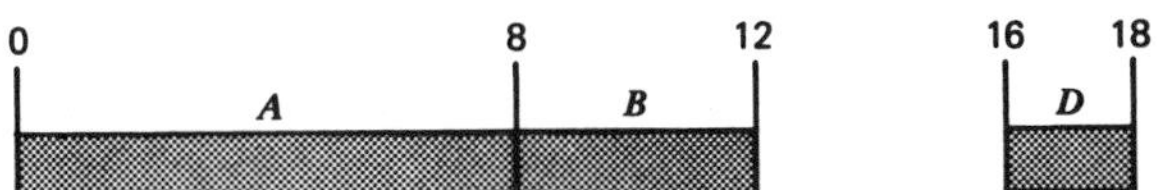

FIGURE 20.4 Claim—Revised Equipment Utilization

also referred to as disruption costs, loss of productivity, or consequential losses. Some basic understanding of construction productivity and the causes of fluctuations in labor productivity is required to assess their impact.

Acceleration can be accomplished by increasing the labor force. This may not be cause for a claim because the work is accomplished at the efficiency anticipated in the cost estimate. However, acceleration is often accomplished by the use of overtime, overmanning, stacking of trades, or multiple shifts, all of which contribute to lower productivity (Dozzi and AbouRizk, 1993).

Disruptions occur when workers are moved prematurely from one task to another, which breaks the job rhythm. Stop and go operations occur because of material shortages or lack of coordination with related activities. Redeployment of materials and labor is required, which can result in a net loss in productivity.

20.5 CALCULATING COST OF CLAIMS

This is often referred to as the cost of damages because one party claims their position has been damaged. Compensation should be based on the amount required to keep the party "whole," that is, compensation should cover costs in excess of what they would have been if the impact had not occurred.

Three methods for quantification of cost are:

1. Total cost
2. Modified total cost
3. Differential cost

20.5.1 Total Cost Method

In this method the claimed cost is the total cost incurred minus the estimated cost. This method may be the only method to calculate damages, but it is usually not acceptable because it turns an otherwise fixed price contract into the cost-reimbursable type. It assumes that the contractor's cost estimate is correct and is a proper measure of the work. It also accepts the contractor's inefficiencies as a valid cost item.

20.5.2 Modified Total Cost Method

In this method the bid estimated cost is validated by making adjustments for contractor errors and the removal of costs for excusable and inexcusable delays. Also excluded are portions of the work that were unaffected by the owner's actions. If in comparison with other bids the difference between the average of the next three bidders and the contractor's bid is within some acceptable range such as 3–5%, then the contractor's bid (less adjustments) is accepted as a reasonable baseline for costs. Comparisons can also be made with published data to achieve similar results. The claimed cost is then the modified total cost minus the contractor's bid price.

20.5.3 Differential Cost Method

This method is also referred to as the measured mile or modified baseline approach. The idea is to select a portion of the project for work similar to that which has been affected by the change or disruption. Comparison of these portions of work will yield the increase in costs due to the owner's actions.

20.6 CLAIMS AVOIDANCE

Avoidance of claims is the best approach for all participants. To avoid claims, it is necessary that thorough consideration be given to the contracting method before proposals are accepted from the contractors. Fixed price contracts attract more claims than cost-reimbursable contracts.

On any project the owner may cause claims to be filed by instituting changes while the project is in progress. This situation occurs when the owner is not sure of project needs at the initial planning stage or when there is insufficient or poor engineering either prior to tender or in dealing with expected conditions. This can usually be avoided if the owner and the engineer undertake lengthy and detailed investigations of the project before the contract is awarded and select the form of contract that suits the situation.

When one of the contractors faults in his or her commitments, all adversely affected contractors are given grounds for claims. Nonperformance of any contractor can generally be avoided if project management takes the time to review thoroughly each contractor's schedule and activity network in view of the contractor's capacity for timely achievement of interface events.

Fair allocation of risk and incentives for cooperation are also good prevention methods.

An owner can prevent claims by creating a climate of cooperation on a project. An increasingly popular execution technique is to adopt the partnering concept where all parties are viewed as key participants and contributors. A modus operandi is established where all parties work diligently to minimize conflict and avoid claims by an open attitude and honest approach toward the solution of problems that arise.

20.7 DOCUMENTATION FOR CLAIMS

In the spirit of partnering, good documentation by both the owner and contractors is essential and avoids unproductive confrontation. An open dialogue and prompt attention to problems will avoid many claims and disputes. Good documentation fosters cooperation.

One of the most inexpensive and efficient methods of protecting against claims is to maintain a diary of the work. When the owner has to refute a contractor's claim, the diary helps to substantiate such refutation. A suggested diary format is given in Figure 20.5.

Documented evidence forms the best basis for a contractor's claim. Careful maintenance of records throughout the project is a must to provide this data. A contractor should record all change orders affecting the contracted work, date of notice to proceed, amount estimated, date of submission, time and amount requested, time and amount approved, and date of approval. The contractor should also maintain a record of his or her claims with their

Location: Chicago Place
Contractors: Ices Ltd.

Date: 5/12/93
Contract No.: 73-08A-14567
Weather: Cloudy
Temp.: Low 10, High 25

Work Category	Work Force	Equipment	Description
ABC Construction		2 cranes	Forms for beams and slab, ninth floor
Operating engineering	2		Masonry for west and south walls, fifth floor
Oiler	2		
Masons	10		
Carpenters	15		
Laborers	12		
Western Electric Co.			
Electricians	5		Rough in ninth floor Conduits, seventh floor
Eastern Plumbers			
Plumbers	8		Waterpipe, fifth floor
Fitters	5		Rough in ninth floor
Southern Earthmoving			
Operating engineers	2	1 backhoe	Excavating in northeast and dumping in southwest
Oilers	2	1 loader	
Truck drivers	5	5 trucks	
Northern Steels			
Ironworkers	6		Placing rebar, ninth floor

Delays and difficulties encountered. Drawings for tenth-floor reinforcement have not been received from the structural engineers. Masonry work on sixth, seventh, and eighth floors behind CPM schedule because of nonavailability of masons.

Work not approved. Masonry on fourth panel, south wall, fifth floor rejected because of unevenness.

Remarks . Jackson, carpenters' foreman for ABC is absent because of illness. Sheet metal work held up because of masonry on sixth, seventh, eighth floors. Teamsters' Union strike begins tomorrow.

Inspection tests performed. Water pressure tested in washroom 27 on third floor - 100 psi. Electrical cables and conduits approved.

Instruction to contractor's representative. Sheet metal contractor advised to resume work. Construction superintendent ABC to demolish rejected masonry.

Prepared by: Alexander Black Phone: 492-1000

FIGURE 20.5 Owner's Diary

description, date of submission, time and amount requested, time and amount granted in negotiated settlement, and documented decision of the owner.

The contractor should keep progress photographs, test reports, shop drawings, and copies of correspondence with the owner in these records. A record of labor, material, equipment, and indirect costs for the additional work should also be maintained by the contractor. This information serves a dual function. It can be used to demonstrate the effects of change and is also available for historic use in estimating. As stated in previous chapters, this kind of information storage is most practically accomplished by means of a computerized information system.

The day to day follow-up and the conduct of the owner or the owner's representative can result in a change of work sequence, disrupting the contractor's original plan. The contractor's daily records, letters, interoffice memos, telegrams, notes of messages and telephone conversations, and other forms of oral and written communication may substantiate subtle owner-imposed changes. Whenever the contractor detects a change, either while work is in progress or after it is done, he or she should promptly inform the owner of an intention to file a claim and then document all pertinent data.

On all projects, but particularly when CPM is a contract obligation, the contractor should keep the network dynamic and useful by regular updating. When this is not done, no current basis exists for settling the claim. A regularly updated network tells the history of the project. The impact of pending time extensions in either the approval or dispute stage also affect the network. An impact analysis sheet with details for each item becomes a handy reference at negotiation time. The sheet should contain descriptions, references, network computations showing the effects of change, and an evaluation of the time and cost additions being claimed.

Changes often affect both the utilization of resources and productivity on a project. The effect on cost sometimes is not explicit and needs more than mere historic records of labor, material, and equipment. If the necessary data are collected, the process affected can be simulated and the computer outputs can be used to document the claim. For instance, a contractor responsible for the boring of a small tunnel wants to use crews on both ends of the tunnel to achieve maximum utilization. Because of interruption by the owner, however, the contractor cannot succeed in this objective. By documenting this claim with simulation of the whole project, the contractor can successfully substantiate the claim for losses.

20.8 SETTLEMENT OF CLAIMS

Claims can be settled by the following methods:

1. Negotiation
2. Dispute review boards

3. Mediation
4. Minitrials
5. Arbitration
6. Litigation

Negotiation involves two parties who agree to communicate with each other and make decisions. The parties reach an agreement which is a modification to the contract.

Dispute review boards (DRB) are formed at the inception of the project and remain throughout construction. Disputes are heard as they arise and resolutions are arrived at in a timely manner. These boards consist of industry experts who make nonbinding recommendations for the settlement of each dispute.

The Dispute Review Board fosters co-operation between the contractor and the owner, and provide a means for prompt and equitable resolution of claims and disputes (ASCE, 1991). The DRB does not supplant the owner's dispute settlement methods, but rather is an intermediate step aimed at avoiding more expensive and less satisfactory procedures. A DRB emphasizes dispute prevention.

The provision for mediation is usually provided for by the contract. A neutral third party acts as a communicator and facilitator as the parties make decisions themselves. An agreement is reached which is nonbinding, but one to which the parties are morally committed. Minitrials are also a nonbinding resolution procedure which follows a structured process similar to litigation and is usually conducted by a judge.

Arbitration is stipulated by contract or legislation or is simply agreed upon by the parties. A neutral third party acts as a decision maker for a panel which consists of representatives from the opposing sides of the dispute. The decision by the arbitrator is final and binding.

The most expensive process for resolving disputes is litigation. There usually are no winners in this process. A decision is rendered by a judge, which is final and binding.

Inspite of the pitfalls and shortcomings of the alternate dispute methods, they are preferred to litigation, which is a lengthy and expensive process. Except for litigation, avoid the presence of attorneys which can inhibit free and open exchange and may create an adversarial mood.

The first method is the least expensive and simplest. A major disadvantage of litigation is that an engineering decision may be placed in the hands of people who have no engineering knowledge. The other alternative dispute methods lie in between.

20.9 ETHICS OF CLAIMS

The owner–contractor relationship has a great effect on the settlement of claims. For an owner, fairness pays off in the long run. If he/she is fair and maintains a reputation for it, the contractors will not add an additional amount to their bids to cover unfairness. Fairness begins with the specification, which must avoid "weasel" clauses. In the long run, fairness will result in lower job costs. If contractors can be confident that they will be treated fairly and paid promptly by owners, they will be willing to submit lower bids because they can be sure that if extra work arises they will be paid for it.

CHAPTER 21

CORRECTIVE ACTION

Although many causes of poor performance on the job are easily discernible (i.e., strikes, weather, price hikes), not infrequently some inexplicable situation will occur that brings performance consistently below par, resulting in persistent schedule slippages and/or cost overruns. Then management must determine exactly why the efficiency of the organization is low and take actions to correct the situation.

In this chapter the task of analyzing a project or past projects to identify the variables that influence performance will be discussed.

21.1 CORRECTIVE ACTION SYSTEM

Performance on any project may be considered as the output resulting from the input into the project of resources including time and money. Information about progress and cost performance (feedback) is received from the work site and is tabulated in the desired progress and cost reports. Measurement of actual achievement through progress reports (the feedback loop) and its comparison with the plan objectives are essential. If, on comparison, the outputs do not conform with the plan objectives, either the inputs must be manipulated or the project plan modified. The project plan must respond to changing conditions in order to meet the project objectives. These concepts, observed in project management, as illustrated in Figure 21.1, are similar to system design concepts.

The first course of action in regaining a desired performance position is by manipulation of available resources within the constraints of the present

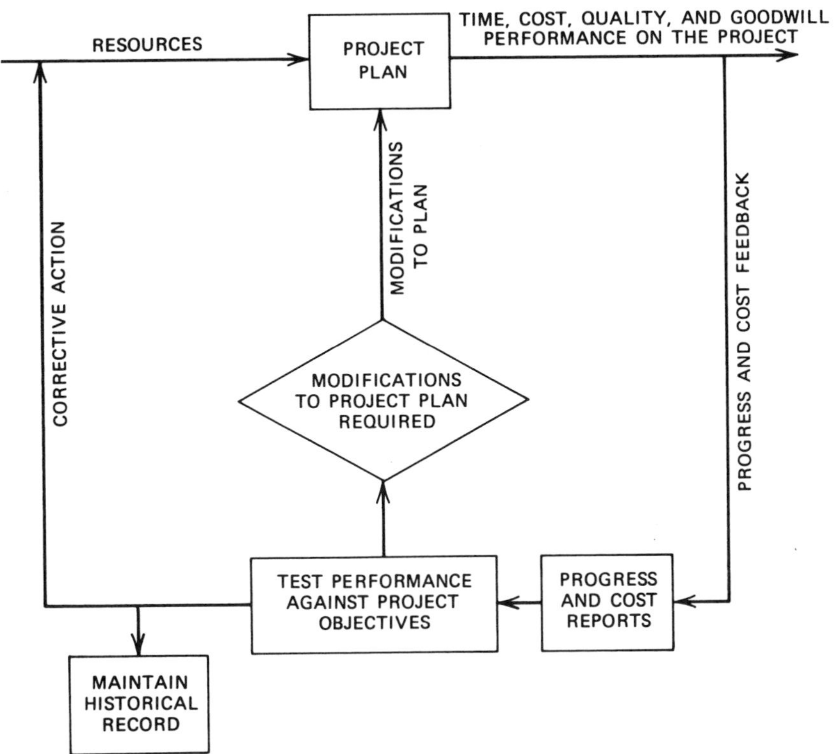

FIGURE 21.1 Corrective Action Process

network. Redistribution of manpower and equipment introduces new characteristics in the project plan and may produce new critical paths. Changes become apparent as soon as the network model is analyzed; thus the proposed remedial scheme must be rescheduled and costed. If this is satisfactory, the project may proceed on this new schedule with new control events; if not, further resource redistribution may be tried. It is possible sometimes to recover performance entirely without recourse to network compression.

The second recourse is to design a completely new network model from the current position to the project completion event, introducing new construction methods and/or equipment, together with other additional resources, to maintain the desired completion date. For this model an estimate of required resources is made, and the net total cost is determined.

After feasibility analysis and optimization, if the plan is acceptable, the project is rescheduled and proceeds according to this new plan. If not, further alternative plans may be devised until, finally, the most economical solution is obtained.

The greater the frequency of feedback and response, the higher will be the probability of attaining the project objectives. The objective is to bring the project plans back into phase with reality. Thus the control of a project requires not only keeping the performance in accordance with the plan, but also keeping the plan updated so that it always depicts the latest strategy to achieve targets as they exist at the time.

A historical record of performance should be maintained so that information is available for use in planning future projects.

21.2 TIME MANAGEMENT PROBLEMS

Completion on time is often difficult to achieve. Although the problem manifests itself as a delay, lost time, or a missed schedule date, the cause usually is not due to the scheduling process itself but by the lack of integrated management of the project. A review of the project management model (Fig. 1.1) provides clues to the source of schedule problem areas. Problems arise because of a lack of integration of "what we manage," that is, quality, cost, time, communications, and risk.

Most problems start in the planning phase, which consists of the **C**onceptual and **D**evelopment stages. Often projects get into trouble because they start in trouble as a result of some planning deficiency. Most of the slack is used up early and the project is encumbered by a tight schedule with little margin for error.

An ill-defined scope will result in changes, which will require a changed schedule. For projects that are similar to a previous project, it may be possible to arrive at a reasonably accurate schedule without a well-defined scope, but most projects are not repeats. Incomplete or inaccurate drawings also lead to changes. There always exists the tendency to make something better and to constantly refine some aspect of a project. It becomes necessary to freeze the scope, which will then fix the schedule duration.

Poor quality management extends the duration of activities or acceleration is required to make up the time to do something a second time.

Poor estimating creates unrealistic expectations and possibly the improper application of resources. If a task is estimated to take 1000 man-hours to complete, the schedule reflects this anticipated work. A crew is assigned to complete this 1000 man-hour task. If the task realistically requires 2000 man-hours, the same crew will require a longer schedule and therefore a cost overrun and a delay occurs.

Uncertainty requires contingency planning. The risk associated with this uncertainty can be narrowed to a few key activities and some contingency must be built into the duration of those activities.

Responsibility for risks should be assigned to the contractual party best able to manage the risk. Often risks are assigned to a contractor, who has less control than the owner over some aspect of the project. A typical example of this is the risk of subsurface conditions being different from those represented by the bid documents. This practice often leads to disputes which disrupt the project and then a delay is likely to occur.

Where a communication system is not structured, project information is probably inadequate. Full and timely information are essential for effective decision making, which is needed for schedule control.

Beside the formal contract, an effective tool to keep the project on track is written internal procedures which clearly outline the authority structure and the operations of each discipline. A project is "never on time" if project information is confusing, late, or incomplete.

The scheduling process must provide realistic schedules and progress reporting. Bimonthly or weekly schedules can be reliable, whereas it is much more difficult to produce realistic long-term schedules.

Completion on time requires integration of the management aspects of "what we manage." It can be seen that many problems manifest themselves as schedule delays.

21.3 REVIEW MEETINGS AND INTERPRETATION OF REPORTS

In practice an effective mechanism used for dealing with overruns and corrective action items is planning and review meetings that are conducted at various levels at regular intervals.

Craft supervisors and resident engineers consult on a daily basis to sort out any changes in the following day's work schedule. Any extraordinary problems or conditions are recorded on the daily progress sheet and are made known to the responsible supervisory personnel.

Weekly meetings are held to assess the progress on work packages which are prepared in compliance with the established work breakdown and cost structure. The review meetings are aimed at translating latest work status and critical problems into a specific action plan. The construction supervisors and the scheduling and cost engineer maintain intimate involvement with the progress of work under daily and weekly schedules. Weekly reports with information on the actual and "forecast to complete" quantities of work serve as the agenda for the weekly meetings. By analyzing the actual manpower distribution and material or equipment usage, the allocation and availability of resources can be adjusted.

The resident engineer, construction superintendents, and safety or quality engineer need only attend these weekly meetings on an as-required basis. They are, however, required to attend the monthly meetings held by the project manager. At these monthly meetings the monthly report, which points up critical matters requiring immediate upper management attention as well as long-run decisions, is reviewed. The monthly progress reports provide a basis for the plan of action for the coming month and quarter. On some projects, the reporting period may be bimonthly or weekly. This may become the norm as information systems improve. Major issues are reviewed for high-impact sensitivity. This type of routine review encourages all levels of the organization to become involved; it is especially useful to executive management in that it permits them to examine the latest trends in physical progress and resource expenditures. Major delays, overruns, and bottlenecks can be analyzed and recommendations for alternate action plans and decision making can be acted upon.

This process does require painstaking effort by the involved team members, but the attention to detail is reflected in a smooth-running project.

If from the progress status reports the project manager finds the project is on schedule and there are no current problems affecting the work in progress, then the effort is to keep the project on course. However, the situation can be misread if the budget is based on an overestimate. A long-term effect of overestimating is that the company is likely to reuse their so-called on-budget figures to bid other jobs with little chance of being successful. If care is taken, the variance reports should indicate to management that something is awry. Table 21.1 shows the inference that management can

TABLE 21.1 Manpower and Schedule Variance/Inference

	Progress		
Manpower Variance	Ahead of Schedule	At Par with Schedule	Behind Schedule
Negative (over budget)	Check percentage progress with the percentage of manpower usage and their increase over the last period.	Check the increase in manpower usage and compare to estimate	Alarming situation; revision to plan is required.
Zero	Performance better than estimate—the latter needs analysis to determine if due to exemplar performance or incorrect productivity used.	Normal—good sign	Excessive manpower usage
Positive (under budget)	Too much overestimating—needs management review	Estimate on the high side—needs correction	Lack of progress—analysis of the situation by management required.

derive from a review of variance in estimated and actual manpower and scheduled and actual progress. It is possible that low actual expenditures result from erroneous charging by employees. If a discrepancy is suspected, the bookkeeping records provide a double check. See also the section on variance analysis in Chapter 14.

If the review of the network plan shows that the project is on schedule and the variances are only small, the project manager must keep in mind that human nature wants the actual and estimated expenditures to agree. Small variances may indicate that a crew is aware of the projections and is making sure that the reported resource usage, if not progress, is on target.

Material variances have their own unique set of problems. The placing of a purchase order represents a commitment for the expenditure of funds, but the expenditure occurs when the company has paid for the material delivered. From the time an order is issued, the material is delivered, an invoice is received, and the material is paid for, several months may have passed. To the accounting department an expenditure is made only after the invoice has been paid. If the material controls are based on expenditures, the project manager may be confused by the apparent discrepancy. To alleviate this vagueness, the project manager should also have a report that indicates commitments made to date. These commitment reports (register) can be made directly from the record of purchase orders.

If the project appears to be running ahead of schedule, it is important that material commitments and manpower requirements also be ahead of schedule. If the project is on schedule with many negative material variances, it could imply that material usage or prices are higher than estimated. If high prices are the cause, the material estimate for the remainder of the project should be reworked. If high material consumption is indicated, the cause(s) should be found immediately, as there may be a need to revaluate the estimate. Check waste allowances.

When the job is proceeding on schedule, but there are positive material variances, either the material requirements were overestimated or better prices were found than were allowed for. Positive variances alone may point to slower progress than expected; in other words, the work has not proceeded at a rate sufficient to warrant the usage of materials according to the original schedule.

Small positive and negative material variance is a good sign if the project is on schedule, indicating the estimates were accurate. However, if the project is behind schedule, it indicates materials are being expended ahead of actual progress. If the job is ahead of schedule, then likely all is well, and chances are that costs are in line with the budget. A material and schedule variance inference table similar to the manpower and schedule variance/inference table can be drawn. A variance/inference table for equipment usage can be used for analyzing equipment usage vis a vis project progress.

21.4 LABOR COST ANALYSIS

One of the more common, and also more complex, causes of schedule slippages and cost overruns can be traced to labor productivity. This is especially true on labor-intensive projects such as building construction. It is important to realize that the factors affecting labor productivity are many and varied. In addition to physical causes for low production rates, a number of psychological factors can lower morale and thus productivity.

Following is a list of factors that can often affect on the job performance. Although it is by no means complete, it should give the reader some insight into the nature of such factors.

1. Acquisition rate of direct labor
2. Attrition rate of direct labor
3. Craft/apprentice ratio
4. Dissatisfaction with manpower scheduling
5. Lack of a steady flow of work
6. Supervisory ratio
7. Overtime
8. Lot size for prefabricated items
9. Time intervals between similar tasks
10. Cumulative output of units produced
11. Out-of-sequence work
12. Elapsed calendar time
13. Engineering changes
14. Rework and scrap rates
15. Equipment changes
16. Percentage of machine control
17. Visible backlog
18. Accident rate
19. Absence of such productivity control techniques as time study, activity sampling, bar charts, and so on
20. Percentage of bad weather

Perhaps the most basic factor affecting on the job performance is one that came into play long before the project plan was implemented, namely, engineering design. Application of value engineering and constructability reviews are important throughout the project, but especially in the early stages. Carelessness in this phase can result in serious consequences once construction has begun. Indeed the consequences may be catastrophic, and the corrective action may be so costly that the project may never recover financially. In this case the only possible corrective action is a program for engineering review at the engineering design phase, not at a point where changes can have detrimental effects. This reviewing procedure should involve specifications, design transmittals, approvals, bid and award of contract. If any errors occur during this review, corrective action may be taken with relative ease. Once this action has been applied the whole reviewing process should be carried out as often as necessary to ensure a sound engineering design.

The feelings and attitudes of the workers affect productivity. An unfavorable response from the laborer is due to lack of genuine interest on the part of management. People like being part of a well-managed team. It gives them a sense of pride to have a good relationship with the organization. It takes money to train people in the organization's methods and standards, and it takes time to integrate them into smooth-running crews. Thus there are sound economic reasons for providing steady work and treating workers fairly. The prospect of ongoing employment will result in higher productivity, and the converse is also true. The workers should know about management's fairness in dealing with them—for instance, in scheduling day and night shifts or holidays according to their preference. Workers can always sense management's genuine concern for them. Hourly employees appreciate the efforts of their management to keep them steadily employed and in return are loyal to the organization. Union workers are known to have worked with one contracting organization for years until the organization could not provide more work.

If workers are not hired and laid off gradually, keeping their number to a minimum, the erratic schedule profile can cause low learning benefits and lead to increased training costs.

Too much or too little supervision can be a cause of low productivity. This is often reflected in overtime work, particularly if people avoid work during their regular hours to get paid a higher wage for overtime. Effective supervision can improve this situation.

By doing a job repeatedly, workers learn about the work, and their productivity goes up. If the work is not spread over a longer period, too many workers have to be engaged to do the job in a short time. Again the benefits from learning are lost. If the work is not continuous but spread over a long period in noncontiguous packages, increase in productivity is not experienced because of lack of continuity. Another cause could be the habit of supervisors to engage workers on one job the first day, send them to another the next day, and then bring them back to the first job the third day, thus breaking the sequence of work. Frustration of workers which leads to low productivity can also be caused by too many changes, resulting in scrapping already completed work. Too often equipment changes can irritate workers and result in lower productivity.

Yet another cause of lower productivity is too little visible backlog of work; work slows down as the end is in sight. This effect is due to workers trying to prolong a job in order to avoid being laid off or transferred. On the other hand, with a backlog there is always an effortless increase in the pace of work in an attempt to catch up.

Workers have a low confidence in management when a project has a high accident record. Good safety inspection and tool box sessions by trade superintendents will enhance safety.

A conscious effort to increase productivity requires the study and implementation of productivity control techniques. These are used to determine the effectiveness of supervision through activity sampling and to test work methods through process charts. The absence of productivity control techniques can be a cause of low productivity.

An effective strategy is to involve the workers in the planning of work and the solution of problems. Brainstorming and quality circles are two techniques that are effective when correctly utilized. This is simply recognizing the fact that those closest to the work usually know how best to correct or improve a situation.

Bad weather, although not controllable by management, can result in many lost work days. Also, when labor has to be paid without actual work being done because of union agreement, the result again is lost production and reduced productivity.

By doubling the crew on an activity, on the job performance may or may not be increased. Care should be taken so as to not create overcrowding. If the action is taken during a period of good productivity, the effect will be to complete the activity ahead of time, and at a lesser cost. Should the productivity be unfavorable, however, the activity will still be accelerated, but the cost may be above that estimated. To offset this, the project manager must look for noncritical activities that can be rescheduled during periods of more favorable productivity. Alternatively, these activities can be spread out in the time available for their completion to minimize the impact of low productivity. An example is scheduling pavement work for summer even when it is a noncritical activity and can be done later in the fall. Scheduling pavement in the summer may require that certain other noncritical bottleneck activities proceed the pavement work—for example, the installation of an asphalt plant. Thus it is necessary to expedite bottleneck activities (not necessarily only critical activities) that open up following activities for exploitation in periods of expected good productivity.

21.5 MATERIAL COST ANALYSIS

Another common cause of excessive costs and delays is the acquisition, transportation, and storage of materials used on the projects. Fortunately, the analysis of material costs requires comparatively little research where adequate material accounting procedures are used. It does, however, require a complete audit of the accounts and a careful scrutiny of the various factors affecting the material costs.

The following are some of the major causes of excessive material costs:

1. Material takeoff
2. Waste
3. Pilferage
4. Inaccurate measurement of quantities delivered
5. Uneconomical lot sizes
6. Poor timing
7. Insufficient provision for escalation
8. Excessive transportation costs
9. Excessive handling of materials
10. Improper selection of materials
11. Excessive storage costs
12. Lack of priority purchasing
13. Improper expediting
14. Late payments for materials (i.e., not taking advantage of discounts)
15. Poor judgment in purchasing

In the analysis of material costs it is vitally important that all purchase orders, material accounts, delivery schedules and receipts, and project estimates be available to the analyst. The preceding list of contributing factors can then be used as a checklist to determine if an item is a factor in the overrun. Each time a question is raised, the available data are analyzed to determine whether the item can be excluded.

The point at which material costs start to become excessive occurs during material takeoff. This may be due to carelessness or correct calculations based on incorrect data, thereby producing false results. If and when these results are later utilized to procure materials, the inherent error surfaces, showing these material costs to be far above those based on the material takeoff. If construction is well advanced, corrective action to deal with this excessive cost is quite often impossible. The only real solution is to abide by the old adage, "An ounce of prevention is worth a pound of cure."

Items 2–4 in the preceding list may not come to pass if the quantities delivered to the site are consistent with the quantities originally estimated as being required and if a record of material usage is available to verify that the quantities delivered are equal to the quantities used. Otherwise, the analyst must find out what happened to the materials that are unaccounted for.

For items 5–7 the analyst should obtain for each type of material a breakdown showing quantities and order dates. By determining whether any orders could have been readjusted to obtain lower prices or lower transportation costs, the analyst can decide whether any of these items contributed to the cost overrun.

The excessive handling of materials is more difficult to ascertain since it is related to the methods employed by the organization. It is often difficult to obtain historical data on the amount of handling that was done. Indeed, if these costs are not charged directly to the materials, the analyst has no means of checking their effects. In this case, however, the added expense would show up in labor, equipment, or overhead costs. If successful in pinpointing such costs, the analyst must consult the site supervisor or project manager to determine if such rehandling is unavoidable. Process flow charts are useful in this endeavor.

The improper selection of materials is often difficult to pinpoint; therefore it should be reserved for those materials that are nonstandard or that are a major expense. When studying this problem, the analyst should obtain opinions from other sources, check the appropriate specifications, and make a list of all possible alternatives and their relative merits.

Items 11–13 are related in that they all require a study of the ordering and expediting of materials. The analyst should determine whether the material storage costs have been justified (i.e., if it was necessary to have these materials on hand) and whether any delays or other problems have developed because of poor expediting, especially on priority items.

If the organization has been lax in the payment for materials, it could mean that it is not getting prime prices from its suppliers. Corrective action would come through establishing and maintaining an alert accounting system capable of tracking all goods delivered and presenting the proper management authorities with correct payment schedules. Money could then be set aside at the appropriate time and late payment penalties can be avoided. Also the carrying charges arising from buying on credit could raise the cost of materials. Again an alert accounting system can provide relief by continuously monitoring the status of all credit transactions and presenting the situation concisely to management.

Finally the analyst should determine whether the materials purchased have been obtained in the most economical way. For instance, the purchasing agent may have bought random length rebar at what he or she thought was a saving compared to buying specified lengths; however, after the field cost of handling, cutting, and wastage are added, the actual cost of random length rebar may well exceed that of uniform lengths. Suffice it to say that any analysis of material costs should involve a search for lower-cost sources of materials, the study of new technology materials, and the serious economic analysis of any bulk purchase plans that may exist.

The reader is referred to Chapter 18 for a more comprehensive discussion on material management.

21.6 EQUIPMENT COST ANALYSIS

Since the plant or equipment costs incurred often comprise a major portion of the budget, the analyst should perform a thorough equipment cost study.

The following are some factors affecting equipment costs:

1. Equipment unsuitable for the organization
2. Uneconomical mode of acquiring and replacing equipment
3. Higher operating and maintenance costs
4. Low productivity of equipment
5. Obsolescence
6. Higher percentage of equipment downtime
7. Lack of preventive maintenance program

Poor selection of equipment is perhaps the single greatest cause for excessive equipment costs. Consider the case of a construction company that obtains a contract to lay a water main. Having already purchased a crane, the management decides to purchase a backhoe to dig the trench and to use the crane for laying the pipe, rather than purchase a side boom. If only the capital outlay is considered, the company has realized a saving by buying the backhoe; however, if the amount of pipe to be laid is very large, it will soon become apparent that the cost of having a large crane tied up is greater than the extra expense of buying the more suitable side boom. Also, the company may have to pass up other jobs in which the crane would be needed. The preceding example is more clear cut than is usually the case: most times the analyst must do a market analysis to find out what alternative types of equipment could have been used and the relative merits of each type.

If records of equipment performance are not maintained and decisions to buy, rent, or lease are not made to achieve economy, the analyst must find out whether any excess costs have occurred in these areas. By studying the equipment usage reports, equipment accounts, cash outlays, and indirect expenses, the analyst can find out in retrospect whether optimum decisions have been made.

Direct and indirect costs of owning and operating the equipment constitute another source of excessive equipment costs. These costs arise from operator expense, maintenance, repairs, fuel, inspections, insurance, and so on.

If all costs incurred are justified and no unsuitable equipment can be found, the analyst should find out whether the actual productivity of the equipment is maximized. If the production rate of the equipment is low, the organization is losing money. Low productivity can be attributed to obsolete equipment, poor maintenance, and many other factors. Thus it becomes a study in itself. Poor maintenance may be singled out as perhaps the most blatant cause of a low production rate. Action to correct this can come about only through the institution of a comprehensive preventive maintenance program. If no such program exists, the analyst may collect relevant data about maintenance in its present breakdown form and then initiate the preventive maintenance program and again collect data. Final comparisons, perhaps after several years, indicate a reduction in excessive equipment cost. On the basis of this study, the analyst can decide what factors are contributing to the excessive equipment costs and how the problem can be corrected.

21.7 OVERHEAD COST ANALYSIS

There are many costs incurred in the day to day operations of the company, including office expenses, management and engineering expenses, taxes, insurance, and other overhead or indirect costs. The following list covers most of the items that contribute to overhead costs. An overrun in any of these items can cause below par performance.

1. Job organization
2. Engineering
3. Testing
4. Supplies
5. Utilities
6. Tools and plants: (a) light trucks; (b) freight and hauling; (c) loading, unloading, erecting, and so on, and (d) maintenance
7. Travel expenses
8. Freight and express
9. Advertising
10. Signs and barricades
11. Photos
12. Legal services
13. Medical and hospitalization services
14. Field offices
15. Permits
16. Insurance
17. Bonds
18. Interest
19. Taxes
20. Cutting and patching (punch list)
21. Winter protection
22. Temporary roads
23. Repairs to adjacent property
24. Pumping
25. Scaffolding
26. Cleanup
27. Contingencies

28. Main office expenses
29. Special items

As can be seen, the number of overhead expense and indirect items is so large that one or more may often be omitted from the items charged to a particular project. Such omissions are very serious since any overhead expenses that are not charged to the project constitute a loss. Thus the analyst should carefully cross check the accounts for overhead against the project accounts.

Although many of the items listed are essential to the daily operations of the company and cannot be lessened, the analyst should check each account thoroughly to find any areas of excessive costs. It would be useful to compare, if possible, the percentage of the total budget spent on each item to the percentage spent by other organizations in similar circumstances, or to that spent by the same organization in the past.

As an example of excessive overhead costs, consider an organization that expanded its main office by hiring five new office employees during the execution of a number of large jobs. Although these jobs are finished and there is little work to be done, the employees are kept on, thereby adding considerably to the office expenses. This type of situation is often difficult to spot since the excess personnel always remain busy, if not on a productive work, perpetuating the type of repetitive effort that normally occurs in a worker-saturated environment.

Another common source of excessive costs arises from poor cash flow policies. Construction companies, in particular, often deal with large amounts of money, which can be very beneficial if the finances are handled properly. Conversely, if the money involved is not put to maximum use, the company can lose substantially. Also, if the progress payments agreed to in a contract are not sufficient, the contractor may have to use borrowed funds, thus incurring interest charges.

It is also important to check the procedure that has been applied to contingencies on a project. If a 5% rate has been assigned to contingencies, the analyst should make sure that the bulk of these funds have not been applied to one portion of the job, thus leaving too little to cover the remaining portion of the work.

21.8 OTHER CAUSES

Besides the direct and indirect expenditure items covered under labor, equipment, material, and overhead, there are other causes that result in underachievement of performance targets:

1. Resistance to change
2. Lack of a management development program

3. Lack of adequate insurance
4. Lack of coordination between various agencies responsible for the job
5. Unplanned and inadequate information
6. Failure to make optimum use of resources
7. Lack of emphasis on purposeful job meetings
8. Overextension of resources and finances
9. Disproportionate capital expenditure, resulting in shortage of working capital
10. Failure to establish a line of credit
11. Bad debts
12. Lack of suitably experienced personnel
13. Unrealistic or inadequate targets of time and cost
14. Failure to predict market fluctuations
15. Unsuitable type of contracting
16. Distance of project from the main office when control is centralized
17. Inadequacy of cost and financial accounting procedures
18. Irregular progress and cost reporting
19. Inaccurate charging of labor, materials, equipment, and overhead costs

Management policy itself can be a cause of performance slippage. Management may resist change. It may be slow to adopt more economical methods and procedures or to turn to new methods and techniques to stay competitive in the market. There may be no management developed program. The management may fail to maintain adequate insurance. All these factors can be cause for trouble to the management.

Ineffective management can be responsible for the failure of an organization to achieve the desired results. Lack of coordination between design disciplines and numerous performing agencies engaged in a project may result in claims from the affecting agencies (Chapter 20). This may be caused by inadequate information flow (Chapter 19) or lack of timely action. Another cause could be delinquency in job meetings or lack of purpose in these meetings, resulting in uncoordinated work. Also, the management may fail to optimize the use of its revenues.

Some organizations overextend themselves, capture as much work as they can, and later find themselves in difficulty because of inadequate capital. Other invest disproportinately in their fixed assets, usually on equipment, which results in a shortage of working capital. These organizations may be solvent but, because they have failed to establish and use a line of credit, in time experience financial difficulties. However, other organizations, though making a good profit, may find, because of a bad choice of clients, a large portion of their good money lost in bad debts. All the preceding cases can cause below normal performance.

It should be remembered that performance control involves comparison with the baseline targets generated by the scheduling and cost engineers. Although the success on a project does depend on efficient implementation, proper supervision, timely coordination, and efficient control, it is essential that the comparison of actual achievement with the planned targets be meaningful. This comparison can be meaningful only if the engineering personnel in the organization are well versed in the use of various techniques explained.

21.9 PERFORMANCE ANALYSIS PROCEDURE

When management realizes the existence of the problem in the performance of the organization, a complete study of the affected areas is definitely needed. The study could be carried out by one of the members of the organization or by an outside agency. Past records can be used to pinpoint the principle areas of concern. For instance, if material costs have been consistently more than anticipated, the analyst should assign priority to a study of the material acquisition procedure (i.e., suppliers, methods of transportation, handling, storage). Likewise, if overhead costs are found to be excessive, it would indicate a need for a complete analysis of the office and management procedure. Whatever aspects of the organization pose a problem, the analyst must perform an in-depth study of all related activities. A useful adage to keep in mind is, "Whatever can go wrong, usually will." No stone should be left unturned in the search for factors contributing to the problem. In the final analysis, however, the analysts should concentrate on the important factors, the principle being that a vital few elements in a business are important, and a great many others are relatively unimportant.

In order to analyze the organization's previous performance effectively, the analyst must find all pertinent sources of information and extract the relevant data. In deciding what information is significant, as well as in actually obtaining such data, the analyst must question all the people involved. Therefore a person in this capacity must never alienate people. Of primary consideration is the potential threat to the careers of managers and executives. It should always be remembered that such people can easily withhold vital information. In fact the analyst should go out of his way to seek the enthusiastic support of people who might have some insight into the problem.

It is apparent then that the analyst should be diplomatic in all relationships with personnel involved in the study. Few people like to be questioned about possible inefficiencies in their work and, if pressed, usually find some plausible excuse by which to shift any profitable blame. This attitude can be overcome in most cases by the analyst's assurances that the study is being conducted in an impersonal manner. Indeed the research should always be impersonal in nature, from the original acquisition of data to the eventual presentation.

Figure 21.2 depicts a methodological analysis procedure in which the analysis is conducted in three phases. The first step consists of an analysis pinpointing the causes of unsatisfactory performance. This is performed by the project management team because of its intimate knowledge of the proj-

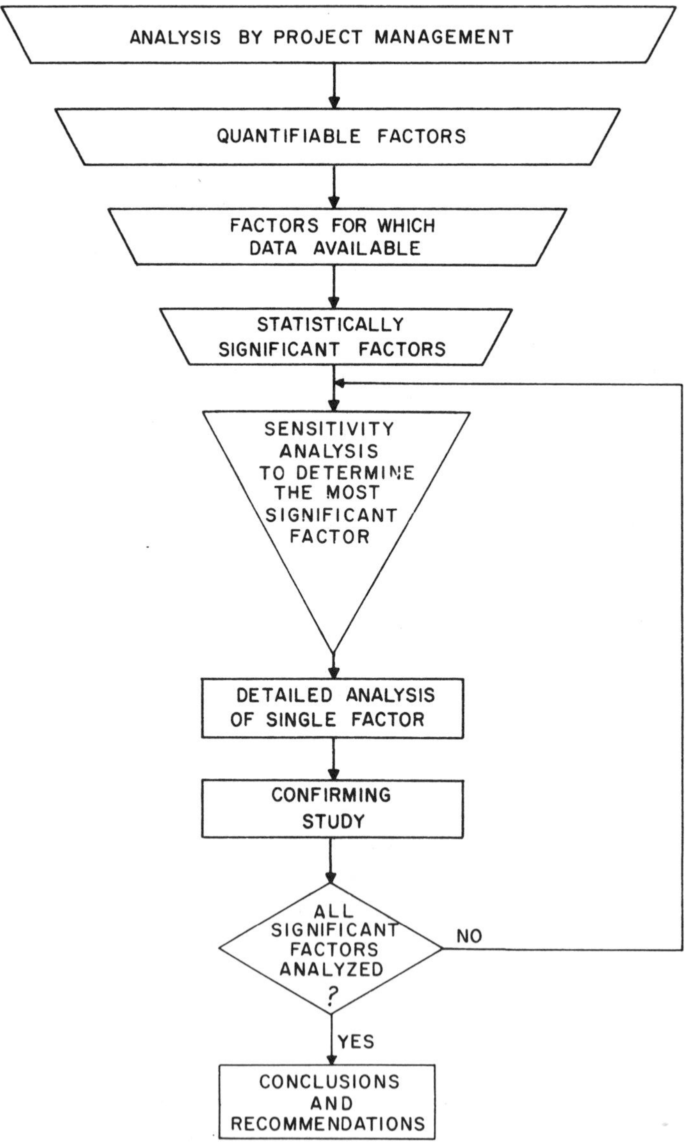

FIGURE 21.2 Performance Analysis Procedure

ect. Starting with the analysis, the analyst can move toward the identification of factors that are quantifiable and perceived to be important.

The search for quantifiable factors is vitally important to a successful analysis; one must be familiar with a broad range of information opinion regarding such factors. In addition to personal interviews and formal questionnaires distributed to members of the organization, there may be related studies in current publications. Based on the literature search and interviews, a list of quantifiable factors can be compiled that represents a consensus.

Most company records, reports, and historic information are insufficient to meet the needs of performance analysis. Although these sources may provide the bulk of the data, the analyst must use ingenuity to track down all possible sources. Another prerequisite to performance analysis is the determination of a realistic baseline by which to measure performance. Original estimates or contract stipulations may provide such a baseline, or the baseline could come from projects of a similar nature performed during the same time period or adjusted to conform with changes in the environment. Having determined a suitable baseline, it is possible to perform sensitivity analysis to find those factors that have the greatest effect on the project. Sensitivity analysis is discussed in the next section.

The significant factors are rated in the order of their influence on the project performance. The analyst then performs a detailed analysis to ascertain the impact of each of these factors, one at a time, followed by a confirming study. This is depicted by the interactive loop in Figure 21.2. As an example of the detailed analysis of a single relevant factor, consider the level of competition among subcontractors as a variable influencing cost overruns. A curve showing anticipated bids as a percentage of the provision in the estimate is plotted. Actual bid prices versus competition are shown on the same graph. The deviation from the projected curve confirms the significant effect that the level of competition has over project cost overruns. The graphical representation is an aid in confirming a study and in presenting the results.

Finally, the results of the study should be presented to both upper and lower levels of management in a series of concise and straightforward briefings, accompanied by all relevant documents and statistical results. These results may be presented to the site supervisor and foremen first; then, based on the responses and ideas of these people, the presentations could be refined and further recommendations added before presenting the report to the middle management. Having obtained their comments and ideas, the analyst can prepare a final presentation for top-level management.

These suggestions can be summarized as follows:

1. Complete study of all related areas.
2. Active promotion of the study at the executive level.
3. Cooperation of data and idea sources.

4. Proper attitude of analyst and involvement of those concerned.
5. Use of impersonal and analytical techniques.
6. Availability of accurate and relevant historical information.
7. Realistic baseline from which to measure performance.
8. Presentation of findings in a carefully structured format.

21.10 SENSITIVITY ANALYSIS

Sensitivity analysis is carried out during the planning and control of a project. During planning, the variables having maximum impact on performance are pointed out to management so that it can take additional care of these items during implementation. This is management by exception. The milestones in time and cost plans reflect the need for additional managerial attention to comparatively sensitive variables in the work preceding a milestone. During control, when management finds it consistently difficult to bring the project performance back to par, sensitivity analysis is an attempt to provide guidance in prescribing suitable remedial action.

Sensitivity analysis points out the variables that are most critical to the proper performance of a project and which values of such variables vary most critically from the performance baseline. It is a broad field, and many different techniques and methods can be applied. Although some problems may be solved more easily than others, it is wise to follow a systematic method in isolating and measuring the effects of factors contributing to late completion dates and/or cost overruns, thereby eliminating many sources of error such as accidentally overlooking important factors or making assumptions before there is any evidence to support them.

Multiple regression is one of the methods for sensitivity analysis. Having obtained the baseline it is possible to perform multiple regression analysis to determine which factors are statistically significant. For example, one of the realistic baselines with which to analyze labor productivity by plotting actual work done is the learning curve. Such curves are especially useful in areas of repetitive work and can accurately predict the ideal dollar or man-hour cost per unit over a period of time, provided that the task is not hindered by some fixed productivity limitation. Learning curves usually follow the general mathematical formula

$$y = ax^{-n}$$

where y is the dollar or man-hour cost per unit, x the number of times the unit has been done, a the value of y for the first unit, and n the exponent that describes the variation.

When a learning curve of this nature is plotted on a log–log scale, the curve follows a straight line, which can be used as the line for regression analysis.

Another method that can be used for sensitivity analysis is linear programming. For example, the management on a project finds it necessary to bring a new pile driver that can be used on this job and later on another job. This step can save them from imposition of a penalty. Normally, pile driving in the present project would have lasted six months, but because the pile driver is needed for another project, the present project must be expedited. Six months' work must be finished in four months by working two shifts. The duration and cost for all pile-driving activities on the first project will change. The management desires to make an informed decision; it wants to determine the sensitivity of the profit on the first project to the proposed corrective action of buying a new pile driver. The right-handed side of many constraints equations changes because of this corrective action, but essentially the additional cost of expediting with the new pile driver is only one parameter. One run with linear programming can thus determine the effect of this parameter on profit. Similarly, other parameters can be changed, one at a time, and the sensitivity of the profit to each parameter can be evaluated. The parameters that can be changed for sensitivity analysis are:

1. Range of the elements for the required vector (one at a time).
2. Range of cost coefficients (one at a time).
3. Complete sensitivity analysis that shows the effect on other parameters of changing the value of a particular parameter.
4. Sensitivity of profit to a change in structural coefficients of a constant.
5. Effect on the solution by the addition of a new variable.

Yet another technique of sensitivity analysis is simulation. As in the case of linear programming, it is necessary to very one variable at a time and observe its effect on the objective function.

The reader must realize now that a number of causes can lead to a slippage in performance, each requiring a type of corrective action that produces different effects on the project performance. It is therefore useful to develop a checklist of these causes and to consider each as a possible reason for below par performance during analysis. It is not possible though to make a list of all remedial actions because they vary greatly with the size and type of project and its environment, financial situation of the company, market conditions, availability of resources, and so on. A thorough study of the causes of malfunctioning as presented in this chapter, must be made first, sequentially for labor, material, equipment, and overhead cost analysis and other causes, and then for the project as a whole.

CHAPTER 22

TOTAL QUALITY MANAGEMENT OF PROJECTS

In the traditional project management environment, the focus has been on the *project* and not on the complete set of *customers*. In today's highly competitive economic environment successful management of projects depends on the synergy of integrating the various stakeholders. Increased costs and lost sales, perhaps even financial failure, may occur if quality is not built into the project from the concept definition stage on through to execution. One way of streamlining operations and directing all efforts of the organization toward serving the customer is the **Total Quality Management** (TQM) philosophy. Total quality management principles can be applied during all the phases of the project life as the "key" to the successful execution of projects from the planning stage through to engineering, procurement, construction, turnover, and startup.

22.1 WHAT IS TQM?

Total quality management is management of quality with the customer as the focus. It is a philosophy of continuous improvement at lower costs that encompasses the entire organization and with total organizational support. It is a framework for excellence striving to achieve zero defects in all endeavors. The technique systematically identifies wasteful efforts, thereby eliminating waste. Total quality management has been considered as a philosophy, but is perhaps better explained as a system and a goal. It is a system in its operation, but a goal in that it is never fully achieved. The concept of continuous improvement, a major feature of TQM, is one key to understanding it. In the current competitive, global marketplace, a firm must constantly improve to survive and prosper.

The underlying feature of TQM is *customer* focus, with the customer being anyone who is next in line in the process. Figure 22.1 illustrates this customer concept, which shows that the output of stage 1 is the input for stage 2.

The client is the final customer, but there are internal and, potentially, external customers. Total quality management of projects involves bringing together successfully a complex mixture of ingredients consisting of teamwork culture, trained personnel, corporate mission, goals and strategies, leadership, organizational structure, and adequate project management tools. It encompasses up-front planning to finalize the scope of work and defines execution philosophy, communication channels to use, and systematic methodology for the project teams to adopt in order to meet project goals and customer objectives.

Total quality management requires that for each functional area the precise goal is identified and objectives are set to maintain quality. It is the extension of quality control techniques into the humanist stage to deploy the "voice of customers" through the organization and to mobilize all employees to focus on continuous quality improvement at lower cost. It also requires team building, and alignment of the efforts of each participant.

It is widely reported that TQM programs have an 80% failure rate in North American organizations. These statistics apply mainly to the manufacturing sector and there are no separate statistics for the construction industry. However, lessons can be learned from the experiences of the manufacturing companies in which the failure of TQM programs occurs, in spite of the educational efforts and other extrinsic stimuli, because there was no meaningful change in the basic attitudes of the individuals involved. Too many organizations try to mimic the successful organizations' effort without understanding the underlying requirements for a successful TQM program. A program must be customized for each organization to be successful.

The traditional project management view has been that compromises are

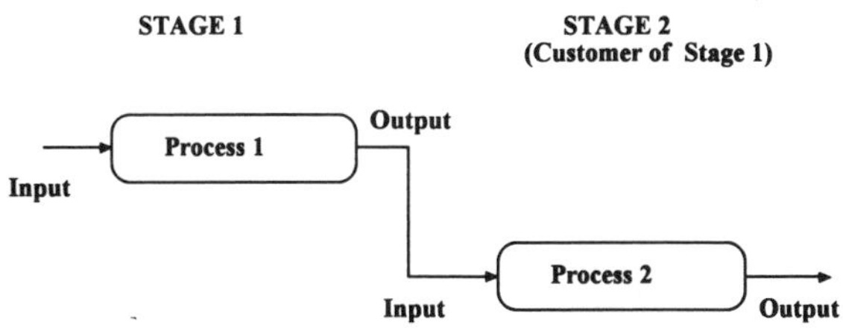

Output of Stage 1 is input for Stage 2

FIGURE 22.1 The Customer Concept

made between quality, cost, and time. In more progressive organizations there is a view emerging that these tradeoffs do not exist. This has led these organizations to strive for excellence in each and every phase of the project to improve quality while reducing costs in a shorter time frame. Continuous improvement of the project management process is possible. Increased effort is needed at the front-end planning stage. Better methods for developing improvements in management processes are required. The status quo is no longer a viable alternative. Project management must address the whole system because improvements in one area may cause problems in another.

Organizations should consider the following issues/questions carefully before starting:

- How does one organize for a TQM program and effectively manage it?
- What organizational structure, skills, and systems must be in place for successful implementation?
- How does one measure improvement?
- Lastly, how does one launch the program to introduce continuous improvement?

22.2 ORGANIZING FOR TQM

Planning for TQM requires a major commitment to quality by top management backed by a plan for action. The system involves three phases which must be pursued simultaneously:

- Planning
- Improvement
- Control

There has been a recognized need for quality control (QC) and this requirement has been met by the implementation of various inspection and measurement procedures. More recently quality has been formally planned into an operation. Quality assurance (QA) is the system which incorporates both the planning and control phases. Quality improvement (QI) is the step that results in major cost reductions. In the TQM philosophy QI is a never-ending action—continuous improvement. Figure 22.2 illustrates the quality management relationship within the context of TQM.

It may be desirable in organizations not accustomed to the philosophy of TQM or to a "continuous quality improvement system" to have some form of quality department or quality council structure.

The quality department is essential in the early stages of the quality program to lend expertise, perform an ongoing quality audit function, and supply training throughout the organization. However, as the program becomes

FIGURE 22.2 Scope of Quality Management with a Customer Focus

more fully entrenched there may be advantages to downsizing this department. It is important that the quality department/council not be seen as being responsible for quality. Quality is everybody's job. The quality council is a cross-functional team which addresses quality issues in the firm.

There are many reasons for a TQM approach, all of which have "bottom line" implications. The appraisal system of inspect, test, and fix is very expensive and destructive. A prevention system is needed in which the requirements are to do things right the first time. This is the zero defect performance standard. Quality is conformance to the requirements, and one of the best ways to avoid litigation is to deliver the requirements of the contract. Many contractors rely on their reputation to negotiate or win contracts on a total evaluation scheme which includes many criteria as well as price. Reputation is a fragile thing and can be maintained by producing quality in all aspects of the project, thereby increasing the likelihood of repeat business.

22.3 STEPS FOR QUALITY PLANNING

Planning for quality is a six-step cyclical process which includes:

1. Identifying the customers
2. Determining the needs of those customers
3. Utilizing management methods which can optimally respond to those needs
4. Developing processes which are optimally able to produce the desired results

5. Communicating the processes to the operating personnel
6. Designing and implementing appropriate measures at each of the stages listed above

The corporate policy required to support the planning process will be concerned with:

- The company mission statement and a clearly defined corporate strategy
- Project organizational structure
- Improving the project generation process
- Project team screening
- Training for projects and operations
- Control measures for projects and operations
- Rewards and recognition
- Communication

A principal ingredient for project planning is a well-defined and communicated corporate policy. This includes team-building strategies, guidelines such as who initiates the project, and project turnover. Organizations should develop a list of critical factors and ask relevant questions that will help explain the relevancy of these factors to the planning process. This exercise will reveal the need for considering these factors in the planning process.

The success of this planning strategy requires a system that will generate a continuous flow of well-planned improvement projects, which is necessary for a successful TQM system. Successful improvements will foster support for a continuous improvement program. Also, the control feedback function insures continuous high-quality improvement in operations as a result of these improvement projects.

Many of these improvement projects will cross several departmental boundaries, which is not conducive to success in the traditional functional organization. Project management techniques and the matrix structure are essential for successful completion of the improvements. This structure allows teams, whose members usually come from various departments, to successfully complete the many and varied improvement projects that will be continuously initiated. All activities within an organization can be viewed as a process. By definition a process is a systematic series of actions directed toward the achievement of a goal. Within the scope of this definition, everything from a single activity to the total operation of the company is a process.

Alignment of the team members is necessary in order to maximize the efforts towards the project goals, rather than allowing divergent agendas to detract from the project focus. As with a group of vectors, their sum will be maximized as the angle of each vector approaches zero, that is, the components at right angles to the resultant approach zero.

As discussed in Chapter One, project management is a system which communicates cross-functionally. It promotes communication and cooperation between functional departments and the hierarchical structure of an organization. This is the process of alignment of the team members at all levels, and if executed in the spirit of TQM, success will be enhanced.

In traditional organizations there usually is a distinct separation between business policy making (done by a single or select few individuals) and project operations. It is important that these two be linked more closely. For TQM to succeed, it is critical that the idea of teamwork be present in the organization.

22.4 THE CONTINUOUS IMPROVEMENT PROCESS

There are three major components to a system for improving project management.

1. A flow chart of the operation
2. The establishment of measures and controls for each step
3. A problem-solving methodology to generate improvements in the process

22.4.1 The Flow Chart

Preparing a flow chart of the project management process is critical to identifying specific stages in a project. In the TQM framework, flow charting identifies customer–vendor relationships. These specify where responsibility lies for particular inputs and outputs. It is important to remember that there are both internal and external customers. A **customer** is the next person in the process. Each stage has its customers. See Figure 22.1 for an illustration of this concept.

Care must be taken to properly identify who the customers are and what specific outputs are required from them. This requires flow charting the whole project from conception to completion. Customers may then understand the vendor's processes in providing their needs. This will facilitate communication and feedback.

22.4.2 Measures and Controls

The flow chart identifies customer–vendor relations. A customer receives inputs from the previous step. The vendor is responsible for meeting the customer's needs. The chart thus establishes responsibility for control purposes.

The first step is to define the product, project characteristics, or service that is desired. Customers will meet with their vendors and establish the desired performance criteria or characteristics. These characteristics are customer defined. More than one iteration will clarify any vagueness of purpose for both the customer and vendor.

The customer will also indicate the relative importance of each characteristic. In addition, the customer will rate each characteristic, reflecting the level of satisfaction with each characteristic as currently delivered. The ratings could be conveniently scaled, such as 1–5 as in Figure 22.3 below. The first column is used to list the characteristics desired by the customer. The next column represent both how important each characteristic or "product" is to the customer (the weight factor) and the rating of the vendor's performance. From the figure, it can be seen that a performance measure may easily be derived by multiplying each performance rating by its corresponding weight and summing the values obtained.

$$\text{Performance measure} = \sum_{i=1}^{n} W_i R_i$$

where W_i is the weighting for each product characteristic and R_i is the current performance rating for each characteristic.

This provides a baseline measurement for the project or product which may be charted and monitored on a monthly basis. This measurement reflects the needs of each customer, the relative importance of each need, and an overall measure of the effectiveness of the vendor or contractor in supplying the desired input. Areas for improvement are indicated by tracking this measurement and determining the target areas which will generate the greatest improvement in performance.

At this point, a method can be established to determine what characteristics of the contractor's or vendor's process may be changed to affect improvement in performance. While still working with the customer, the vendor/contractor will prepare a list of performance variables which may affect the customer's desired characteristics. This process may also take more than one iteration to firmly establish the true performance variables required to meet the customer's needs.

Characteristic or Performance Criteria	Weight Rating
	1 2 3 4 5
	1 2 3 4 5
	1 2 3 4 5
	1 2 3 4 5
	1 2 3 4 5

FIGURE 22.3 Customer Characteristics and Measures

Control and improvement can only be achieved if appropriate measures are in place. It is critical that the customer establishes and clearly communicates what is required of the vendor. In project management, this may have to be done on a project by project basis. When the requirements are clearly defined, measures of effectiveness are then developed. It is critical that this process be in place in the front-end planning phases.

Following are examples of performance criteria that were used, in a more expanded format, on an actual project:

A. Quality performance
   1. Overall plant
   2. Project management
      - Budget and time control
      - Scope and change management
      - Innovation in execution
   3. Construction management
      - Contracting strategy
      - Productivity
      - Safety
      - Coordination
      - Claims
   4. Procurement
      - Cost effectiveness
      - Procurement and delivery schedules
      - Equipment quality
   5. Engineering
      - Design errors and rework
      - Innovation
      - Productivity

B. Plant operability
   - Ease of startup
   - Reliability of systems and installation
   - Performance testing

C. Modules
   - Errors and rework
   - Growth of scope
   - Delivery
   - Fit-up and field problems

Each of these items (A–C) can be expanded in more detail.

The customer should also have firm accept/reject/recycle criteria for inputs. Control limits for acceptance are normal quality assurance procedure. If the inputs are not satisfactory for proper continuance of the project, they should be rejected.

A key ingredient of this procedure is flexibility. The personnel involved must react to the needs as indicated by the measures to meet common goals and objectives. This method of developing measures and relationships also indicates who should improve what.

22.4.3 Improvements

Improvement projects can and should be generated from every level in the organization. Regular meetings by small groups such as quality circles or quality councils, analyzing customer (external and internal) feedback information, can generate project ideas for improvement through the use of simple quality tools like Pareto's law and cause and effect analysis. It is important that every individual in the organization continuously be on the lookout for improvement projects. Some of the best ideas come from line workers who are often adverse to paperwork.

The initial list of possible candidates for improvement projects should be screened to provide a final list of project candidates. The key factors which should be addressed at this phase are:

- Which potential project candidates are most important? The potential candidates in the initial list need to be ranked in order of importance.
- Are the component factors of each candidate controllable? If not, the candidate may have to be removed from the list. Most candidates will have both controllable and uncontrollable components which should be identified. The decision process to determine the final project list requires information on how much room for improvement exists.
- Based on the importance and the degree of controllability, determine which potential candidates will generate the best benefit to the firm. The list is thus reordered to reflect the possible benefits to the firm.
- What resources are required to successfully plan, implement, and complete the potential project? Projects must be feasible.

22.4.4 Systems Approach to TQM

The high failure rate of traditional TQM implementation has frustrated and discouraged many organizations that are planning to streamline operations. This failure occurs because many organizations still approach TQM as a quick-fix tool without completely understanding the complex interrelationships that exist in any organization. More recently, a total "systems" approach seems to offer some hope for successfully applying TQM methods. Understanding the systems effects will lead organizations to a better appreciation of the total picture. This can lead to proper planning, design, measurement, feedback, and control mechanisms that must be in place to assure success.

Any process can and should be viewed in the systems framework as shown in Figure 22.4. A system has a set of inputs, a transfer process that operates

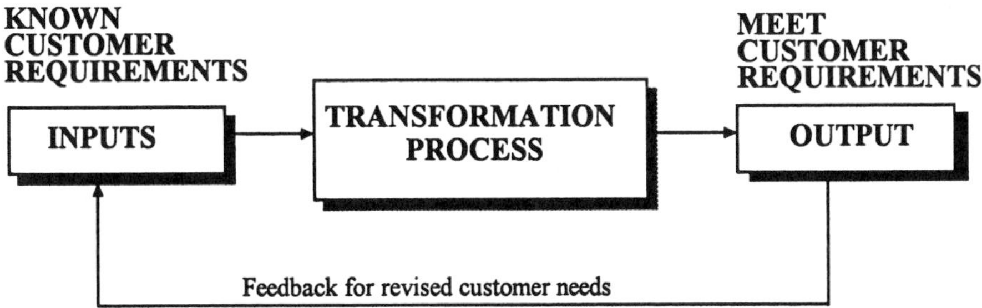

FIGURE 22.4 A Typical Processing Stage

on the set of inputs to deliver a predetermined set of outputs. The system should also have some form of feedback mechanism in place to track and guide the performance toward the target of meeting customer requirement.

Characteristically, systems never fail—they work as designed. This is probably the most critical idea in understanding systems. Systems operate as designed, translating the given set of inputs to outputs as per the set of relationships provided for the transfer process. When any deviation is observed in any input, the system can only deviate from its predicted behavior. It still operates on the given set of inputs as designed.

CHAPTER 23

CONSTRUCTABILITY

The concept of constructability is an old one, but the terminology is relatively new. It is the early involvement of construction thinking; it is simply common sense.

Constructability is the optimum use of construction knowledge and experience in planning, engineering, procurement, and field operations to achieve overall project objectives (CII, 1988). It is the effective and timely integration of resources and technology into the early phases of the project, and then maintaining the involvement of all the stakeholders including the owner and contractor.

Maximum benefits occur if all stakeholders with construction knowledge and experience become involved early in the project. This is illustrated in Figure 23.1 which shows that decisions in the early stages have greatest cost impact.

Constructability enhances the effectiveness of construction. It is a macro-productivity factor that should be the way of thinking of the entire project organization. It is management (at all levels) action that creates this culture, not as a separate function, but an ongoing process. It can be a motivator to the worker when the "smart" details or methods are used.

For example, Figure 23.2*a* shows an arrangement at a steel column splice in a multistory building. The weld preparations are partial penetration welds which receive their maximum loads during construction. In this real example, the preparations were made so that welding was done from the outside faces of the column member. This is shown in Figure 23.2*a*. In a multistory building, access must be provided to each outside face. On the inside of the building, the permanent steel decking provides the working platform and plenty of

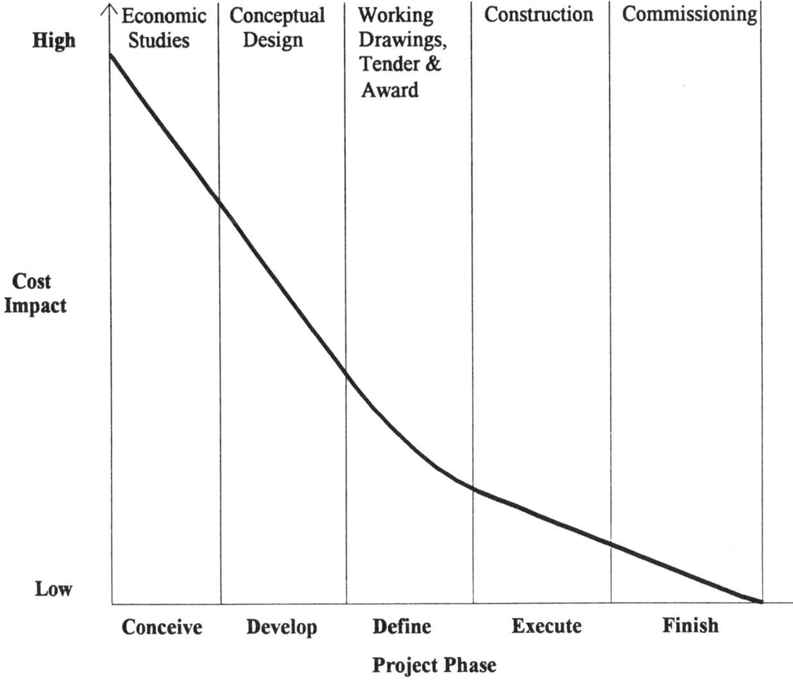

FIGURE 23.1 Cost Influence as a Function of Project Phase

access for the welder. On the outside face, scaffolding would be constructed to provide access and working space for the welder.

A simple revision, as shown in Figure 23.2*b* would reduce both fabrication and field welding costs because the weld preparation is made as before, except on the outside flange the weld preparation is made on the inside of

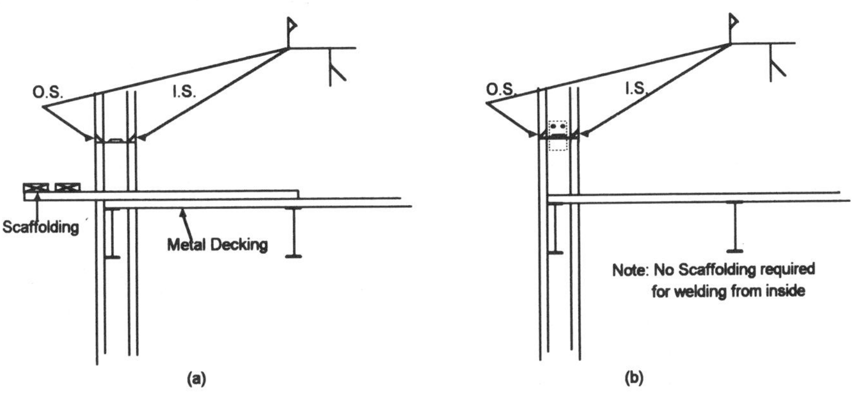

FIGURE 23.2 Field Welding of Steel Column

the outer flange. This eliminates the need to turn the member over in the fabrication shop in order to make the weld preparation. Both flanges are prepared from the same side. There is a more significant cost impact for the field welding operation because no scaffolding would be required for the welder. Access and the work platform are provided by the steel deck as for the inside column flange. Also, safety is enhanced because the welder is not required to work on the outside of the building several floors above ground. This idea was provided by the welder who was faced with the welding task.

23.1 A TRADITIONAL PROBLEM

As the construction process has evolved and become more sophisticated, separation of the design and construction functions has increased under the traditional form of construction procurement. Traditionally the owner hires an engineer/architect who designs the facility. Construction is awarded to a contractor who procures material, labor, and equipment and executes the contract requirements. This method, which results in the separation of functions, is primarily responsible for the lack of constructability as it may exist. A return to the historical master builder concept is heralded to be a step toward more efficient and cost effective projects. In the master builder concept, the architect was the engineer and the contractor also, that is, the responsibility remained with one person.

In modern construction, this concept translates into an integrated design–build process where there is no division of responsibility and the constructor's input is more readily accepted.

23.2 CONSTRUCTABILITY CONCEPTS

The concepts of constructability can be applied during all phases of a project. These concepts are simply applications of constructability.

During the conceptual and definition stages construction involvement is required to develop a contracting strategy, establish project objectives, and select major construction methods.

Overall project schedules need to be construction sensitive. It is necessary to establish a sequence of activities with realistic durations to prevent costly overtime, schedule acceleration, or counterproductive high levels of craft labor.

Major construction methods are considered during the conceptual stage. Special methods include prefabrication, preassembly, and modularization.

Effective site layouts can facilitate construction activities and reduce costs. There must be provision for adequate storage spaces, access, and roads with particular emphasis on clearances for operating equipment and traffic flows. It may be cost effective to use permanent facilities and utilities during construction. Environmental concerns are becoming increasingly important and the disposition of construction waste must be carefully considered in the plan.

To enable efficient construction, designs must avoid complex details and shapes. Designs must also permit flexibility in construction methods and material substitutions. The design schedule must support the construction fieldwork sequence. Good quality drawings, specifications, and site information will improve productivity.

Drawings are frequently criticized for lack of clarity and content. Unfinished drawings or those that lack details require the field crews to devise their own solutions. This transfers part of the design function to the site, which is costly, disruptive, and inefficient. Dimensioning should consider construction needs and should not be scattered over several drawings.

With vendors and suppliers, constructability is enhanced by timely engineering data, preassembly, shop testing of components, and lifting lugs.

Constructability is enhanced by standardization, such as the use of manufacturers' standard dimensions, steel connection details, piping assemblies, and off-the-shelf electrical and mechanical equipment. Designs can be standardized in order to realize the benefits of duplication, symmetry, and repeatability. For example, if the formwork for every member is a new experience, costs will sky rocket. The Montreal Olympic Stadium and the Sydney Opera House in Australia are classic examples of large projects where cost of formwork went out of control. Numerous examples exist where the use of modular design (4-ft or 1200-mm modules) can reduce costs. Concrete formwork and residential construction are examples where wastage could be minimized using modular design.

Structural constructability considerations include the use of elements such as precast staircases in hi-rise cores. Straight reinforcement bars, prefabrication of cages, and detailing of reinforcement to suit pour heights are also cost effective steps.

Effective design and construction requires that construction expertise is utilized early in the project shedule. Constructability improvement is possible with construction-sensitive design and construction-driven schedules.

CHAPTER 24

COMPUTERS IN CONSTRUCTION

Computers are the backbone of today's information systems which form an integral part of project management. Personal computers are almost everywhere and their use covers all applications from word processing and electronic spreadsheet to sophisticated 3-D animation, simulation, and walk-through applications. Once only reserved for the very large EPC contractors and design firms, computers are now found at nearly every desk and on every construction site. Computers are in essence data and information* processors. If properly applied and managed they can greatly enhance productivity and improve quality, providing a company with a competitive advantage.

The introduction of inexpensive personal computers (PC) greatly affected the role of information technology in construction as well as other fields. The PCs provide better access, less centralization of information, and less central control. This can work to a company's advantage or disadvantage.

In the construction industry, management resides in the hands of those who graduated decades ago with little exposure, if any, to computer technology. Such individuals are faced with the difficulties of understanding the value and use of computers in the daily project life. They have to manage the existence of the computer as an integral part of their project or office, make decisions regarding what hardware and software to acquire, who gets the training, and how this whole phenomenon affects the bottom line, the cost per unit of construction. Although new graduates possess the appropriate computer knowledge, it will be some time before they become decision

* Information in this chapter generically refers to data of all sorts that can be processed by a computer (activity information, payroll data, etc.).

makers. This results in a state of *transition* which must be properly managed. This will eventually subside at a time when computers truly become integrated with everyone's daily work routines.

At another level, decentralization of information technology boosted productivity of those armed with proper knowledge in their fields and in the field of computing. Proper applications of remote computing makes is possible to solve problems thousands of miles away from the project site without any loss of detail or input from the project management personnel. Advanced data collection techniques provide better control of cost, time, and quality as well as improved future estimates. Scheduling exercises of CPM or precedence take very little time to process once the schedule is set. Estimating can be more reliable with the proper use of software that facilitates reuse of unit costs, crew composition, and package or work assemblies. Payroll has always been a prime application for computers in mid- to large-size companies. Because running software on a PC cost less than $500, it became accessible to smaller firms, providing comprehensive accounting features for better company-based financial planning and control.

Computers have become an integral part of project management and the selection and use of computers requires an understanding of certain issues. These can be summarized as (1) understanding computers and what they can offer, (2) understanding software applications, (3) appreciating what is involved in managing this new resource, and (4) understanding the direction that new technologies will be taking in the coming years. In the remainder of this chapter we discuss these issues.

24.1 MANAGING THE COMPUTER SYSTEM

Computers have many applications in construction as will be discussed in the following section. The issues dealt with in this section cover strategic decisions regarding managing the computer-based information system.

In larger companies a formal computer information system setup is desired where a specialist—a system engineer (or a whole department)—looks after the entire information system from hardware to software to user training. For small and mid-size construction companies this is not feasible because the costs cannot be justified. It is however desirable to instate the same concepts of the formal computer information system found in large companies. The guidelines for a simple information system can be outlined as follows (Mensching & Adams, 1991):

1. Set up a simple and flexible master plan for the computer information system.
2. Establish standards and procedures for acquisition of computer hardware and software in line with the master plan.

3. Streamline hardware and software acquisition (or development) by adopting a system that is likely to grow with the company and that will be there in the future. This simply implies that while purchasing a hardware or software item make sure that the vendor will be around in the future or that someone else can provide the service.
4. Appoint specific individuals to look after the information system or individual components of it.
5. Revise the master plan periodically to reflect new technological changes and new company goals and directions.
6. Establish a plan for data security and integrity.

24.1.1 Adopting a Master Plan for the Computer System

A plan for the system should be simple, flexible, and straightforward since computer technology becomes obsolete in very short periods of time. A detailed, inflexible plan may backfire. The following issues should be addressed in the plan:

Computer System Platform. What computer system should the company adopt—a personal computer of the IBM-compatible family, a Macintosh-based system, an inexpensive SUN system, and so on. This should be part of the master plan as it will affect every other aspect of the information system from upgradability to data interchange to user-friendliness and software purchases. This should also include the operating system (e.g., for PCs DOS, Windows, OS/2, or UNIX). This, of course, depends on many factors including the primary uses of the system (word processing, presentation graphics, estimating, scheduling, etc.).

Determine the Prime Applications for Computerization. Many companies may have been successfully performing various tasks manually. How could productivity be improved by automating the appropriate tasks? For example, to which extent would you automate estimating while still improving productivity of the estimators. Should one embark on automating the entire estimating process from CAD files to bid preparation? Or alternatively, maintain a manual quantity take-off approach and support it using an electronic spreadsheet or estimating program that will enhance the arithmetic processing, productivity analysis, and extension of the estimate? Many off the shelf products like Precision Estimating from Timberline, MC^2, and others may be used in the second scenario, while much investment in software development will be required for the first.

Determine a Reasonable Budget for the Computer System. A reasonable budget should be allocated for the computer system while considering hardware and software upgrade requirements, maintenance, and replacement costs. This aspect is of prime importance as the market trend in information technology is such that new software requires more capable hardware and while the software upgrades in operating systems as well as applications takes place on the average at least once every year, this might become a significant expense. Hardware is also becoming obsolete in very short periods of time. A laser printer with a resolution of 300 × 300 dpi may no longer produce the competitive reports that new printers do. Likewise, the fast, powerful 386 PC with an 80-MB hard drive can no longer keep up with software applications written for faster computers and larger drives. To maintain a competitive advantage such equipment replacement must be considered periodically, with the average life of a PC being less than three years.

Determine Whether the Company will be Involved in Software Development and for Which Applications. This aspect may have a significant influence on productivity. Developing in-house software may consume large sums of money, while professional software developers have produced an equivalent system selling for a fraction of the development cost. Many argue that the off the shelf systems are not appropriate for what they do. Developers have enhanced their systems and many allow programming inside their applications, thus providing an alternative.

Set Up a "Disaster Recovery Plan." This plan should address issues like periodic backup of data on a regular basis (e.g., important files daily, others weekly or monthly). Two backups should be maintained simultaneously at two different locations and personnel should be trained to operate more than one system in case a particular individual is no longer available to run a specific software. Finally, what to do when the whole system is not operational must be addressed (e.g., finalizing an estimate that was prepared using a computer system when the system is not operational).

24.1.2 Standards and Procedures for the Computer System

A simple set of standards will keep the system in check and insure compliance with the overall strategic plan of the company. The following issues should be addressed when setting standards and procedures:

Guidelines for Hardware and Software Acquisition (Mensching & Adams, 1991). A simple system may be established to help the end user define his needs and assist him in ranking available alternatives in the market. The system could be as simple as defining the following:

1. Define mandatory attributes of the required system (hardware or software). Appendix 24A provides a set of criteria for a generic software system for illustration.

2. Identify the possible competing systems (i.e., alternative candidate for the system).
3. Eliminate the alternative candidates that do not meet all mandatory attributes.
4. Define a weighting scheme and a ranking system that reflects the preferences for each of the desirable criteria and weigh the alternatives accordingly.
5. Select the candidate with the highest rating.

An illustration of how to set up procedures for selecting a software system is provided in Appendix 24B.

24.1.3 In-House Program Development

The discussion in this section is limited to in-house program development for in-house use, because development of a program for commercial or wide public use is a major undertaking often associated with major risks. In addition, the decision to develop a program to perform a particular task or to acquire an off the shelf package to solve the same problem is not dealt with in this discussion. In general terms the decision should be based on a formal analysis of the pros and cons of the development process and all risks associated with it. Many of us get carried away with a piece of work we develop to solve our own problems and embark on marketing it. It should be noted that producing a commercial package is a major undertaking requiring continual support and upgrading of the package as well as addressing the concerns and attending to the problems of its users.

The development of computer programs for construction and project management applications used to be reserved to the very large companies with in-house computing staff. With the advancement of software technologies, program development is getting more and more into the hands of most computer users. Popular development environments include database programs (e.g., DBASE IV, PARADOX, ACCESS), electronic spreadsheets (e.g., LOTUS 123, EXCEL, QUATRO-PRO), and more recently rule-based expert system shells. What makes these development environments unique is their user-friendly programming structure where very little is required on the user-interface programming part of the program. Many short cuts are also preprogrammed to facilitate creating and manipulating data structures, arithmetic operations, and program control constructs.

Traditional programming environments may be also used for in-house program development. This has many advantages and disadvantages. Among the advantages are the flexibility of the programming environment where the programmer controls most of the program structure, user interface, and implementation details. Portability of the product is also easier as there would be no requirements for running the applications once compiled except having the required operating system environment. In contrast to this, developing a spreadsheet using EXCEL, for example, is dependent on having the EXCEL spreadsheet whenever the application is to be executed. The disadvantages are the environment is usually more difficult to learn and the productivity is low as the programmer will have to write all details from error trapping to user interface. Many new systems have attempted to address this problem. VISUAL C++ and VISUAL BASIC are recent implementations of C and BASIC from Microsoft with many tools that will cut program development time and enhance productivity.

In general, most nonprogrammers will opt for the first development approach with more going with the spreadsheet environment. The low learning curve and easy development environment comes at the expense of flexibility, portability, and expandability, however. A rule of thumb is to select the application environment with which one is familiar and comfortable. If more than one environment is possible, a balance between development time, flexibility, and reuseability is advisable. Data interchange is very important in construction as we normally deal with excessive amounts of information. That should always be a consideration when selecting a programming environment.

Regardless of what approach is taken, certain rules for program and application development must be observed. In general if the application is developed for use by more than one individual it must contain some documentation regarding its operations. This could be a combination of the following:

1. A user manual showing the intended user how the application would run and a sample application.
2. Description of the scope of the program and its limitations.
3. Input/output formats and a data flow diagram.

The application must also possess adequate internal documentation and proper programming practices to facilitate further development and work by others on the same programming code. This may be achieved through (1) proper remarks associated with the written code or next to a spreadsheet macro, (2) a detailed procedures manual, (3) an operations flowchart, (4) a description of internal data structures and, (5) input/output interface specifications.

24.2 COMPUTER-PROCESSING APPLICATIONS IN CONSTRUCTION

Computer-processing applications in the construction industry are increasing as more and more companies come to realize the benefits inherent in such methods. In the highly competitive construction field the need to reduce costs is paramount.

Computer applications, which have been used for increased economy in the construction industry, may be broadly classified under the following four major categories: Administration, Engineering, Project Control, and Plant and Equipment. The various processors used for computation are listed separately for each category.

Administration
Payroll
Accounts payable
Accounts receivable
Inventory control and order processing
Billing
General ledger
Government reporting
Union reporting
Document control
Insurance reporting
Correspondence control and word processing

Engineering
Specifications
Cost data base
Estimating
Takeoff and billing quantities
Surveying aids and plotting
Temporary work design
Technical computations and analyses

Project Control
Scheduling, rescheduling, and updating
Calendar data activity schedule
Optimum time/cost schedule
Progress reporting
Time study and activity sampling programs
Purchasing/procurement (material espediting)
Bid vs. estimate analysis
Contract administration
Worker scheduling
Manpower budgeting and control
Cost control
Design and field change administration
Performance review
Simulation

Plant and Equipment
Mix analysis
Job scheduling
Maintenance scheduling
Equipment down time analysis
In-house equipment availability accounting
Equipment replacement program

Unfortunately, these applications are mainly isolated tools that are useful at only one stage of the project. The main reason for the lack of comprehensive computer-aided project management system has been that no two projects are alike. Because they differ greatly in size, location, constraints, costs, organization, analytical methods in use, degrees of specialization, and computing facilities, each project uses its own selection of computer programs. Any uniform system that could be applied to such a broad range of job conditions would be so general as to be impractical in most cases and would still be insufficient to handle many situations.

Finally, such a system would have to be capable of being integrated with present management functions within the various departments of an organization. Again the task would be difficult to say the least.

24.3 INTEGRATED COMPUTER SYSTEM MODELS

An "integrated" computer system would reduce the data from the individual processors within the system to one common level. The system would effectively relate these processors by means of the information passed from one to the other.

The processors can be standard programs available through various sources of supply or, alternatively, new programs can be designed by, and for, a specific organization. As stated, the integration of these programs is achieved by changing the format of the information passed from one program to the other so that the format of the output from one program is acceptable as input to the next.

The errors are eliminated in an integrated system because an entry made only once is passed on to all appropriate processors after being subjected to an extensive analysis of its correctness. The connection among processors eliminates transcribing and multientry errors, reduces omission errors, and keeps all processors current while reducing time and effort by eliminating the need for separate entry for each processor.

The actual transmission of data is optional as well. It may be automatic in certain parts of the system, whereas in other cases the output will be punched on cards, stored on tape, disk, or diskette, or screened to allow management to study the information before it is passed on to the next procedure. Consider, for example, the material purchasing and expediting procedure. The inventory files may be updated automatically as each material is assigned to the project. The accounts payable file, on the other hand, is updated through cost data transmitted by means of data entered by the purchasing staff.

The degree of integration among the data-processing functions of a construction organization must be decided by the management, depending on the requirements of the company, facilities available, existing techniques, and so on. In many cases management may want to decide interactively at one stage whether any changes in the project plan should be made before proceeding to the next stage. It is imperative that overall control of the project remain with management at all time.

The model presented in this section follows the conceptual approach to the integration of the data-processing functions of a construction company.

It is a model in that it shows the type of integration that can be achieved. An alternative conceptual model is described at the end of the section; however, many other designs could be used to provide equally good, or perhaps better, processing of data.

The system is represented in Figure 24.1 by an information flow diagram depicting the logical position of each program within the total management setup of the company and showing the flow of data between the various elements of the system. This figure considers only a few of the processors from the exhaustive list given in the preceding section. Note that the diagram shows the logical relationship of elements in the system rather than the chronological sequence. For instance, network scheduling continues throughout the life of the project.

The following paragraphs explain in detail the design concept and input–output detail of a few of the major processors used in the construction industry.

Payroll Accounting. The payroll procedure calculates the payments to be made to all the employees working either in the central and branch offices or on the projects currently in progress. It can be related to the project cost control program. Labor time cards can be fed in the payroll procedure to

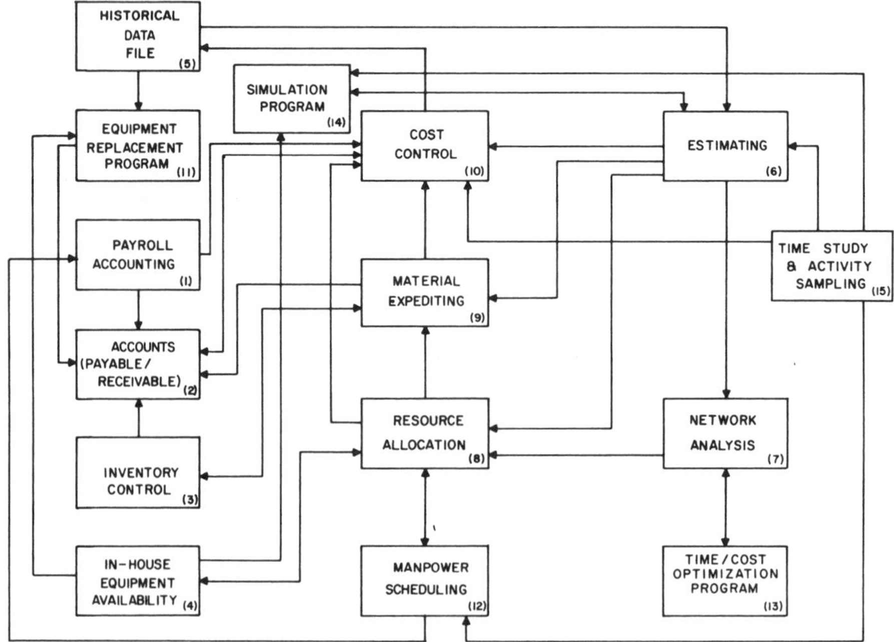

FIGURE 24.1 Information Flow Diagram

calculate the earnings of each employee and print checks. These costs can then be transmitted directly to a file of actual costs on the project, which the cost control program can use to convert the incurred costs to actual costs.

Accounts (Payable/Receivable) System. The accounts system aids in the financial accounting for the company. In addition to keeping track of labor costs and costs incurrred in the purchase of equipment, it can be linked with the material-expediting procedure. As materials are purchased for a particular project, costs are applied to each purchase, and the amounts payable to the supplier are handled by the accounting system. Also, progress payment data for contract work calculated by the cost control program can be used by the accounting system to update the accounts receivable.

Inventory Control System. The inventory control system handles the necessary calculations and filing for all data related to the inventory. All materials used on a project are noted and the stocks replenished accordingly. Thus the data obtained from the material-expediting program would be used as input to the inventory system. On the other hand, the expediting program would need to assess the inventory files to determine whether the required materials are available.

In-House Equipment Availability Accounting. In-house equipment availability accounting keeps track of the equipment owned by the company and updates its files regularly as equipment is assigned to projects in progress. The equipment files provide data for the resource allocation procedure, which in turn reports the usage schedule for each piece of equipment used on the project to the equipment-accounting program. The equipment file would also be updated as each new piece of equipment is purchased.

Historical Data File. The historical data file contains the information derived from the previous projects executed by the company and is updated at the conclusion of each job. It includes unit costs for equipment, materials, and labor, as well as elemental costs for standard elements. These data are obtained from the cost control program, but such programs as the equipment replacement dynamic programming procedure would also be designed to access this file.

Estimating. The estimating procedure reads quantity data cards prepared by the estimator, locates the appropriate formula for each item, and computes the quantity estimates. These quantity records are then processed with historical elements of cost, and the rate table file, to produce detailed cost estimates. Output from the estimating procedure is used as input to the resource allocation, material-expediting, and cost control procedures. Input to it also originates from the simulation and time-study procedures.

Network Analysis. The network analysis procedure involves the initial scheduling and simulation of the project by means of a network plan. Input consists of the duration and relationship of each activity in the project. Either the input is given directly or standard subnets are retrieved and integrated to form a project network. A plotting program is used to draw a network. The network data are updated regularly as the project progresses, and the resulting project schedule is used by the resource allocation procedure.

Resource Allocation. The project schedule produced in the network analysis program is combined with the estimated equipment and labor requirements for the resource allocation procedure.

Any restrictions on the resources result in the schedule being revised accordingly. Equipment data are received from the in-house equipment availability procedure, and labor availability from the worker-scheduling program. The revised project schedule is ultimately used by the material-expediting and cost control programs. The resource allocation procedure should also incorporate a leveling subroutine.

Materials-Expediting Program. From the bill of materials and the revised project schedule, the materials-expediting program assesses the inventory file and determines delivery requirements for materials and equipment. The inventory file is updated, and information for accounts payable is produced. Also, materials-expediting schedules are generated.

Cost Control. The cost control program keeps track of all costs incurred on a project, tabulates them into formats useful to different levels of management, and puts out various reports on the status of the project. It also reports original and revised budgeted costs, appropriations, committed and actual costs, and forecasts of cost to complete. Cost estimates are obtained from the estimating procedure, and actual costs incurred as the project progresses are received from the payroll, accounting, and materials-expediting procedures. Schedule and progress reports are obtained from the network schedule and resource allocation procedures. Output from cost control is used to update the accounting and historical data files and to provide cash flow forecasts to management, as well as for progress payment certification.

Equipment Replacement Program. The equipment replacement program provides the analysis to decide when each piece of in-house equipment should be replaced. It gathers the pertinent equipment costs from the historical data file, exposes them to the dynamic programming comparison process, and provides the resultant statistics to management personnel for a final decision. This program also decides the acquisition mode (i.e., rent, buy, or lease).

Manpower Scheduling. When the manpower requirements for a project have been determined from the resource allocation procedure, management may want to schedule the workers on the various shifts as fairly as possible to avoid conflicts. The worker-scheduling program can aid in this process.

Optimum Time/Cost Schedule. To find the minimum-cost project schedule, it will be necessary to make time/cost tradeoffs for the activities involved. A heuristic procedure compares the time and cost estimates for various alternatives and determines the minimum-cost project duration. This information can then be used to reschedule the project.

Simulation Programs. Simulation programs such as SLAM II or CYCLONE are used to aid in determining the most economical equipment combination and selecting the most economical construction methods. Simulation allows several alternatives to be tested; in other words, alternatives formed by equipment to be matched, alternative construction methods, alternative strategies to mitigate the effects of weather, and so on, from the points of view of cost, time, and resources required. Thus the programs provide useful data for the estimating processor. Input originates from the estimating, time study, and equipment availability modules.

Time Study and Activity-Sampling Programs. This package is available to perform statistical analysis of the data collected in the time study and activity-sampling procedure. The output is used in estimating, cost control, simulation, and possibly worker scheduling.

An Alternative Concept. An alternative concept for integrating all the modules described earlier is to organize them around a central database management system and to connect each module through an interface program to the system, as shown in Figure 24.2. The modules are completely independent of each other, and each receives its input from the central database management system through an interface program. The individual programs would retain their configurations, thus preserving the modular aspect of the data-processing system. Such a design permits a step by step development of an overall scheme of management control. The modular concept also offers the flexibility that is so important in the highly diverse construction industry and permits the user to tailor a system best suited to his or her particular needs.

Implementation of The Integrated System. It is strongly recommended that the implementation problem be approached systematically. The first step in any system analysis is to establish the need or requirements that must be met. The present data processing organization of the company must be subjected to an in-depth analysis. From this study will emerge the current and future requirements of the company and a list of priorities for implementing the solution.

The next consideration is the cost of the computer system. This is no easy task because the costs vary among different computer configurations,

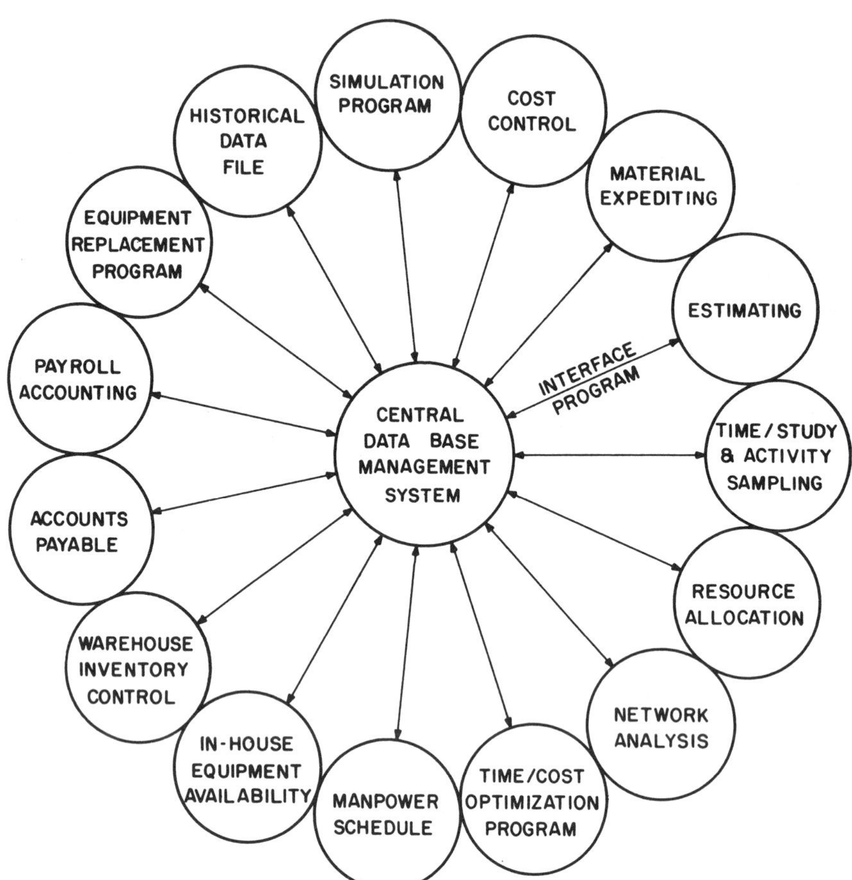

FIGURE 24.2 Integrated Computerized System

different methods of acquiring computer processing, the amount of processing required by types of software, supporting staff, and so on. To determine whether an integrated computer-processing system is economically feasible, it will be necessary to compare the potential value of timeliness of the information to be generated by the system with the additional costs of computer equipment, software, supporting staff, and initial setup costs. If the functions are already being performed manually in the organization, their cost can be compared with mechanized processing.

Having determined that an integrated computer system is justifiable, management must determine the most practical means of attaining the required facilities. The most suitable type of cumputer configuration has to be determined. The final system must reflect the needs of the company and its ability to support the costs of operating such a system. Decisions must be made whether (1) to buy, rent, or lease a computer system for use within the company; (2) to buy machine time to process data in the batch mode on an outside main frame computer; (3) to hire a service bureau; and (4) to buy a minicomputer, microcomputers, or a combination of computers. Similarly the software to be used to provide the necessary programs can be obtained from computer manufacturers or software consultants, and others can be obtained from professional associations, technical colleges, and universities. Since these programs seldom provide an integrated package, it will be necessary to have other interface programs developed by in-house personnel or software consultants.

After the preparatory planning is complete and a specific system design has been completed, management can begin the process of incorporating the system into the existing organization. A carefully thought out implementation plan should be developed and followed. Instead of attempting to mechanize all aspects of the system at one time, the plan should provide for the initial implementation of one major area, followed by a continuing development of processors covering other major areas. For instance, management may want to mechanize the accounting procedures as the first step; install an inventory control system; and finally incorporate a project scheduling and cost control system.

At least initially, each processor should be designed and operated as an entity rather than developed as part of a single integrated system. This will allow the personnel using the programs to become acquainted with the idiosyncrasies of each. This initial period of adjustment could prove beneficial once integration has begun. Possibilities that had not been considered before may be brought forth by personnel who have become familiar with the individual programs. These changes may thus be incorporated into the system during the initial installation.

The processors that are integrated need not be limited to the ones shown in Figure 24.1. All the processors listed in this chapter can be used by a construction organization. They can be gradually added to build a mammoth system that can meet all the needs of the organization.

It must be remembered that implementation of a system means having the required ability, appropriate hardware, suitable software, competent personnel, and constant management support. Computer capability is generated by an interaction of all these factors and a proper environment. This capability does not come with the purchase of a mainframe, mini, or microcomputer nor is it part of any computer program library; it must be developed through planned implementation. Paramount to its development, all individuals involved in the construction management team must thoroughly understand the processes of their respective work, as well as the computer's role in it. This may require that a training program is put in place to explain the importance of computer processing as well as to enhance the facility to analyze for corrective action the various computer reports.

24.4 APPLICATION DEVELOPMENT ENVIRONMENTS

Development of software for in-house use has been gaining momentum since the introduction of the PC. Small programs have proven to be very effective in improving productivity, streamlining the management process, and improving the quality of work. This section reviews some of the most widely used tools for such application development. Two categories of software can be identified: traditional and advanced. Traditional tools focus on spreadsheets and database management programs and are very popular amongst many PC users for their simplicity. Advanced tools include expert systems, neural networks and other recent software development technologies that are capable of complementing the traditional tools.

24.4.1 Traditional Tools

Traditional tools include spreadsheets, relational databases, and procedural programming languages. Our discussion is limited to the first two; an examination of the procedural programming languages is beyond the scope of this book.

Spreadsheets. Spreadsheets are simple user-friendly programs that provide the user with a work space (called a spreadsheet) divided into an arbitrary number of cells. Each cell is capable of accepting (1) data as input, (2) formulas for arithmetic calculations, and (3) programming instructions (decision structures or looping mechanisms), as well as providing output results. Most spreadsheets have integrated many features into their systems to provide integrated "programming" environments. Examples include database management features, statistical analysis, advanced programming instructions, and graphic utilities. A spreadsheet is normally divided into columns with alphanumeric headers and rows with numeric headers providing a grid for cell reference. Table 24.1 shows part of the range-estimating spreadsheet developed in Chapter 16. Columns A–D are used to enter the line item, description, quantity range, unit price, and subtotal for an estimate. The table shows only seven rows for brevity. Cell A1 is occupied with a descriptor that has no function in the spreadsheet other than to identify the use of the column. Cell A2 specifies the line item and is also a descriptor. Some of the cells in Column C are constant numbers representing the estimated quantity, whereas other cells like C3 include a formula [e.g., = Risk Uniform (2200, 2500)]. This formula generates a uniformly distributed number within the indicated range, as described in Chapter 16. Column D contains the unit cost which is either a constant or a formula reflecting a distribution range while column E is a simple formula indicating that the cell contains the result of the multiplication of columns C and D. Finally, cell E8 is a formula indicating that the cell contains the sum of all cells from E2 to E7. This would be the total estimated cost.

TABLE 24.1 A Portion of the Range Estimating Spreadsheet

	A	B	C	D	E
1	Item	Description	Quantity Range	Unit Cost Range	Subtotal Range
2	1	Excavation m3	2300	11.33	= C2*D2
3	2	Backfill m3	= RiskUniform (1900, 2200)	10.67	20,800.00
4	3	Piling 300 dia m	160.00	= RiskUniform (28, 32)	4,640.00
5		Piling 750 dia m	510.00	182.67	93,160.00
6		Bells 1500 dia ea	42.00	393.33	16,520.00
7		Bells 1200 dia ea	16.00	340.00	5,440.00
8					= Sum (E2, E7)

Spreadsheets are very effective tools that are widely used. Their applications in estimating, equipment management, and miscellaneous data analysis are numerous. Applications include bid preparation and risk analysis (e.g., using @RISK from Palisade), simple arithmetic operations, statistical analysis such as regression models for estimating purposes, analysis of productivity data and productivity trends, and sophisticated implementations of add-ons (e.g., simulation, neural networks, and genetic algorithms).

Relational Databases. Relational database management programs (RDBM) are in wide use in construction management application owing to the nature of construction data. In construction we deal with large amounts of information which must be manipulated, combined, and presented in many different ways. Relational databases are very effective for such applications. Although many new advancements in database management took place (e.g., object-oriented programming), the relational model is still dominant in the construction software market. A database management system is simply a computerized record-keeping system. For the system to be useful it must provide effective data manipulation features addressing issues such as: organization, efficiency, flexibility, security, and consistency. The relational model of data management is separated from all other models in that the data organization may be conceptually perceived as being organized in tables. Tables are composed of rows (also referred to as records) and columns (also referred to as fields). The RDBMs provide the user (or the programmer) with the ability to (1) create tables that will house pieces of related information and (2) manipulate the data within the table. Creating a table involves defining the fields that compose a table and providing it with a reference name. Data manipulation provides for adding, deleting, and editing individual entries within a record, whole records, or whole tables.

The architectural model of databases can be divided into three levels—internal, external, and conceptual. The internal level addresses interfaces between the computer system and the database engine (e.g., SQL) and provides the ability to communicate, physically store, and retrieve information in the computer system. The conceptual level does not truly exist, but provides the mechanism of visualizing the organization of data in a simplistic way, thus enabling the programmer to control the data structure more effectively. Data in the RDBM system is visualized as being part of a table, and the system composed of many tables. The external level is what the user of the program sees (e.g., input forms and output reports).

Although we mostly interact with relational databases at the external level, visualizing information at the conceptual level is very helpful for understanding how the data handling within the program is carried. To illustrate the conceptual level, consider the range-estimating example previously shown in Table 24.1. With a database program the data is visualized as being entries within a table. First, define the table by specifying its fields. Performing this includes specifying the name of each field and the type of data it contains [e.g., integer, double precision numbers, or string (specified collection of characters), etc.]. Generically we can define the table as follows:

Define (Table) RangeEstimate
    Item AS integer
    Description AS String, maximum 100 characters
    Quantity AS double precision number
    Unit cost AS currency number
    Subtotal AS double precision number
End of Table definition

Building a RDBM involves creating tables, defining indexes to sort the data within the tables, and filling the tables with appropriate data. Although information will be inserted sequentially as the user is specifying the records, sorting, searching, and presentation of information is internally handled without physically relocating any records. The database management system maintains a set of indexes that will show the logical location of each record according to the index specified. For the same example define an index (a desirable way of sorting the records at some point of time) based on the field "Item" and call it "ItemIndex." This will enable the RDBM to present the table to the user sorted by item number. It would also allow searching the table for any item number specified. Likewise we can specify an index for description and the RDBM will be able to sort all items by their description (alphabetically) and so on. What distinguishes the RDBM from a spreadsheet is its flexibility. The presentation of information does not have to be done through cells as with a spreadsheet, for example. Moreover individual tables may be defined and linked with each other, thus providing more structured data representation than with a spreadsheet. Only information that the user is interested in may be displayed, with all conceptual tables being handled in the background by the RDBM. Many scheduling software programs currently use relational database management systems to handle data. Primavera Project Planner relies on BTRIEV (Novel) to perform all of its RDBM functions. An activity in a scheduling network may be represented in the following simplified RDBM table:

Define (Table) ActivityTable
    Activity ID   AS Long Integer
    Activity Preceder   AS Long Integer
    Calendar ID   AS Integer
    Original Duration   AS Single precision number
    Remaining duration   AS Single precision number
    Actual start   AS Date
    Actual finish   AS Date
    Percent complete   AS Integer

```
        Early start        AS Date
        Early finish       AS Date
        Actual start       AS Date
        Actual finish      AS Date
        Late start         AS Date
        Late finish        AS Date
        Cost code          AS Long Integer
        Total float        AS Long Integer
    End Table
```

Any time an activity is added to the network it occupies one new record, allowing it to hold information defined by the fields (a description, procedures, etc.). The information is entered to the program (like P3) in different representation at the external level and is presented differently as well. The handling of information is done through manipulation of the information in the table-like structure.

The advantages of a RDBM compared to other forms of data representation (e.g., linked lists and arrays) are numerous. The organization of information and compactness of representation make the RDBM superior to other forms of data management systems. The speed of access is another advantage as many reports are normally presorted when an index is defined, thus providing speedy search and delivery of information on demand. The RDBM is easier to handle and maintain than other forms of data representation since the information is logically presented by tables. These systems also reduce redundancy in the information since information available in a table but needed by another need only be represented once as it can be quickly accessed from its table when required. Finally RDBM systems provide a medium for shared data as the information may be reused in other forms with minimum overhead.

24.5 ADVANCED SOFTWARE/APPLICATION DEVELOPMENT TECHNOLOGIES

Many new software technologies have emerged over the past few years. Few have survived the useability test. This section limits the discussion to rule-based expert systems and neural networks with the intention of providing some insight into such technologies, but by no means providing a comprehensive list of new software technologies.

24.5.1 Artificial Intelligence and Rule-Based Expert Systems

Artificial intelligence (AI) may be defined as a branch of computer science that is concerned with the automation of intelligent behavior. Artificial intelligence is one attempt to make computers process in an intelligent way. This is accomplished by analyzing and emulating the thinking nature of humans. This involves understanding how humans make decisions and solve problems, breaking the thought process down into its fundamental steps, and designing a computer program that simulates these processes.

Expert systems (ES) constitute one application of artificial intelligence. Expert systems are intended to solve problems in a particular domain while simulating the performance of a human expert in that domain.

For an ES, facts are an essential ingredient. To reach a specific goal, supporting facts specific to that goal are needed. Rules are the evaluation tool in any ES system. Facts and rules define the general solution strategy. The human mind has an enormous and dynamic set of facts and rules related to all aspects of life. In a domain-specific expert system, the domain-related knowledge (rules) are stored in an area of the computer called *the knowledge base*.

The portion of the intelligence that helps arriving at new facts is called *inference mechanism*. It is central to the ability to learn from experience because it enables the generation of new facts from existing ones.

An expert system is generically composed of the following parts (1) knowledge base, (2) working memory, (3) inference engine, and the (4) I/O user interface.

Knowledge Base. This contains the knowledge of a particular application. In a rule-based expert system, the knowledge base contains a condition–action, or IF–THEN set of rules that enables the system to test facts, assert their compliance with facts within the KB, and take actions accordingly.

Working Memory. This contains an updated description of the "current state of the world" in a reasoning process. The working memory is initialized with data given in a problem instance, keeps track of partial conclusions and new facts as the solution process goes on, and finally ends with the conclusion if found. All of this information is separate from the knowledge base.

The Inference Engine. The inference engine applies the knowledge (rules from the knowledge base) to the particular problem instance in the working memory. The inference engine matches the data in the working memory against the condition (premise) part of the rule (case of data-driven ES) or against the part of the rule (case of goal-driven ES). The rule that matches is then fired and its action part is added to the working memory (data-driven) or its premise is added to the working memory (goal-driven). If more than one rule match (a conflict set), a conflict resolution strategy should be employed to choose one to be fired. The inference mechanism continues with this recognize–act cycle until the goal (conclusion) is reached.

The User Interface. This allows the user to interact with the expert system. The user interface facilitates the dialogue between the user and the computer (the artificial expert). Normally the user interface is also supplemented with a help-style explanation feature which allows the program to explain its reasoning to the user. These explanations include justification for the system's conclusions or "HOW" conclusions were reached, and explanation of "WHY" the system needs a particular piece of data. The explanation can take the form of displaying the rules used in reaching a conclusion, or the rule(s) that need the piece of data. The capability to answer "WHY" and "HOW" queries is a major feature of expert systems. Users accept a recommendation from the computer when they are satisfied the solution is correct and when they understand the reasons behind the conclusion, especially if it is based on heuristics.

One point to emphasize is that, among the component parts of an expert system (knowledge base, inference mechanism, user interface, etc.), only the knowledge base is domain specific. All other components are parts of a general purpose expert system framework applicable to other application domains. This means that the system developer (the engineer concerned with building the knowledge base pertinent to a specific domain, or in short, the knowledge engineer) can use any of the available expert systems "shells" to build the ES. A shell is a complete expert system framework with an empty knowledge base, which is to be completed by the system developer using the system's support tools. Examples of microcomputer-based shells include: Exsys, Insight 2+, M4, Personal Consultant Plus, and 1st Class.

For engineers using computers only as a tool and not as an end in itself there is seldom the need to build a complete shell. Use of one of the available shells is often recommended with most of the effort then going into building the domain-specific knowledge base. This is by no means an easy task. Eliciting rules from experts is difficult because often experts do not really know how to reason about solving problems. In many cases they have not tried to quantify the steps to a decision. Another difficulty is that experts often disagree on the causes of problems and acceptable solutions, and many problems do not have clear-cut right and wrong answers.

The role of the system developer is to capture the best available experience and expertise, and not just a general consensus among different practitioners. Creating such a rule-based knowledge base from divergent positions requires a very experienced system developer, and a means of identifying and resolving conflicts in rules.

24.5.2 A Simplified Expert System Example

Roadways often need repair because of distresses caused by moving traffic, environment conditons, and age. Distresses may include cracks, potholes, and so on. For a highway maintenance crew to fix these distresses, the causes should be diagnosed and treated. If only the symptoms are treated, distresses will shortly reoccur. An expert system can be built to help solve the problem. The following set of simplified rules may be part of the ES:

RULE #1
IF distress is pothole
AND moisture level is high
THEN the cause is poor drainage
RULE #2
IF distress is transverse cracking
AND an old concrete pavement exists below the asphalt surface
THEN the cause is reflection of old cracks and joints
RULE #3
IF distress is transverse cracking
AND an old concrete pavement does not exist
AND temperature is below −20 C
THEN the cause is low temperature shrinkage
RULE #4
IF water table is near the surface
THEN moisture level is high

When this knowledge base is executed (run) under a goal-driven system, the top level goal "the cause is ?" is placed in working memory. The inference engine finds three rules in the knowledge base that match the contents of the working memory, rule 1, rule 2, and rule 3. If the inference engine resolves the conflict between the three rules in favor of the lowest-numbered rule, then rule 1 will fire. This causes ? to be replaced by "poor drainage" and the premises of rule 1 to be placed in the working memory. Note that there are two premises to rule 1, both of which must be satisfied to prove that "the cause is poor drainage" is true. The problem is decomposed into two subproblems which are knowing the distress type and finding whether the moisture level is high. The conclusion of rule 4 matches the working memory. Rule 4 is fired and "water table is high" is added to the working memory. Now the system needs answers to two questions: "is distress type potholes ?" and "is water table high ?" As there are no more rules in the knowledge base that can answer these questions, the expert system will turn to the user to get more information. If the user confirms that distress is pothole and water table is high then the system concludes that the cause is really poor drainage.

24.5.3 Feasibility of Expert System Implementation Within a Construction Domain

Expert system feasibility studies may utilize any of a number of approaches to determine whether a particular operation is suitable for consideration as an expert system application area or not. However, the following two issues should be addressed.

Suitability of the Domain. The tasks to be automated within the expert system should be highly dependent on an expert's knowledge or experience and requiring mostly cognitive skills to accomplish. It must be representable using symbolic logic. In addition, the task must be well scoped and properly defined as a system with known boundaries, input, and output. Finally, experts should be available to supply the knowledge for the system development, literature, or documentation of the system to be automated and test cases for validating its operations.

Benefit to the Users. Once a domain has been determined as suitable for applications of expert systems technology, a determination is required of the benefits which might accrue to the organizaton if an expert system were put into place. These benefits may include: (1) widening the availability of rather scarce expertise amongst the organization and at the same time relieving the few experts from excessive demand on their expertise work; (2) standardization of approaches to solving related problems with consistent results; (3) retaining the valuable knowledge of proven experts when they are no longer part of the organization; (4) improving productivity and consistency as their output and availability are significantly faster and more consistent than the human experts; (5) training of new members of the organization by providing them with an alternative solution based on expert's knowledge.

Construction includes many areas where expert systems may be applied. A noncomprehensive list is presented in Table 24.2. It should be mentioned that many of these applications may require major undertakings. New expert system shells provide room for smaller applications within the everyday construction project requirements, however.

TABLE 24.2 Sample Area of Application for Expert Systems

Risk assessment on large construction projects (e.g., Hitachi developed a system that identifies risks and prevents them in advance)
Safety issues (HOWSAFE system provides a mechanism for checking the safety program within a construction company)
Evaluation of construction schedules
Contractor prequalification
Construction claims prevention
Welding procedure selection, welder qualification test selection, weld defect diagnostics
Equipment diagnostics
Scheduling
Plant operation

24.5.4 Neural Networks and Their Applications

Neural networks (NN) present another AI model for emulating the human expertise. They are drastically different in their approach than rule-based expert systems and are well suited to cover a broad range of applications requiring pattern recognition, learning, recall by association, and fault tolerance. Without addressing the issues of biological resemblance between the human brain and NN, an attempt is made in this section to give the reader an overview of what NN are and their possible applications in construction.

Neural networks, as the name implies are networks consisting of nodes, referred to as processing elements (PE), which are connected through arcs with an associated weight. A node receives algebraic input from incident arcs. Input along each arc is weighted by the weight associated with the given arc and the total summed at the node. The sum is then transformed using a normalization function and an output sent through the output arc to other nodes. A schematic representation of a PE of a NN is given in Figure 24.3.

What makes the NN work is the combination of many PEs connected by arcs in different ways. In other words, one PE by itself is nothing but a tranformation function of limited value, but the connection of large number of PEs in multiple layers makes the NN useful and functional.

Neural networks are normally presented in layers conveniently divided into an input layer, and output layer, and one or more intermediate layers (referred to as hidden layers). In the NN shown in Figure 24.3 the output is flowing from bottom to top. It should be noted that the connections need not be from each PE to every other PE in the succeeding layer. Connections are problem dependent, just like the weights associated with the arcs.

To construct a NN one simply decides how many layers are required, the number of PEs for each layer, the connections between the PEs, the NN model (e.g., feed forward), and the learning algorithm (e.g., Delta rule) to

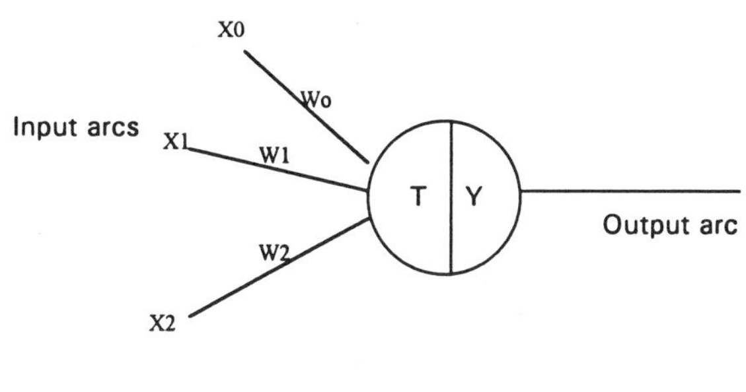

$$T = \Sigma\, Wi \,.\, Xi$$
$$Y = f(T)$$

FIGURE 24.3 A Schematic Representation of a NN Node

be applied as there are a variety of NN implementations. Although one can conceptually program a NN using a programming language or a spreadsheet, dedicated NN programs like NeuralWorks designer pack from NeuralWare make the work easier and more professional. Such systems provide many of the implementations and facilitate the user input and output.

Once a NN such as the one shown in Figure 24.4 is defined on a program like NeuralWorks, it is has to undergo what is referred to as *learning* before it can be used. In the learning phase the NN is subjected to a set of examples of the problem to be solved so it would learn how to solve similar problems. In essence this is reduced to finding out the weights of the various connections that would produce the desired output. The NN program normally uses a learning algorithm that will specify how the weights are to be adjusted if the results differ from those expected. Recently Genetic Algorithm has been used for this purpose. Once the network is run many times for the learning to take place and the weights are adjusted to reflect the desired output for the given input, the NN may be used to solve similar problems as those used in learning. The difference between this approach and the rule-based expert system approach is that with the NN, even if a case was not observed during learning, the output may still be interpolated.

Using the NN is referred to as the *recall* phase. A new problem is presented to the NN; the weights across the connections are used to provide the output. The suitability of using NN lies in reducing the problem input and output into individualized entries on the nodes. To illustrate building a NN consider the problem of forecasting productivity factors for construction activities as a function of weather.

A neural network was developed to generate construction productivity values, given factors which influence the productivity that can be achieved during construction operations. For illustration purposes only, productivity data was obtained from Koehn and Brown (1985) where construction produc-

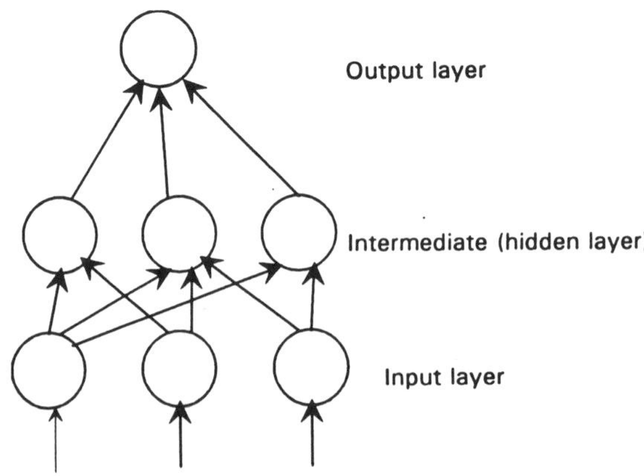

FIGURE 24.4 Generic Feed-Forward NN Structure

tivity is related to temperature and relative humidity. The Koehn and Brown data is derived from empirical equations based upon actual productivity data. The model is simple to construct as two factors determine the productivity factor, temperature, and humidity. Therefore two nodes are at the input layer. The output will consist of one node representing the productivity factor used. A simple feed-forward back propagation NN is used with one hidden layer composed of three nodes, as shown in Figure 24.5.

The NN was developed using the NeuralWare Explorer software. The user provides the network with temperature and relative humidity values and the corresponding productivity factor is generated.

Training the Network. The network is trained by providing a set of learning examples where the input values are presented along with the corresponding desired output values. The network begins the learning process by randomly selecting connection weights, joining each node in the network. For each example provided to the network a result is calculated. The calculated result is then compared to the desired result. The error is then back propagated through the network using an error propagation formula which essentially adjusts the connection weights such that a more agreeable result will be obtained. The process of adjusting connection weights is in fact the learning procedure for the network. By adjusting these weights the network is able to train itself to provide the correct output for a given set of inputs.

In this case the network was provided with 75 learning examples by which it was to train itself. The training set is given in Table 24.3, with the temperature in the first column, the humidity in the second, and the expected productivity according to the tables of Koehn and Brown (1985) in the third column.

After approximately 15,000 training cycles the network was able to predict quite accurately the desired output for each set of inputs.

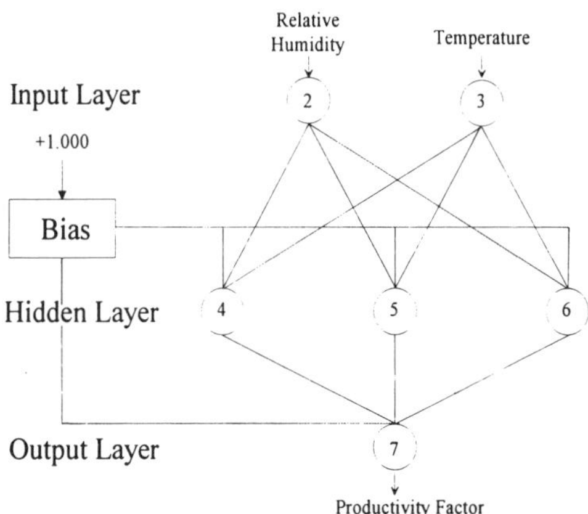

FIGURE 24.5 Productivity Factor Neural Network

TABLE 24.3 Training Set for NN

Humidity	Temperature	Productivity Factor	Humidity	Temperature	Productivity
5	−20	0.28	95	50	1
25	−20	0.25	5	60	1
45	−20	0.18	5	20	0.81
65	−20	0.05	45	60	1
85	−20	0	65	60	1
15	−10	0.43	85	60	1
35	−10	0.4	15	70	1
55	−10	0.34	35	70	1
75	−10	0.21	55	70	1
95	−10	0	75	70	1
5	0	0.59	95	70	1
25	0	0.57	5	80	1
45	0	0.54	25	80	1
65	0	0.49	45	80	1
85	0	0.36	65	80	0.98
15	10	0.71	85	80	0.95
35	10	0.7	15	90	0.95
55	10	0.67	35	90	0.93
75	10	0.62	55	90	0.9
95	10	0.5	75	90	0.85
25	20	0.81	95	90	0.78
45	20	0.81	5		0.81
65	20	0.79	25	100	0.8
85	20	0.75	45	100	0.77
15	30	0.9	65	100	0.71
35	30	0.9	85	100	0.61
55	30	0.89	15	110	0.58
75	30	0.89	35	110	0.57
95	30	0.87	55	110	0.51
5	40	0.96	75	110	0.41
25	40	0.96	95	110	0.21
45	40	0.96	5	120	0
65	40	0.96	25	120	0.28
85	40	0.96	45	120	0.25
15	50	1	65	120	0.15
35	50	1	85	120	0
55	50	1	25	60	1
75	50	1			

An additional 75 test examples were provided to test the network output to the output that is expected. The results are illustrated in Figure 24.6. Covariance and correlation values relating the original data set to the network results indicate a high degree of correlation. As can be seen the patterns are very similar.

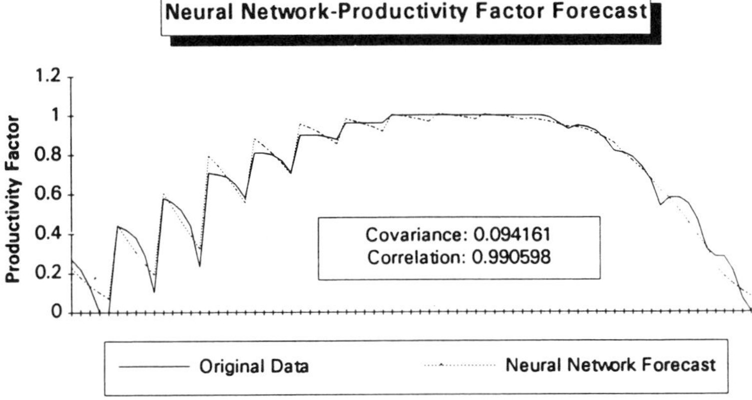

FIGURE 24.6 Forecast Results vs. Original Data

Internal Neural Network Values. For the training set of data provided to the network the following internal factors were generated by the network during the training period.

Connection	Weights
2–4	+0.131061
2–5	+0.097813
2–6	+0.257702
3–4	−2.025511
3–5	+2.277357
3–6	+0.429409
Bias-4	−1.228370
Bias-5	−2.052341
Bias-6	+0.177736
4–7	−1.261046
5–7	−1.508096
6–7	−0.333804
Bias-7	−1.332220

The trained network may be run within NeuralWare by simply specifying the new input. The program returns the output automatically. For example, for the input Temperature = −20 and Humidity = 15%, the output is found to be 0.232253, which is in agreement with the values published in Koehn and Brown (1985).

This information may also be used independently of the NN program as a set of equations with the appropriate weights and summation values. It can be implemented as part of spreadsheet for estimating purposes on any program.

APPENDIX 24—A SAMPLE CRITERIA FOR SELECTING A GENERIC SOFTWARE SYSTEM

The criteria for selecting a software system could include the following:

Accessibility. Access to the particular computer system required by the program must be available. Also, of course, the relative advantages and limitations of the system and its suitability to the organization's needs should be noted.

Simplicity. The software must be relatively easy to install and operate. The input data must be easy to prepare, and the output reports must be understandable.

Data Sequencing. Data sorting is one of the basic uses of computers. The degree of capability for the various sorts of data can be the crucial factor in determining whether a computer program should be used.

Documentation. Many potentially powerful programs have documentation that is difficult to read and understand. The use of such programs proves costly in the long run. Many trials may result for each step. Such programs should be avoided.

Reliability. It must be a fully tested system and should have a proven record. Advice should be sought from past users of the program.

Maintainability. The user must have access to programming personnel familiar with the system, to correct latent program bugs and implement any modifications and/or extensions deemed necessary.

Flexibility. The program should have the capacity for many types of applications. For instance, a scheduling program must be able to handle both small and large networks. It should have the capacity to store standard subnets which can be retrieved and modified to suit the new project needs and integrated to form the project network. A satisfactory way of incorporating modifications into the network must be available. Such operations as multiple starts and ends should be possible. The program should be capable of converting calendars from five- to six-day weeks. It should have the ability to compensate for different workweeks for different activities, which allows processing of international projects involving countries with different national calendars. It should accept additions and deletions of holidays and use variable time units for activity durations. It should have the ability to vary the number of time units per calendar week and generate schedules that have all starts scheduled for the morning and all completions scheduled for the afternoon. The program should have the ability to schedule imposed dates.

The program should have the ability to specify activities in terms of actual start, time remaining, work done, or percentage complete. It should have the ability to report current data only and suppress already completed activities. The program should be capable of analyzing several interfaced projects concurrently in progress. It should also integrate subnets within a project network. The program should have a problem-oriented language that can be used to modify the format of reports to meet the needs of different levels of management.

To facilitate the selection of software, the flexibility criterion for the programs of interest should be stated by the prospective buyer in a manner similar to the scheduling program.

Efficiency. The program must be written in such a manner as to take advantage of current programming technologies and must make efficient use of computer resources. The operation and file-handling procedures must avoid complications in the computer room and allow for adequate turnaround.

Database. The database must contain all the necessary elements so it can be managed to generate the desired information reports and to retain historical information in a suitable format.

Controls. The system must have adequate editing capability for the detection of errors in the input data and must contain controls that ensure that all input data are accounted for. A sufficient trail must be provided as well as a restart of recovery procedures.

Compatibility. The program should be compatible with other programs and systems in use in the company. The program should be written in modules so that the output from a module can be written on a data output device. This may save input–output time for another program. The program should also be capable of retrieving output data sets from other programs and processing them as required.

Costs. The system must be economical in terms of installation, operations, and maintenance.

The selection criteria discussed here should be considered whenever choosing a computer program. A task force in the prospective software buyer's organization should determine the needs, evaluate alternative software against predetermined criteria, and request the selected contenders to demonstrate the use of their software in meeting the needs of the organization. It is important that the program provides for all the requirements of the organization since changing from one system to another is usually quite costly and time consuming.

The American Society of Civil Engineers, the Canadian Society of Civil

Engineers, and many other societies and associations have standing divisions on computer applications in engineering. These divisions can be extremely helpful in directing the prospective user to the most knowledgeable person in his or her area of interest. The prospective user may thus have the benefit of expert opinion based on experience before being committed to a program.

APPENDIX 24.B—EXAMPLE OF SETTING UP PROCEDURES FOR SELECTING A SOFTWARE ITEM

An expert system shell is to be acquired within an organization. Various expert systems shells were identified by referring to the *AI* magazine and *Expert Systems Strategies* magazine as follows:

1. **NeuralWare Professional II Plus**
   By: NeuralWare
2. **KBMS**
   By: AICorp Inc.
3. **GoldWork III**
   By: Gold Hill Inc.
4. **Mentor**
   By: ICARUS Corporation
5. **Level 5 Object**
   By: Information Builders' Inc.
6. **ART-IM**
   By: Inference
7. **Nexpert Object**
   By: Neuron Data
8. **Guru**
   By: Micro Data Base Systems Inc.
9. **Kee**
   By: IntelliCorp

These systems are high-end expert system shells frequently used by various expert system developers. A comparative statement about these software systems has been prepared based on the information collected from the developers of these shells. The expected features of the expert system to be developed were identified as follows:

1. The system should run on an IBM PC or compatible, preferably under the Windows environment.
2. The system should be developed based on the object oriented programming concept in order to fully exploit the electronic data integration concept of programming.

3. The system should call external progam for various purposes including algorithmic functions, data manipulation, and utilities.
4. Since the proposed system is developed to support both the expert and novice user, it should have an extensive help feature including graphical representation.
5. In order to make the system available to wide users, the cost of the system should be as low as possible.

These expected features of the automated system were translated into systems requirements, in order to carry out the comparative study of the expert systems shells available in the market. The systems requirements identified are as follows:

1. The selected expert systems shell should support object-oriented concept.
2. The selected expert systems shell should support advanced graphic features.
3. The selected expert systems shell should support integration with external programs.
4. The selected expert systems shell should provide hypertext features.
5. The selected expert systems shell should provide extendibility.
6. The selected expert systems shell should work in Windows environment.

Table 24.B1 shows the comparative study of the expert systems shells listed above.

Mandatory Attributes

- Hardware compatibility
- Software compatibility
- Purchase price
- Speed
- Capacity
- Maintenance contract
- Delivery date
- Availability of training
- Installation time
- Guarantees and warranty
- Operating cost
- Documentation

Desirable Attributes

- Upgradability
- Electrical power needs
- Free trial period
- Local agents
- Trouble shooting team
- Retention of market value
- User group
- Past record of the company

TABLE 24.B1 Comparative Study of the Available Expert Systems Shells

Expert Systems Shell	OOP	Graphics	H.T.	External Programs	Windows Driven	Extendibility
1. NeuralWare Professional II Plus	Neural Net	Yes	No	No	Window-like interface	No
2. KBMS						
3. GoldWork III	Yes	Yes	Yes	Yes	Yes	Yes
4. Mentor	Yes	Yes	No	Yes	Yes	Yes (procedural programming)
5. Level 5 Object						
6. ART-IM						
7. Nexpert Object	Yes	Yes ToolBook	Yes	Yes	Yes	Yes (C, Fortran, Exe)
8. Guru	Yes	Yes	No	Yes	No	Yes (KGL)
9. KEE 4.0	Unix based	Check information for DOS based				

- Number of users
- Learning time
- Run time feature
- Documentation

Additional Requirements of the Hybrid System

1. Frame- or object-oriented representations of domain objects.
2. Rules for representing heuristic knowledge.
3. Support for a variety of search strategies.
4. Definition of demons to implement interactions and side effects.
5. Rich graphics-based interfaces.
6. Provides extendibility features by providing access to the underlying language.
7. The ability to compile a knowledge base for a faster execution or delivery on a smaller machine.

Features of the Selected Shells

Guru 3.0: KGL (KnowledgeMan/Guru Language)—SQL, Fuzzy Logic, Certainty Factors, Algebra Choices, BLOB (Binary Large OBjects technology), Not Windows based.

ART-IM/Windows: CBR Reasoning, Hypothetical Reasoning with Consistency Management—Supports what–if explorations, Supports Asymetrix ToolBook

NEXPERT Object: Hypermedia, *NEXTRA*

KAPPA-PC 2.0: KAL–KAPPA-PC Application Language

Level5 Object: Hypertext, Incremental Compilation, Supports BMP files

Mentor: Bayesian logic supported: Dynamic modification of confidence factors

GoldWorks III: Fully integrated into Window 3.0 interface use of virtual memory, Dynamic Graphics: dials, guages, *x-y* plots, and pop-up canvases; Foreign Function Interface; Device Independent Bitmaps (DIBs); Dynamic Data Exchange (DDE); Golden Common Lisp Integration; Certainty Factors; Explanation Features; Multiple Inheritance; Dependency Network

Other issues that came up during selection:

- Run-time version
- Number of users
- Software/hardware lock
- On-line help
- Legal issues

GLOSSARY

AACE American Association of Cost Engineers.

Activity A basic element of work which consumes time and resources, and is represented by an arrow in AOA and a node in AON.

Activity duration Time required to complete an activity or task.

Activity network A graphical means of showing the interrelation of a group of activities.

ACWP Actual cost of work performed, usually includes direct and indirect costs.

As-built-schedule Final project schedule which depicts actual start and completion dates and durations.

Benefits and burdens Costs of employee benefits (vacation pay, life insurance, health insurance, pension, etc.) and statutory burdens (Federal Unemployment Tax, Workmen's Compensation Insurance, Statutory Holiday Pay, etc.)

Budget Resources assigned for the accomplishment of a scope of work; becomes the baseline for comparison of performance.

BCWP Budget cost for work performed is the value earned based on the budget for the actual work completed (i.e., physical % complete x approved budget).

BCWS Budget cost for work scheduled is value allocated in the budget for work that has been scheduled up to a particular calendar date.

Budget estimate An estimate used for appropriation of funds for a defined scope of work. AACE accuracy range is −15% to +30%.

Capital cost Total cost including direct and indirect costs; costs for engineering, home office support, equipment, material, labor, and financing; does not include operating materials or spare parts.

Cash flow The steady stream of money, in the form of payments and receipts to and from an individual or organization.

Change order A written authorization to perform additional work or omit specified work on an existing contract.

Charge number The accounts code number used to identify the work package or summary cost category for estimating and accumulating costs.

Chart of accounts A list of cost account numbers for gathering and reporting of cost.

Claim A request for compensation whenever any cost is incurred by the contractor which he feels is not covered by the contract agreement.

Commissioning Activities that are performed that substantiate capabilities of equipment or systems to function as specified.

Commitment A transaction which constitutes a firm open order placed for work to be accomplished, goods or materials to be delivered, or services to be performed, and for which the ordering source is thereby obligated to disburse funds.

Constraint Any factor that affects the schedule of an activity.

Constructability This involves experienced construction input throughout all phases of a project to ensure a feasible and economic construction.

Construction management contract Contract for management of contractors and construction contracts.

Contingencies Costs that are unforeseen but are expected based on experience and risk analysis. These costs do not include for scope changes, escalation due to inflation, or allowances which are known but undefined.

Control estimate An estimate used as the basis for control.

Control level schedule The most detailed schedule which is task oriented. This is a foreman/supervisors daily schedule and is known as the daily or weekly work schedule.

Control planning The planning stage at which plans for the eventual monitoring and control of the project are prepared.

Control schedule Milestone schedule used for monitoring progress.

Control team Organizational group responsible for estimating, scheduling, and cost control.

Cost code This is the code which identifies a work item.

Cost plus contract A contract on which the contractor is reimbursed all his or her costs and is paid a profit in addition.

Cost reimbursable contract *See* Cost plus contract.

Cost slope The amount of the funds required to reduce the duration of an activity by one day.

Crash cost The cost of completing an activity by crash duration.

Crash duration The less than normal time required to complete an activity with extra funds or resources.

Critical activity An activity on the critical path.

Critical path The longest chain of activities that determines project duration and has no float.

CPM Critical path method. An activity on arrow type network planning method, using deterministic durations, developed by James F. Kelly of Rand Corporation in 1956.

Design–build contract Contract for all aspects of design, engineering, procurement, and construction for a project.

Detailed cost estimate (definitive) A detailed estimate based on a complete scope definition (80%+) and used for bidding and allocation of budget.

Dummy An imaginary activity, in CPM with no duration, used either to clarify identification or show an indirect relationship between two activities.

Earliest finish Earliest start plus activity duration.

Earliest start For an activity, same as the early event of the preceding node.

Early event time For a given node, it is the earliest time an activity leaving that node can begin.

Earned value The budget cost of the work actually completed.

Economic feasibility The process of selecting the minimum cost or maximum profit solution out of several acceptable alternative solutions.

Efficiency A measure of productivity compared with an established norm.

Elemental cost estimate Estimates based upon individual costs of the various functional elements of the project. These costs are obtained from an analysis of previous projects.

Equipment factored estimate An estimate based on equipment costs multiplied by factors which make allowance for all other costs associated with the installed piece of equipment. Other costs not attributable to equipment must also be included.

Escalation Provision for change in costs from a base value due to price level changes over a time period.

Estimate The compilation of the bid in which the quantity survey materials are priced and the labor for installing them is added. To these figures are added the job overhead costs and profit to make up the estimate.

Event That point in time when all preceding activities have been completed and all immediately succeeding activities can begin.

Expected elapsed time The expected duration determined in PERT from the pessimistic, optimistic and most likely time of an activity using a formula to find weighted average.

Expediting schedule A schedule designed generally to assist the expediter in expediting delivery of materials and equipment to coordinate the deliveries to their use. Expediting schedules also help monitor progress in design and contract packages.

Facility Major division of a project into which it is divided for cost capitalization.

Feedback The flow of project data from the site of management needed to reassess and reappraise project plans so that efficient control is maintained throughout the project duration and future plans can be improved.

Final funding approval Approval of funds which officially permits project to advance to the execution stage.

Financial feasibility The process of estimating the inflow and outflow of funds to determine if the project can be funded.

Finish float The difference between an activity's earliest and latest finish time.

Fixed cost contract *See* Lump sum contract.

Float The spare time between the predecessor and the successor nodes of an activity which can be used as required. The variations of float are; start float, finish float, total float, free float, independent float.

Forecast to complete When a project is in progress, the sum of actual cost, commitments, obligations and the amount required to complete work not covered by these three costs.

Forward Pass A computational process used to determine the early event time for the events.

Free float The early event time of the following node minus the early event time of the preceding node minus the duration of the activity identified by these nodes. Its use does not affect the downstream activities.

Front-end schedule Usually a bar chart for the early work or tasks in a project.

Implementation Execution of work on a project according to the plan through orientation, good human relations, cooperation and enthusiasm of the project personnel.

Independent float The early event time of the following node minus the late event time of the preceding node minus the duration of the activity identified by these nodes. Its use does not affect either the preceding or the succeeding activities.

Indicated total cost (ITC) A forecast of revised total costs which includes all approved changes.

Indirect cost Salary of supervisory personnel, costs of camp facilities, temporary roads, transportation and site office which are neither part of direct expenses on trade costs nor attributable to headquarter's expenses. Indirect cost is charged to the project as a percentage of direct expenses.

Initial planning The planning stage at which the conceptual design effort on a project is outlined.

Interface event An event which is common to two or more subnets.

Labor cost The salary plus all fringe benefits of construction craftsmen and general labor on construction projects which can be definitely assigned to one work item or cost center.

Lag The logical start/finish relationship, and time, between preceding and succeeding activities.

Late event time For a given node, is the latest time an activity entering that node can finish without delaying the project.

Latest finish For an activity, same as late event time of the following node.

Latest start Latest finish minus duration of an activity.

Lead The lead time assigned to the preceding activity before the succeeding activity can begin.

Levels of schedules The three main levels of scheduling which are differentiated by the degree of detail—Management Level, Project Level, and Control Level.

Learning curve effect The effect on productivity resulting from the fact that workers will be able to produce more quickly once they become used to the work. A learning curve is a graphical representation of the progress in production effectiveness as time passes.

Life cycle cost The total cost for the entire life of a facility which considers planning, financing, construction, operating, environmental, and closing costs.

Linear programming A mathematical technique for solving a general class of optimization problems through minimization (or maximization) of a criterion function subject to linear constraints.

Lump Sum Cost Contract A contract on which fixed price is determined for a well defined project. Additional cost is involved only when the scope of the project changes.

Management level schedule Schedule that sets out milestones used for monitoring by upper management levels.

Master plan The management's plan that shows how they can achieve their targets, it is not composed of the action plans of the contractors or the departments of the organization who are responsible for physically building the project.

Master schedule A bar chart prepared early in the project and used to establish major requirements for engineering, procurement, and construction.

Material expediting The coordination of the delivery of materials to the site so that all deliveries are on time.

Material management Management of all aspects of procuring materials and services.

Milestone An important event in an activity network, specified as target for the subcontractors, requiring special managerial attention for control and is specially marked.

Milestone network It is the intermediate stage in network development which succeeds the WBS but precedes an activity network. It is useful for information summarization for various management levels.

Modular construction A construction technique in which large sections of a facility are assembled off the site for placing in position later.

Monte Carlo technique A simulation technique using random sampling of variables to approximately evaluate probability of occurrence.

Multiple resource allocation A procedure to assign more than one type of resource per job.

Multiproject resource allocation A procedure to make maximum use of available resources from a single resource pool on two or more independent projects.

Negative cash flow The excess of money paid out over the money received.

Nodal activity An activity described on a node as in precedence networks.

Node A point in time where an activity or activities start or end.

Normal cost The cost of completing an activity in its normal duration.

Normal time The estimated duration of an activity needed to perform it with the resources normally used on such activities in the organization under normal conditions.

Optimistic duration The estimated duration of an activity if everything goes exceptionally well, for which there is one in a hundred chance.

Optimization Choosing the best solution to meet the specified criteria from a number of feasible alternatives.

Order-of-magnitude estimate This estimate is made when the initial concepts are formulated, and is based on a small amount of engineering work. AACE categorizes the accuracy as −30$ to +50%.

Organization resource chart (ORC) A diagram which sets out the structure of an organization responsible for a project, including subcontractors.

Overhead cost The costs of personnel in the office, travelling, office rent, depreciation on furniture and equipment, printing and stationary, postage and stamps, etc.

Overrun The actual cost for the work performed to day, minus the value for the same work. When actual cost exceeds value there is an overrun. When value exceeds actual cost, there is an underrun.

Owner The individual who supplies the funds and receives the end product of a project.

Parallel method A heuristic procedure for resource allocation in which resources are allocated one day at a time.

Paretto's law 80 : 20 rule—a small fraction of elements account for a large fraction in terms of effect, that is, 20% of the elements require 80% of the total effort.

Performing agency A contractor or in-house department of the organization responsible for performing work on the project. It is an element in the OAT.

Pessimistic duration The estimated duration for an activity when everything goes wrong, for which there is one in a hundred chance.

PMI Project Management Institute

Positive cash flow The excess of money received over the money paid out.

Precedence diagramming method (PDM) A logic network using nodes to represent activities and connecting them by lines showing their dependencies.

Program evaluation and review technique (PERT) A planning method, using probabilistic durations developed by the U.S. Navy Bureau of Ordinance in 1957.

Progress reporting An evaluation of actual physical progress presented in various formats.

Project calendar A calendar that shows all project dates in terms of calendar dates.

Project clock A device to denote the time when resources are available, activities can be started and therefore priorities can be assigned for allocation of resources.

Project controls Functions of estimating, scheduling, and cost controls.

Project duration The earliest time by which a project can be completed and is determined by the chain of activities which takes the longest time to be completed.

Project level schedule An activity schedule which consolidates activities from a more detailed control level schedule.

Project manager The owner, or individual or organization hired by the owner, who awards contracts and oversees the execution of the project.

Punch list A list of uncompleted or corrective items of work to be done to complete the contract. These lists are prepared by the Project Manager after an inspection of the project at substantial completion.

Purchase order (PO) A document of an agreement for the acquisition of materials, equipment, or services.

Queue A number of sequential events that are waiting to be serviced. Examples are a waiting line, a machine assignment, etc.

Real time dummy Represent the subnet and its effects on the main network, they may be inserted in the main network to replace a subnet. They act as positioning restraints either to delay the interface event until the correct calendar time after the start of the project, or to ensure that particular interface events are achieved in time for the remainder of the project to finish on time.

Relaxation The process of increasing the duration of an activity so that its cost is reduced.

Request for quotation (RFQ) A bid document listing item quantities and descriptions which is sent to vendors who indicate their interest in bidding on the items by quoting prices.

Resource allocation A method of scheduling work by balancing resource need with availability of resources at a given time.

Resource pool Quantity of resources available for activities which can be started but have not been assigned resources until now.

Resource profile A graph of the resources required for each day of the project.

Series method A heuristic procedure for resource allocation in which an activity is allocated resources for its entire duration.

Simulation A term generally given to mathematical representations that take random samples from a probability distribution curve in order to simulate a real-life situation.

Skeletonization The replacement of a network by the smallest number of real-time dummies to show correctly the network start and finish date and all interface events.

Slack The spare time between the predecessor and the successor nodes of an activity which can be used as required. Synonym for float. Used in PERT.

Source node A node which does not have any branch incident to it. Activities emanating from it are started at time zero. Mainly used in GERT.

Square meter method Used to prepare estimates for building projects, it involves measuring the gross floor area and applying a price per square meter obtained from previous projects.

Standard deviation For an activity, the amount of spread about the mean in the many time estimates used for one activity. For an event, the uncertainty of a calculated start for an event, obtained by summing the standard deviations of the activities on the most critical path to the event.

Start float The difference between an activity's earliest and latest start times.

Status reporting A report which compares planned to actual accomplishments as of a specific date.

Subcontract A contract between the prime contractor and a specialty trade or service.

Subcritical path The second longest chain of activities between two nodes in a network. Although this path has a float, it is so small that a delay on any activity can make it closer to critical.

Subelement The progressive breakdown of a functional element into smaller and smaller items for which costs are determined either as a percentage of total cost, or as an amount of money per unit area, volume, or production as required for the elemental cost estimate.

Subnet Short for sub-network, is a section of an activity network, representing a distinct part of the project for which a separate performing agency is responsible.

Subsystem The major items of hardware making up the total system. Subsystems may occur at several Work Breakdown Structure levels.

Systems approach The systematic application of tools and procedures for planning and control to achieve optimum results on a project.

Take off (material) List of quantities taken from drawings or other engineering documents.

Task A work element which is part of an activity.

Task force A team assembled for the purpose of executing a specific project.

Total quality management A program for planning, controlling, and improving quality with the focus being on the customer.

Turn-key contract An agreement whereby one contracting entity is responsible for all project services and work.

Underrun The amount by which the current approved estimate exceeds the sum of the actual costs and the estimates-to-complete.

Unit price contract A contract on which the contractor is paid for the measured quantity of work at a fixed price for each item.

Updating The constant rescheduling and/or restructuring of project plans brought on by reports from the project site.

Value (work performed to date) The planned cost for completed work, including the part of work-in-process which has been finished. This value is determined by summing the planned cost for each completed work package. If a work package is in process, the part of its total planned cost which applies to work completed is approximated by applying the ratio of actual cost to the latest revised estimate for that work package.

Variance The sum of the squared deviations from the mean.

Work breakdown structure The progressive breakdown of the project into smaller and smaller increments, to the lowest practical level to which cost is to be applied. The Work Breakdown Structure graphically depicts the summary cost categories applicable to each major subdivision of the subsystem.

Work item The progressive breakdown of work into smaller and smaller items for which unit costs can be determined to compare with unit prices.

Work package (as is) This is the lowest level within the WBS.

APPENDIX A

MASTERFORMAT, BROADSCOPE SECTION TITLES*

Bidding Requirements, Contract Forms, and Conditions of the Contract

00010 Pre-bid information
00100 Instructions to bidders
00200 Information available to bidders
00300 Bid forms
00400 Supplements to bid forms
00500 Agreement forms
00600 Bonds and certificates
00700 General conditions
00800 Supplementary conditions
00900 Addenda

Note: The items listed above are not specification sections and are referred to as "Documents" rather than "Sections" in the Master List of Section Titles, Numbers, and Broadscope Section Explanations.

Specifications

Division 1—General Requirements
01010 Summary of work
01020 Allowances
01025 Measurement and payment
01030 Alternates/alternatives
01035 Modification procedures
01040 Coordination
01050 Field engineering
01060 Regulatory requirements
01070 Identification systems
01090 References
01100 Special project procedures
01200 Project meetings
01300 Submittals
01400 Quality control
01500 Construction facilities and temporary controls
01600 Material and equipment
01650 Facility startup/commissioning

* This listing has been reproduced with the permission of the Construction Specifications Institute, 601 Madison Street, Alexandria, Virginia, 22314-1791.

01700 Contract closeout
01800 Maintenance

Division 2—Sitework
02010 Subsurface investigation
02050 Demolition
02100 Site preparation
02140 Dewatering
02150 Shoring and underpinning
02160 Excavation support systems
02170 Cofferdams
02200 Earthwork
02300 Tunneling
02350 Piles and caissons
02450 Railroad work
02480 Marine work
02500 Paving and surfacing
02600 Utility piping materials
02660 Water distribution
02680 Fuel and steam distribution
02700 Sewerage and drainage
02760 Restoration of underground pipe
02770 Ponds and reservoirs
02780 Power and communications
02800 Site improvements
02900 Landscaping

Division 3–Concrete
03100 Concrete framework
03200 Concrete reinforcement
03250 Concrete accessories
03300 Cast-in-place concrete
03370 Concrete curing
03400 Precast concrete
03500 Cementitious decks and toppings
03600 Grout
03700 Concrete restoration and cleaning
03800 Mass concrete

Division 4—Masonry
04100 Mortar and masonry grout
04150 Masonry accessories
04200 Unit masonry
04400 Stone
04500 Masonry restoration and cleaning
04550 Refractories
04600 Corrosion resistant masonry
04700 Simulated masonry

Division 5—Metals
05010 Metal materials
05030 Metal coatings
05050 Metal fastening
05100 Structural metal framing
05200 Metal joists
05300 Metal decking
05400 Cold formed metal framing
05500 Metal fabrications
05580 Sheet metal fabrications
05700 Ornamental metal
05800 Expansion control
05900 Hydraulic structures

Division 6—Wood and Plastics
06050 Fasteners and adhesives
06100 Rough carpentry
06130 Heavy timber construction
06150 Wood and metal systems
06170 Prefabricated structural wood
06200 Finish carpentry
06300 Wood treatment
06400 Architectural woodwork
06500 Structural plastics
06600 Plastic fabrications
06650 Solid polymer fabrications

Division 7—Thermal and Moisture Protection
07100 Waterproofing
07150 Dampproofing
07180 Water repellents
07190 Vapor retarders
07195 Air barriers

07200 Insulation
07240 Exterior insulation and finish systems
07250 Fireproofing
07270 Firestopping
07300 Shingles and roofing tiles
07400 Manufactured roofing and siding
07480 Exterior wall assemblies
07500 Membrane roofing
07570 Traffic coatings
07600 Flashing and sheet metal
07700 Roof specialties and accessories
07800 Skylights
07900 Joint sealers

Division 8—Doors and Windows
08100 Metal doors and frames
08200 Wood and plastic doors
08250 Door opening assemblies
08300 Special doors
08400 Entrances and storefronts
08500 Metal windows
08600 Wood and plastic windows
08650 Special windows
08700 Hardware
08800 Glazing
08900 Glazed curtain walls

Division 9—Finishes
09100 Metal support systems
09200 Lath and plaster
09250 Gypsum board
09300 Tile
09400 Terrazzo
09450 Stone facing
09500 Acoustical treatment
09540 Special wall surfaces
09545 Special ceiling surfaces
09550 Wood flooring
09600 Stone flooring
09630 Unit masonry flooring
09650 Resilient flooring
09680 Carpet
09700 Special flooring
09780 Floor treatment
09800 Special coatings
09900 Painting
09950 Wall coverings

Division 10—Specialties
10100 Visual display boards
10150 Compartments and cubicles
10200 Louvers and vents
10240 Grilles and screens
10250 Service wall systems
10260 Wall and corner guards
10270 Access flooring
10290 Pest control
10300 Fireplaces and stoves
10340 Manufactured exterior specialties
10350 Flagpoles
10400 Identifying devices
10450 Pedestrian control devices
10500 Lockers
10520 Fire protection specialties
10530 Protective covers
10550 Postal specialties
10600 Partitions
10650 Operable partitions
10670 Storage shelving
10700 Exterior protection devices for openings
10750 Telephone specialties
10800 Toilet and bath accessories
10880 Scales
10900 Wardrobe and closet specialties

Division 11—Equipment
11010 Maintenance equipment
11020 Security and vault equipment
11030 Teller and service equipment
11040 Ecclesiastical equipment
11050 Library equipment

11060 Theater and stage equipment
11070 Instrumental equipment
11080 Registration equipment
11090 Checkroom equipment
11100 Mercantile equipment
11110 Commercial laundry and dry cleaning equipment
11120 Vending equipment
11130 Audio-visual equipment
11140 Vehicle service equipment
11150 Parking control equipment
11160 Loading dock equipment
11170 Solid waste handling equipment
11190 Detention equipment
11200 Water supply and treatment equipment
11280 Hydraulic gates and valves
11300 Fluid waste treatment and disposal equipment
11400 Food service equipment
11450 Residential equipment
11460 Unit kitchens
11470 Darkroom equipment
11480 Athletic, recreational, and therapeutic equipment
11500 Industrial and process equipment
11600 Laboratory equipment
11650 Planetarium equipment
11660 Observatory equipment
11680 Office equipment
11700 Medical equipment
11780 Mortuary equipment
11850 Navigation equipment
11870 Agricultural equipment

Division 12—Furnishings
12050 Fabrics
12100 Artwork
12300 Manufactured casework
12500 Window treatment
12600 Furniture and accessories
12670 Rugs and mats
12700 Multiple seating
12800 Interior plants and planters

Division 13—Special Construction
13010 Air supported structures
13020 Integrated assemblies
13030 Special purpose rooms
13080 Sound, vibration, and seismic control
13090 Radiation protection
13100 Nuclear reactors
13120 Preengineered structures
13150 Aquatic facilities
13175 Ice rinks
13180 Site constructed incinerators
13185 Kennels and animal shelters
13200 Liquid and gas storage tanks
13220 Filter underdrains and media
13230 Digester covers and appurtenances
13240 Oxygenation systems
13260 Sludge conditioning systems
13300 Utility control systems
13400 Industrial and process control systems
13500 Recording instrumentation
13550 Transportation control instrumentation
13600 Solar energy systems
13700 Wind energy systems
13750 Cogeneration systems
13800 Building automation systems
13900 Fire suppression and supervisory systems
13950 Special security construction

Division 14—Conveying systems
14100 Dumbwaiters

14200 Elevators
14300 Escalators and moving walks
14400 Lifts
14500 Material handling systems
14600 Hoists and cranes
14700 Turntables
14800 Scaffolding
14900 Transportation systems

Division 15—Mechanical
15050 Basic mechanical materials and methods
15250 Mechanical insulation
15300 Fire protection
15400 Plumbing
15500 Heating, ventilating, and air conditioning
15550 Heat generation
15650 Refrigeration
15750 Heat transfer
15850 Air handling
15880 Air distribution
15950 Controls
15990 Testing, adjusting, and balancing

Division 16—Electrical
16050 Basic electrical materials and methods
16200 Power generation–built-up systems
16300 Medium voltage distribution
16400 Service and distribution
16500 Lighting
16600 Special systems
16700 Communications
16850 Electrical resistance heating
16900 Controls
16950 Testing

BIBLIOGRAPHY (REFERENCES AND ADDITIONAL READINGS)

AACE, *Cost Engineer's Handbook.*

AbouRizk, S. M. "Input Modeling for Construction Simulation." Ph.D. Dissertation, School of Civil Engineering, Purdue University, West Lafayette, IN, 1990.

AbouRizk, S. M. "Statistical Considerations in Simulating Construction Operations," Independent research study report, Division of Construction Engineering and Management, Purdue University, West Lafayette, IN, 1988.

AbouRizk, S. M., Halpin D. W., and Wilson J. R. (1991), *"Visual Interactive Fitting of Beta Distributions,"* Journal of Construction Engineering and Management, ASCE, Vol. 117 No. 4, Dec. 1991, pp 589–605.

AbouRizk, S. M. and Dozzi, S. P. (1993). *"Applications of Computer Simulation in Resolving Construction Disputes,"* Journal of Construction Engineering and Management, ASCE, -Special Issue: Applications of Microcomputers and Workstations in Construction. Vol. 119 No. 2 June 1993. pp 355–373.

AbouRizk, S. M. and Sawhney A. (1993), *"A Subjective and Interactive Distribution Estimation System,"* Canadian Journal of Civil Engineering, CSCE, Vol 20. pp 457–470.

Adeli, H. *Expert Systems in Construction and Structural Engineering,* Chapman and Hall Ltd., London, 1988.

Ahuja, Hira, N. *Construction Performance Control by Networks,* Wiley, New York, 1976.

Ahuja, Hira, N. *Successful Construction Cost Control,* Wiley, New York, 1980.

Ahuja, Hira, N. and Walsh, Michael, A. *Successful Methods in Cost Engineering,* Wiley, New York, 1983.

Ahuja, Hira, N. and Campbell, W. J. *Estimating: From Concept to Completion,* Prentice Hall, Englewood Cliffs, NJ, 1987.

American Society of Civil Engineers, "Construction Cost Control." *ASCE Manuals and Reports of Engineering Practice,* No. 65, Rev. Ed., New York, 1985.

American Society of Civil Engineers. (ASCE) *Avoiding and Resolving Disputes During Construction,* New York, 1991.

Amos, John M. and Sarchet, Bernard, R. *Management for Engineers,* Prentice-Hall, Englewood Cliffs, NJ, 1981.

Barrie, Donald S. and Paulson, Boyd, C., Jr. *Professional Construction Management,* McGraw-Hill, New York, 10992.

Bell, L. C. and Stukhart, G. "Cost and Benefits of Material Management" ASCE Journal of Construction Engineering and Management Vol 113, No. 2, June 1987.

Brately, P., Fox, B. L., and Shrage, L. E. "A Guide to Simulation," Springer-Verlag, 1983.

Box, G. E. P., and Muller, M. E., "A note on the generation of random normal deviates." Ann. Math. Stat. 29, pp. 610–611.

Business Roundtable Report A-3. "Improving Construction Safety Performance," May 1988. Business Roundtable Report A-6. "Modern Management Systems," November 1982.

CII Construction Industry Institute Publication Nos: 17-1—*In Search of Partnering Excellence;* 3-1—*Constructability: A Primer,* July 1986; 2-3—Productivity Measurement: An Introduction, October 1990; 7-2—*Project Materials Management Primer,* 1988.

SD-1 Attributes of Material Management, 1985.

Chang, Y., and Sullivan, R. Quant Systems (QS) Version 2. Users Manual, Prentice Hall, Englewood Cliffs, N.J. 1991.

Clark, C. E. "The PERT Model for the Distribution of an Activity Time." *Oper. Res.,* **8,** 405–406 (1961).

Clark, F. D. and Lorenzoni, A. B. *Applied Cost Engineering,* Marcel Dekker, New York, 1988.

Clough, R. R. *Construction Contracting,* 4th ed., Wiley, New York, 1981.

Construction Specifications Institute. "MASTERFORMAT—A Master List of Section Titles and Numbers." The Construction Specifications Institute, Alexandria, VA, 1983.

Dabbas, M. A. and Halpin, D. W. "Integrated Project and Process Management." *J. Construction Div., ASCE* **103**(3), Col:361–374(1982).

Dozzi, S. P. "Estimating and Controlling the Cost of Capital Projects." Project Management for Developing Countries, International Seminar, Oxford and IBM Publishing Co., PVT, Ltd., pp. 219–240, 1991.

Dozzi, S. P. and AbouRizk, S. M.

Feigenbaum, A. V. *Total Quality Control, Engineering and Management,* McGraw-Hill, New York, 1983.

Fishman, G. S. Concepts and Methods in Discrete Event Digital Simulation,'' Wiley, New York, 1973.

Fisk, E. R. *Construction Project Administration,* Wiley, New York, 1992.

Grubbs, F. E. ''Attempts to Validate Certain PERT Statistics or 'Picking on PERT.' '' *Oper. Res.,* **10,** 921–915 (1962).

Hahn, G. H. and Shapiro, S. S. *Statistical Models in Engineering,* Wiley, New York, 1967.

Halpin, D. W. *Financial and Cost Concepts for Construction Management,* Wiley, New York, 1985.

Halpin, D. W. and Riggs, L. S. *Planning and Analysis of Construction Operations,* Wiley, New York, 1992.

Halpin, D. W. and Woodhead, R. W. *Construction Management,* Wiley, New York, 1980.

Harris, R. B. *Precedence and Arrow Networking Techniques for Construction,* Wiley, New York, 1978.

Juran, J. M. *Quality Control Handbook.* McGraw-Hill, New York, 1974.

Kepner C. C., Tregoe B. B. *The New Rational Manager,* Princeton Research Press, Princeton, NJ, 1981.

Kerzner, H. *Project Management: A Systems Approach to Planning Scheduling and Controlling,* 3rd ed., Van Nostrand Reinhold, New York, 1987.

Koehn and Brown.

Law, A. M. and Kelton, W. D. *Simulation Modeling and Analysis,* McGraw-Hill, New York, 1982.

Lewis, T. G., and Payne, W. H. Generalized Feedback Shift Register Pseudo-random Number Algorithm, J. A. C. M. Vol 20, pp 456–468.

MacCrimmon, K. R. and Rayvac, C. A. ''An Analytical Study of the PERT Assumptions,'' *Oper. Res.,* **12**(1), 16–37 (1964).

McCaffer, R. and Harris, F. *Modern Construction Management,* BSP Professional Books, Oxford, 1989.

Mensching, J. and Adams, D. ''Managing an Information System,'' Prentice Hall, Englewood Cliffs, NJ, 1991.

Meuller, F. E. *Integrated Cost and Schedule Control for Construction Projects,* Van Nostrand Reinhold, New York, 1986.

Moder, J. J., Phillips, C. R., and Davis, E. *Project Management with CPM, PERT and Precedence Diagramming,* 3rd ed., Van Nostrand Reinhold, New York, 1983.

Newman, 1951.

O'Brien, K. E. *Improvement of On-Site Productivity,* K. E. O'Brien & Associates Inc., Toronto, 1989.

Pritsker, A. A. *Introduction to Simulation and SLAM-II,* Wiley, New York, 1985.

Project Management Institute, *Project Management Body of Knowledge.* Special Summer Issue, Vol. XVII, No. 3, Drexill Hill, NJ, August 1986.

Royston, J. P. ''The W. Test for Normality,'' *App. Statistics,* **31,** 176–180 (1982).

Peterson, C. and Miller, A. ''Mode, Median and Mean as Optimal Strategies.'' Journal of Experimental Psychology'' Vol. 68, pp 363–367.

Sprague, J. C. and Whittaker, J. D. *Economic Analysis for Engineers and Managers,* Prentice Hall, Englewood Cliffs, NJ, 1986.

Stukhart, G. and Bell, L. C. *Attributes of Material Management,* Construction Industry Institute, 1985.

Szongi, A. J. et al. *Principles of Engineering Economic Analysis,* Wiley, Canada, 1982.

Tardif, L. "Snapshort Technique." *The Revay Report,* Vol. 7, No. 1, Revay and Associates Ltd., Montreal, 1988.

Welch, P. D. "The Statistical Analysis of Simulation Results." In Lavenberg, S. S., Ed., *Computer Performance Handbook,* Academic, New York, 1983.

Willis, E. M. *Scheduling Construction Projects,* Wiley, New York, 1986.

USA Navy, "PERT Summary Report, Phase 1, Special Projects Office, Bureau of Naval Weapons, Department of the Navy, Washington, D.C., 1958.

Wideman, R. M. "Cost Control of Capital Projects and the Project Cost Management System Requirements." AEW Services, Vancouver, Canada, 1983.

Wilson, J. R. "Statistical Aspects of Simulation, Proceedings, IFORS, 1984, pp 825–841.

INDEX